RED WINE ENOLOGY

ADVANCED WINEMAKING STRATEGIES FOR FINE RED WINES

Tannin and Redox Management of Red Wines

Volker Schneider
Mark Tracey

WINE APPRECIATION GUILD PRESS
SAN FRANCISCO

Red Wine Enology

Wine Appreciation Guild Press
an imprint of
Board and Bench Publishing
www.boardandbench.com

Editorial direction by Annika LD Imelli
Design and composition by PerfecType, Nashville, TN

ISBNs
978-1-935879-15-2 (print)
978-1-935879-16-9 (ePub)

Library of Congress Cataloging-in-publication is on file with the Library of Congress

Preface

This book is dedicated exclusively to the enology of red wines. They differ from white wines by their high levels of tannins and colored compounds, which are extracted from grape solids during the red winemaking process and constitute their very essence. These tannins and the associated color-active anthocyanins are phenolic substances that are largely responsible for the distinctive mouthfeel properties of red wines. This makes phenol chemistry and tannin management the central topic of red wine enology.

Whilst everybody in the wine industry has an understanding of what tannin is, there is much less knowledge about what redox means. Indeed, red wine phenolics easily react with oxygen. In this process, they change their chemical structure and consequently their sensory properties. This explains why oxygen is deliberately used to further shape red wine mouthfeel properties during what is commonly designated as aging or maturation. However, when there is oxidation, there are also reactions of reduction. Maintaining an appropriate balance between oxidation and reduction is the purpose of redox management. Ultimately, phenols, oxidation, and reduction must be seen in context.

Having in mind the importance of tannin and redox management in red winemaking, we identified the need for a special book that could demonstrate to the reader how to use basic knowledge of phenol chemistry to rationally explain or even predict the sensory outcome of wine treatments and winery processes. Rather than providing an extensive description of red wine constituents, we tackle the types of chemical reactions triggered by these processes, the ways to control them, and their sensory outcomes. In particular, we highlight how decisions made during winemaking favor or disfavor chemical reactions leading to differences in wine composition and sensory properties.

Seeking to guide readers towards judiciously employing redox management to better exploit the full potential of their grapes and thus enhance their wines, we discuss in particular the importance of tailoring the post-fermentation redox regime to the phenolic composition of the respective wine. Thus, examples of low-phenol wines being subjected to excessive oxygenation are examined as well as wines with high phenol levels obtained from premium grapes or with unique phenolic profiles requiring particular care in redox management such as Pinot noir and similar challenging varieties.

Red Wine Enology

This book has been written for professional winemakers, newcomers to the wine industry, and students of enology both seeking to deepen their knowledge and appreciate the practical application of what they have learnt in their courses. Instead of targeting a merely scientific audience, it tries to find the right balance between theoretical expertise and hands-on issues. We assume basic chemistry knowledge including organic chemistry, albeit we acknowledge that some lessons might have fallen into oblivion and seek to support the reader in this regard. Having backgrounds in both practical winemaking and enological chemistry, we are privileged to synthesize considerable knowledge and experience that has been acquired from the crush pad to the bottling line under the most variable climate and winemaking conditions.

The authors

Contents

1. Introduction

This book is more than a compilation of the most recent scientific findings on red wines because winemaking is not only science. Contemporary winemaking rather takes an engineer's approach, which involves applying science as much as possible but also completing the process with intuition and experience. Thus, in many circumstances, winemaking strives to go beyond the facts that are really known about the complexity of wine and yet to integrate science, intuition and experience together to yield a system offering all the best individual winemaking steps. This is due to both the limits of scientific knowledge and the complex human sensory response to wine composition.

Since wine has a material basis that is chemistry, we start with a review of the phenolic substance classes found in red grapes and wines, their typical concentrations, their basic reactivities, and their sensory characteristics. The infinite number of possible tannin structures becomes obvious. This first chapter is also designated to be used as a reference for subsequent sections.

Thereafter, analytical measures are proposed to quantify the phenolic strength of red wines and some of the most relevant phenolic fractions. This has a simple reason: Serious red wines have become much too expensive to evaluate their basic mouthfeel features elicited by tannins and related compounds only by sensory means. However, instead of presenting the whole array of analytical approaches, only standard methods, their pitfalls, and the interpretation of their results are described from a winemaker's perspective. These results also explain the significant differences between cool and hot-climate red winemaking.

Endowed with the basics outlined in the first two chapters, the reader will better understand the great section on fermentative maceration and other ways of extracting the phenolic material from the fruit during the winemaking process. The importance of maceration duration and temperature is emphasized, whilst simple total phenol measurements are presented as a means to optimize the timing of devatting.

Before moving on to post-fermentation operations sharpening mouthfeel perceptions, the sensory properties of tannins and related compounds are described in detail, in particular the astringency as the primary sensory expression they elicit, the pitfalls of its sensory evaluation, and the way it is modulated by wine compounds other than tannins. Since post-fermentation operations associated with aging and maturation rely largely on deliberate oxygen supply as during micro-oxygenation and barrel aging, emphasis is put on the various oxygen acceptors. Among the latter, phenolic compounds can be the most important ones, but other wine constituents compete with them for oxygen to a variable extent, depending on the individual wine and its current stage. Hence, winemaking outcomes can be interpreted through the lens of competitive reactions, of which the oxidation and subsequent polymerization of phenolic substances are in the center of enological interest. Excessive oxidation is always harmful and is evidenced by a temporary accumulation of hydrogen peroxide and acetaldehyde. Therefore, the art of red wine maturation lies in keeping reactions of oxidation and reduction in balance by means of redox management. These means are explained in detail and summarized under the notion of oxidations under reductive conditions.

From the perspective of applied technology, measures of quantifiable oxygen supply by micro-oxygenation, macro-oxygenation and diffusion through container wall materials are also presented as are the use of sulfur dioxide, commercial tannins, oak barrels, and oak alternatives. Lastly, particular attention is paid to the often neglected steps of preparing wines for bottling. These final steps comprise ultimate taste corrections, measuring and stabilizing free SO_2 as a function of post-bottling oxygen uptake, and filtration. With regard to the growing interest in non-interventionist or organic winemaking, bottling of wines without added sulfites or of unfiltered wines is also discussed.

The authors have sought to write a book whose style lies between that of a text book and a treatise on the management of phenols and redox potential in red wines. Accordingly, some readers may only ever refer to specific sections to address specific requirements, whilst others may wish to read the book fully to appreciate the story that it tells: The authors encourage this! For those intending to read fully we are aware that the ideas within the book are very interrelated and therefore we suggest that readers should allow adequate time for their reading and not expect to fully appreciate the arguments within a first reading. With this in mind, we will suggest a slightly shorter initial reading sequence for readers who, whilst perhaps experienced winemakers, do not have a formal enology training. Such a first reading could then be followed by fully re-reading the book equipped with an awareness of its contents, philosophy, and style.

Our suggested shorter initial reading sequence is as follows: Chapter 2, outlining the basics of phenol chemistry, can be 'lightly read' so as to refamiliarize the reader if necessary with the basic classes and properties of wine phenols, however it may well be necessary to refer back to chapter 2 whilst reading later chapters to clarify phenol chemistry specifics. Reading of chapter 3 should only comprise 3.1.1 and 3.1.4. Chapter 4 can be limited to 4.1. In chapter 7, section 7.2 can be omitted.

There is no doubt that many questions remain unanswered at the end of this book. For this reason, we have included a detailed and carefully selected bibliography that allows to explore certain points in greater depth. At the same time, we are aware that no one can yet answer all questions that might arise in the context of tannin and redox management of red wines, one of the most complex and fascinating subfields of enology. As a field of ongoing research, it promises many additional insights in the coming years.

2. The phenolic make-up of red grapes and wines

Tannins and colored compounds extracted from grape solids during the winemaking process are the very essence of red wines and differentiate them from white wines. Amongst all wine constituents, they are the most reactive and the most variable in both quantitative and qualitative terms. Their amounts and composition shape the wine style and are in the focus of almost all decisions taken during red winemaking and aging. In chemical terms, they are all phenolic compounds. This chapter gives a short introduction into the chemical fundamentals of red wine phenolics. It deals with flavan-3-ols, tannins, anthocyanins, polymeric pigments, and the way they react one with another. Respecting a logical sequence, it is placed at the beginning of this book for the benefit of those who do not have a solid grasp on organic chemistry. Of course, you are not required to bear all details in mind to be a good winemaker, but they will help you better understand what is happening in your wine when your college chemistry lies a long while back. Readers willing to move directly to more practical hands-on issues might prefer to start reading with chapter 4, but they are invited to refer to the chemical basics outlined in the following whenever they deem it necessary to do so.

2.1. Phenols are the very essence of red wines

The typical taste and color properties of red wines are derived from tannins, colored anthocyanins, and interrelated compounds. These compounds or their precursors are extracted to a variable extent from the grape skins and other grape solids during the processing steps inherent to red winemaking before pressing. By far the most widely used procedure for their extraction is maceration, commonly referred to as skin contact, of variable technical implementation and duration, whilst other techniques such as thermovinification, carbonic maceration, and flash détente are used much less frequently.

In contrast to red winemaking, minimal or restricted skin contact is the general rule in modern methods of white vinification. Therefore, white wines usually do not contain appreciable amounts of tannin, if any. When they do so, measures are taken to lower them. The presence of fairly high amounts of tannins, anthocyanins, and the reaction products they generate during winemaking and storage are the very essence of red wines and responsible for their most obvious and admired sensory features. They differentiate them from white wines.

On the other hand, these compounds are also the most reactive and the most variable of the total red wine constituents in both quantitative and qualitative terms. The only common feature they display is their belonging to the large chemical group of non-volatile phenols, which are secondary plant metabolites. This explains why irrespective of whether they are called phenols, polyphenols, or phenolics, they make phenol chemistry and, ultimately, the phenolic composition of red grapes and wines the key to understanding the behavior of red wines during aging and their response to enological interventions.

2.2. Flavonoid phenols

2.2.1. General properties

A phenol is a molecule that consists of what is known as a benzene ring, a structure consisting of six carbons joined in a ring with a hydrogen atom (-H) attached to five of them and with a hydroxyl (-OH) group attached to the sixth. Thus, its chemical formula is C_6H_5OH.

There are no true phenols in either grapes or wine, but only polyphenols. There are several definitions of polyphenols, but the most common one is that they have more than one hydroxyl group attached to one or several benzene rings. Polyphenols occurring in plants and wine are modified in various ways, often joined together and characterized by the presence of large multiples of phenol structural units.

Flavonoid phenols are phenols that possess a three-ring system composed of 15 numbered carbon atoms and share a general C6-C3-C6 structure, consisting of two benzene rings (A and B) and a central, heterocyclic ring (C) (Figure 1). The C ring contains one oxygen atom and can adopt various oxidation states. The B ring is fused to the C ring at position 2. The A ring is the least variable portion of the C6-C3-C6 backbone.

Figure 1: Basic flavonoid ring structure and numbering of carbon atoms.

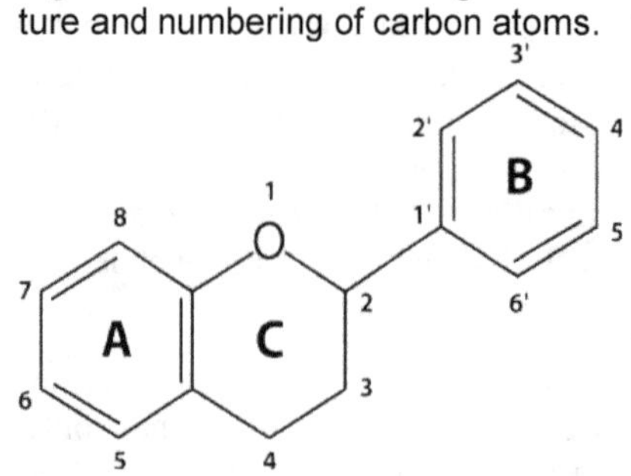

All flavonoids found in grapes and wine have the same substitution pattern of hydroxyl (-OH) groups at carbons 5 and 7 of the A ring. Furthermore, differences in the oxidation state and substitution on the C ring define the different classes of flavonoids, for example flavanols, flavonols, flavones, etc. On the other hand, different substitutions in the B ring define different compounds within the same class.

Flavonoid phenols are extracted from the solid parts of the grape cluster. They are present in all grapes regardless of their being red or white. It is only the additional presence of anthocyanins which distinguishes red cultivars from white ones. The wine type, determined by the quality and quantity of flavonoids, is then a function of their extraction during the wine making process. White wines with limited skin contact after crushing display very low levels of flavonoids.

The extent to which flavonoids accumulate in red wine depends on fruit composition and maturity, the length of the skin contact, the temperature regime and mechanical action during that period, and matrix variables such as SO_2 additions, pH, and alcohol content. They are the building blocks of tannins and colored pigments responsible for the sensory features of matured red wines. Hence, their amount and chemical composition substantially control the sensory outcome of enological interventions during pre-bottling storage and of post-bottling aging.

From a quantitative, chemical, and sensory perspective, the three most relevant polyphenolic classes within the flavonoid family extracted from grapes are flavan-3-ols, anthocyanins, and true tannins. A heterogeneous family of reaction products resulting from the interaction between flavan-3-ols, tannins, and anthocyanins, the so-called polymeric pigments, is not originally present in grapes but only formed during winemaking operations and wine aging. Minor levels of flavonols complete the picture but are not considered in this survey due to their limited importance.

2.2.2. Flavan-3-ols

Flavan-3-ols, sometimes also referred to as flavanols, are flavonoids that are characterized by a hydroxyl group attached to carbon 3 of the C ring and various degrees of hydroxylation at their A and B rings. The hydroxyl group at C3 of the C ring can be esterified[1] with gallic acid, thus forming a gallate. The main structural difference from other flavonoids is the absence of a double bond between C3 and C5 in flavan-3-ols and no carbonyl[2] at C4 in ring C.

In grapes, flavan-3-ols occur mainly in the solid tissues as five monomers[3]: (+)-catechin, (-)-epicatechin, (+)-gallocatechin, (+)-epigallocatechin, and (-)-epicatechin-3-gallate (Figure 2). These monomers display different isomeric[4] configurations having some impact on their sensory properties.

In contrast to many other flavonoids, flavan-3-ols are not bound to a sugar molecule. Furthermore, they are not to be confused with flavonols, another flavonoid subgroup which contains a keto group and plays a subordinate role in wine. In earlier times, there were also reports on flavan-3,4-diols, the existence of which in wine is now doubted.

---

[1] Esterification takes place when a hydroxyl (-OH) group reacts with the carboxyl (-COOH) group of an organic acid with splitting off of water.

[2] A carbonyl is a functional group of a carbon double-bounded to an oxygen: -C=O.

[3] A monomer is a chemical substance whose basic molecules can join together to form polymers.

[4] Isomers are molecules with identical formulas and the same number of atoms of the same elements, but with a different arrangement of the atoms in space.

Figure 2: Structures of monomeric flavan-3-ols found in grapes and wines.

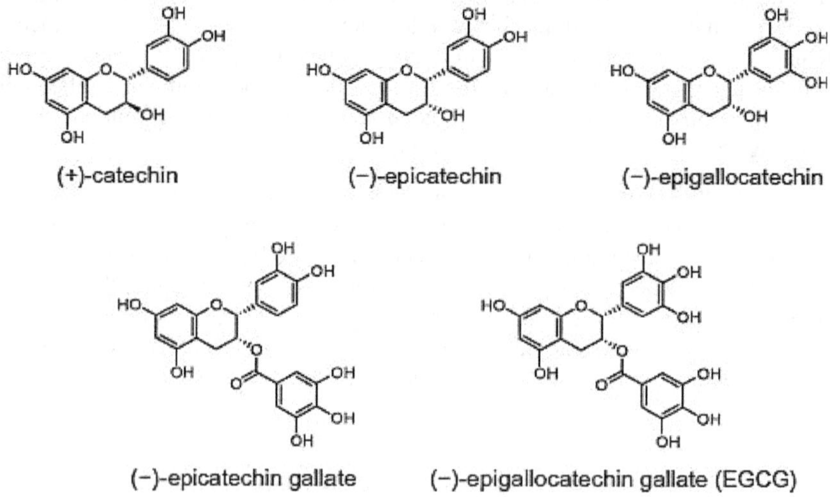

In the skins, flavan-3-ols occur in the vacuoles of the subepidermal cells. In seeds, they are located in the cells between the external cuticle and the inner lignified layers (Kennedy et al. 2000) and comprise only (+)-catechin, (-)-epicatechin, and (-)-epicatechin gallate. Seeds concentrate the vast majority of flavan-3-ols of the berry (Mattivi et al. 2009, Obreque-Slier et al. 2010 a, Rodriguez Montealegre et al. 2006), and differences as compared to skins can be more than tenfold, depending on grape variety and ripeness. Stems are not considered in this context because they are removed by destemming to avoid the release of undesirable flavonoid phenols they contain in high amounts.

Flavan-3-ols contribute to astringency and have a marked bitter taste, whereby (-)-epicatechin is significantly more bitter than (+)-catechin (Robichaud and Noble 1990, Kallithraka et al. 1997). They are poorly soluble in water but rapidly dissolve in alcohol.

Release of flavan-3-ols from skins occurs during the first few days of maceration, whilst the extraction of flavan-3-ols from seeds requires longer maceration times (Gonzáles-Manzano et al. 2006). When maceration is extended over several weeks, the seeds contribute to almost 90 % of the total flavan-3-ol content of the final wines (Gonzáles-Manzano et al. 2004).

Clearly, flavan-3-ols from seeds are more slowly extracted than those from skins and primarily after the end of the regular maceration period of one to two weeks. This delay is caused more by the time required for seed hydration indispensable for flavan-3-ol release than by the low alcohol levels at the initial stages of primary fermentation (Singleton and Draper 1964). Thus, maceration length is the main driver of seed flavan-3-ol and also tannin extraction under conditions of typical final alcohol concentrations (Casassa et al. 2013 a, 2013 b).

2.2.3. Tannins

Grape-derived tannins, also designated as proanthocyanidins or condensed tannins, comprise oligomeric (degree of polymerization[5] = 2 to 5) and polymeric (degree of polymerization > 5) flavan-3-ols formed from the five monomeric flavan-3-ols shown in figure 2. The oligomeric and polymeric tannins in grapes and wine are characterized by a mean degree of polymerization (mDP), which refers to the average number of constitutive flavan-3-ol subunits in the oligomer or polymer.

When polymerization occurs, flavan-3-ols can be connected between the carbons at positions 4 and 8, between the carbons at positions 4 and 6 with and without an ethyl bridge[6], respectively, and between the hydroxyl groups at C5 and C7 forming an ether bond[7]. Figure 3 gives examples. In all these possible connectivities, each flavan-3-ol subunit can display any of the isomeric forms of its corresponding monomer.

Figure 3: Flavan-3-ol subunits connected between positions C4→C8 in a trimer (A), and between C6→C8 including an ethyl bridge in a dimer (B),

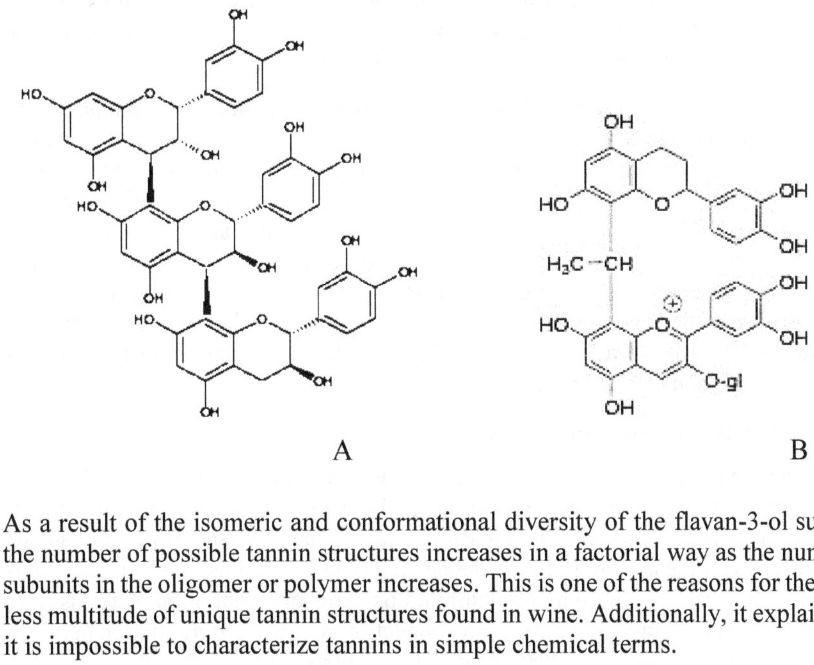

A B

As a result of the isomeric and conformational diversity of the flavan-3-ol subunits, the number of possible tannin structures increases in a factorial way as the number of subunits in the oligomer or polymer increases. This is one of the reasons for the countless multitude of unique tannin structures found in wine. Additionally, it explains why it is impossible to characterize tannins in simple chemical terms.

Furthermore, neither the structure nor the mDP of these tannins is constant over time. On the contrary, they undergo continuous changes. Under the mild acidic conditions of wine and under exclusion of oxygen, tannins are subject to acid-catalyzed C-C

[5] Polymerization is a process of reacting single monomeric molecules, called subunits, together in a chemical reaction to form polymeric chains or three-dimensional networks.

[6] An ethyl bridge is any part of a molecule with the formula =CH–CH_3 and connected to two other distinct atoms in the rest of the molecule.

[7] An ether bond is an oxygen atom that connects two molecules by a C–O–C linkage.

bond-breaking and bond-making reactions (Haslam 1980). This process impacts the sensory perception of wine astringency (Chapter 5.3.1). Which of the two types of reaction is predominant depends on pH, the ratio of polymeric to monomeric species, and the availability of SO_2 and acetaldehyde. Even when no changes in the mDP occur, structural changes can be observed (Dallas et al. 1995, Vidal et al. 2002). This shows that oxygen exclusion during storage cannot prevent the tannin of red wines from undergoing permanent changes (Somers and Evans 1986). However, when oxygen uptake is allowed to take place, these changes are being forced and driven in a different direction. Under these conditions, tannins can strongly polymerize and reach molecular weights as high as 5,000 and more than 15 subunits. This increase of the mDP is all the greater the higher the initial tannin concentration, leading to the formation of longer linear chains or branched polymers (Vernhet et al. 2014).

Clearly, under oxidative storage conditions, high-tannin red wines more easily generate high-molecular weight tannins than low-tannin wines. The reason is easy to understand: In thin wines, molecules have less chance to react one with another than in concentrated wines. Last, but not least, tannins other than only C-C bound ones can be formed, for instance ethyl-bridged tannins (Figure 3 B) and anthocyanin-tannin adducts leading to polymeric pigments (Chapter 2.2.5).

Tannin macromolecules resulting from excessive polymerization have practical enological implications: They become less soluble and can precipitate when their mDP exceeds a critical limit. Furthermore, they easily react with proteins, whether protein-like fining agents (Chapters 12.1.3) or salivary proteins (Chapter 5.2). This characteristic interaction differentiates them from other wine phenolics and causes the sensation of astringency. They can also react with grape-derived or yeast-derived polysaccharides and thus even pass into a colloidal state[8]. This reaction modulates their sensory properties. In particular, it can lower their astringency (Chapter 5.3.3). The role of colloids during filtration, and the potential sensory consequences resulting therefrom are discussed in chapter 12.2.

Differences between skin and seed tannins

The simultaneous designation of all these tannins as proanthocyanidins results from the fact that they are converted in colored anthocyanins such as cyanidin or delphinidin (Chapter 2.2.4) when heated in dilute mineral acid. Proanthocyanidins yielding cyanidin under these conditions are designated as procyanidins, whilst those yielding delphinidin are referred to as prodelphinidins. This differentiation of terms may seem excessive, but it gains importance in connection with the differentiation between skin and seed tannins and their specific sensory qualities.

Similarly to flavan-3-ols, the flavonoid-based proanthocyanidins occur in both the skins and the seeds, with the vast majority of them located in the seeds. Amongst them, procyanidins are found in both skin and seed tissue and prodelphinidins restricted to the skin tissue. Hence, skin-derived tannins are characterized by the presence of prodelphinidins. Additionally, they display a higher mDP, a lower percentage

[8] Colloids are nano-sized particles that are generated by a physical association of macromolecules. They have a diameter of 1 to 1,000 nm and a glue-like character.

of galloylation[9], and noticeably less astringency per concentration unit than seed tannins (Hanlin et al. 2011).

The stronger astringency of seed tannins as compared to skin tannins can have a far-reaching impact on quality. In general, the concentration of seed tannins decreases during grape ripening, while skin tannin concentration increases. Thus, technological seed maturity, which occurs when grape seeds are brown, can become a critical factor and limit general wine quality in bad harvests or cool climates. Tannin extraction techniques (Chapter 4) strive to favor skin tannin extraction and reduce seed tannin extraction.

In wine, oligomeric tannins are the major constituents of total tannins (Gonzáles-Manzano et al. 2006). During maceration, they follow an extraction kinetics similar to that of flavan-3-ols, with their concentration increasing with increasing maceration length. They can be extracted to a limited extent in the absence of alcohol as it occurs during pre-fermentation cold soak (Hernández-Jiménez et al. 2012). As these oligomers increase in size becoming polymers, their extraction into must progresses more slowly.

During real world macerations, total tannin extraction proceeds almost linearly over the first 10 days, then reaches a plateau lasting one to two weeks, and is followed by another linear increase up to the peak of extraction. While the plateau is consistent with the time required for seed hydration and swelling, the subsequent linear increase reflects the extraction of seed tannins. After one month of maceration, the amount of seed tannins is appreciably higher than that of skin tannins in most of the finished wines, with the exact ratio depending on the variety (Casassa et al. 2013 a, 2013 b). This behavior suggests that the maceration period should not exceed three weeks if one wishes to limit the extraction of astringent or aggressive seed tannins. More details are given in chapter 4.1.7.

2.2.4. Anthocyanins

Anthocyanins are the red pigments present in the vacuoles of the berry skin tissue of red grapes and also in the pulp juice of the so-called teinturier varieties. They represent the essential difference between red and white grapes as all other phenols occur in comparable amounts in grapes regardless of their color, though they are scarcely extracted in white winemaking. Pure anthocyanins are tasteless or indistinctly flavored, but they are able to lower astringency upon addition to wine as they easily interact with tannins (Chapter 5.3.2).

Depending on the substitution pattern of their B ring, anthocyanins in grapes occur as five glycosylated[10] monomers, namely the glucosides of malvidin, cyanidin, petunidin, peonidin, and delphinidin (Figure 4). The glucose moiety is bound at their C3 position and makes them water-soluble. Although the distribution of anthocyanins varies between cultivars, malvidin-3-glucoside is invariably the most important of

[9] Galloylation is the attachment of gallic acid.

[10] Glycosylation is the attachment of a sugar molecule, here glucose.

them, generally accounting for approximately half of the total anthocyanin content of grapes.

The glucose is further bound at its C6 position with an acid, primarily acetic, p-coumaric, or caffeic acid. This attachment of an acid is called acylation and improves the chemical stability of the anthocyanin molecule. Pinot noir is exceptional in having no acylated anthocyanins, which explains why its wines are usually paler in color than most other red wines.

Figure 4: Basic anthocyanin structure, substitution patterns of the main five anthocyanins associated with red color in wine, and their acyl groups.

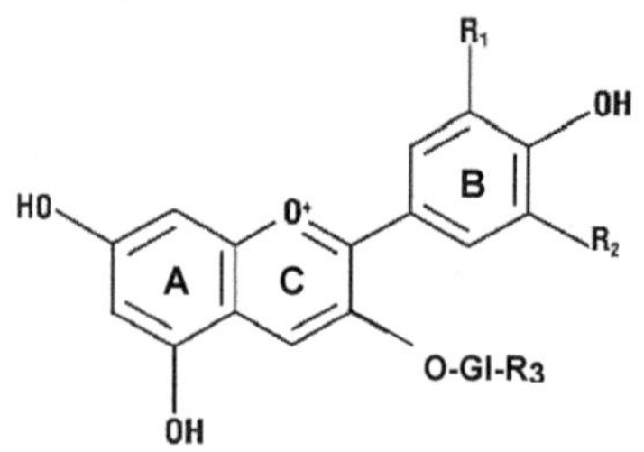

Anthocyanin	R_1	R_2
malvidin-3-glucoside	OCH_3	OCH_3
cyanidin-3-glucoside	OH	H
petunidin-3-glucoside	OCH_3	OH
peonidin-3-glucoside	OCH_3	H
delphinidin-3-glucoside	OH	OH

R_3 = acetic acid, p-coumaric acid, or caffeic acid
Gl = glucose

Anthocyanins are the only class of flavonoid phenols able to carry an electric charge, more precisely a positive charge on their C ring. However, at wine pH, only a limited percentage of the anthocyanins has that charge. The ion resulting from that behavior is called the flavylium cation, which is responsible for the chromatic properties of young red wines. It causes anthocyanins having a maximum absorbance between 475 and 540 nm with a mean of 520 nm, conferring upon them a red to blue color that also depends on their substitution pattern.

The pH value exerts a strong impact on the color provided by the native, monomeric anthocyanins, particularly in young red wines in which they are the most responsible for total color. Only at pH lower than 0.5, they occur exclusively in the colored red flavylium form. As pH increases, they are gradually and reversibly converted into other structures, namely a colorless carbinol pseudobase, a nearly colorless chalcone, and a blue quinoidal base. These four anthocyanin structures exist together in a dynamic equilibrium. In the wine pH range 3.0 to 4.0, the colorless carbinol pseudobase is the major form present. This means that if red wines had a substantially lower pH than they have, they would display several times more color intensity than they actually do.

The pH-dependent behavior of anthocyanins is displayed in figure 5. It can easily be observed when titratable acidity is measured using potentiometric titration. It also explains why color losses occur upon pH increase when young red wines are deacidified regardless of the means employed for that purpose.

Figure 5: Molecular structure, charge, and color of anthocyanins as affected by pH.

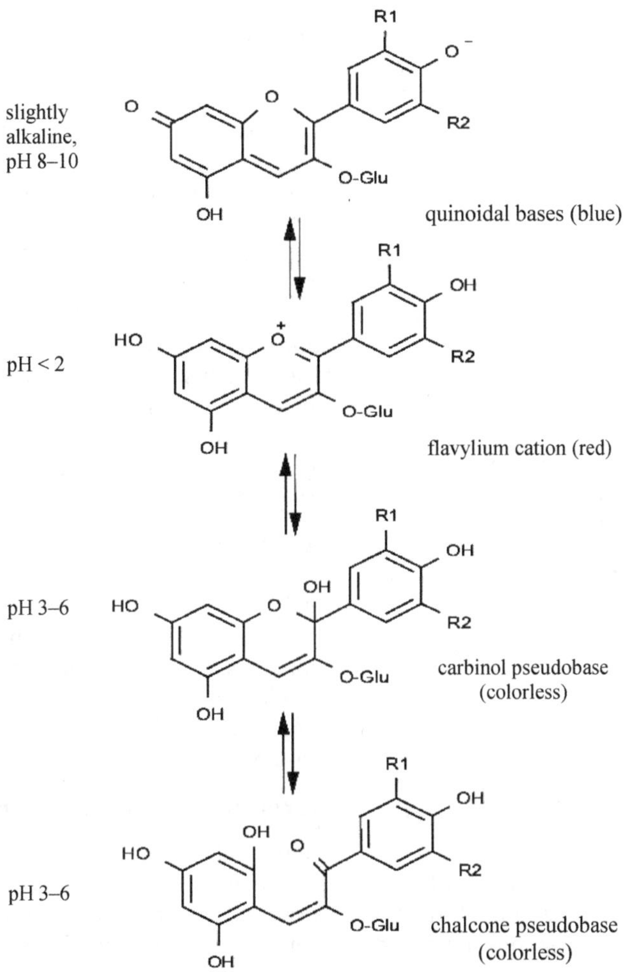

Anthocyanins also react reversibly with bisulfite (HSO_3^-) generated by SO_2 when this is added to must or wine, thus forming a colorless anthocyanin-4-sulfonate adduct shown in figure 6. This behavior explains the partial bleaching of young red wines upon any SO_2 addition. However, this bleaching effect is moderated by the presence of other SO_2-binding compounds. It largely disappears when anthocyanins undergo reactions with tannins to form polymeric pigments (Chapter 2.2.5) during wine aging.

During the skin contact period after crushing, anthocyanins are rapidly extracted from the skins by a diffusion process that is favored by their water-soluble nature, resulting in maximum extraction rates and a peak of extraction within the first 3 to 7 days of maceration at standard temperature under commercial winemaking conditions. However, their extraction is never complete and depends on a number of technical variables the winemaker decides to put in place during maceration. A variable remainder of anthocyanins inevitably remains in the pomace after pressing.

Figure 6: Formation of the anthocyanin-4-sulfonate adduct with bisulfite.

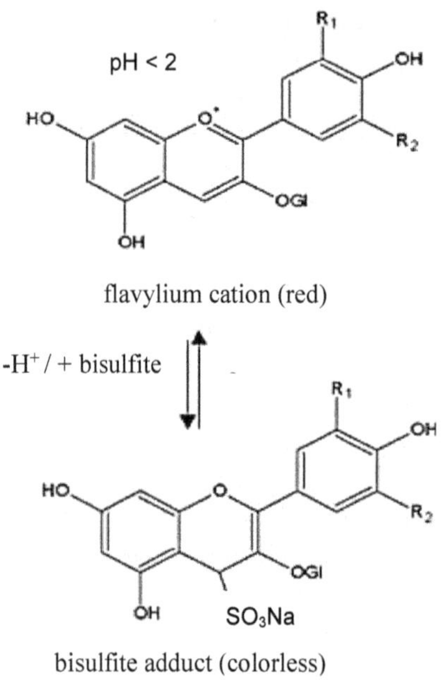

After anthocyanins reach a peak concentration in fermenting must, a variable drop in concentration is typically observed. These losses can be largely explained by adsorption onto the negatively charged yeast cell walls, adsorption onto bitartrate crystals or any other particulate matter, and oxidative cleavage of the C ring leading to irreversible color losses. More important, however, is the chemical incorporation of anthocyanins into polymeric pigments, which start being formed from the very early beginning of maceration onwards. Furthermore, direct anthocyanin degradation by oxidation may also occur during extended macerations.

The decay of anthocyanins is a continuous process with a half-life period of approximately one year. In general, after four years of aging, insignificant quantities of colored anthocyanins remain in the wine. Non-bleachable polymeric pigments (Chapter 2.2.5) play a much larger role in long-term color than anthocyanins.

The copigmentation phenomenon

Colored anthocyanins tend to undergo a reaction called copigmentation, which designates their interaction with colorless so-called cofactors leading to the formation of loose molecular associations with these cofactors. These associations are called copigments. They consist of weakly bound, sandwich-like stacked molecular structures of up to eight molecules each that place the sugar moieties of the anthocyanin towards the external part of the complex, thus protecting them from decoloration by water. Their stacked structure in three dimensions is still subject to speculations, but has been shown to cause two main effects, which are a strong increase of wine color intensity, and a shift of their color from red towards purple (Boulton 2001).

It is important to underline that copigments must not be confounded with polymerized anthocyanins or polymeric pigments generated by a polymerization process and displaying much stronger covalent[11] bonds (Chapter 2.2.5).

[11] Covalent bonds occur when two atoms combine by sharing electrons. They are strong and difficult to break. In contrast, a non-covalent bond is weaker and does not involve sharing atoms, but it is a

Copigmentation effects are reversible by the action of the solvent. Their structural stability depends on the highly ordered, lattice structure of water, and is decreased by any factor affecting that structure, such as an increase of temperature or alcohol content. This can be easily observed when a young red wine is alternately heated and cooled in a test tube over about a 50° C range. Thereby, color is instantly decreased by heating and restored by cooling. A similar effect appears when alcohol is removed from young red wine and replaced by an equivalent volume of water adjusted to wine pH: Alcohol removal significantly enhances color density. Alcohol is the fermentation factor which reduces hydrophobic effects, thus declining color density in must and young red wines (Somers 1998).

Comparable effects can also be observed under real world conditions: Thus, many red wines display less color during the first weeks after primary fermentation than at the halfway point of fermentation. Higher initial levels of copigments in cold-soaked or heated musts are also partially broken down by the alcohol produced during fermentation. However, within the typical alcohol range of wine, the impact of alcohol on the copigmentation phenomenon should not be overestimated. In the long term, other reactions involving anthocyanins are more important for changes of visible color.

The effect of alcohol on copigmentation, however, is important when comparing wine color to results obtained in aqueous media, and especially so for analytical methods for anthocyanins and copigmentation involving buffered solutions for sample dilution. While it is clear that wine dilution with water (pH 7) does not generate a linear decrease of color due to the pH impact on colored anthocyanins (Figure 5), the equilibrium nature of copigment formation also leads to a nonlinear relationship between color and anthocyanins' concentration upon dilution at constant pH and ethanol level. Therefore, measuring anthocyanin concentration by simple assessment of their color is not feasible as long as copigmentation effects are not cancelled out by appropriate dilution. It is not exactly known how many times red wines must be diluted to completely overcome copigmentation and transfer all anthocyanins into their free forms, but a dilution factor of approximately 50 appears reasonable for that purpose. This behavior has implications in anthocyanin concentration measurements, which are discussed in chapter 3.2.2.

Copigmentation only occurs with monomeric anthocyanins in the must and at early stages of a wine's age. It accounts for approximately half of the visible color in young red wines. It causes anthocyanins to exhibit more color than expected from their concentration. This color enhancement by copigmentation is 2 to 10 times that expected from the free anthocyanins alone (Boulton 2001).

Copigments are precursors of the more stable and covalently bound polymeric pigments (Chapter 2.2.5). As there is a progressive displacement of the anthocyanins by polymeric pigments, copigmentation effects diminish during storage and aging. However, as long as they exist, they are one of the main causes explaining why there is no good correlation between red color intensity and anthocyanin concentration in young red wines.

---

common way of joining macromolecules together, for example through hydrogen bonding and hydrophobic effects.

Copigmentation largely depends on the concentration of specific, uncolored phenols acting as cofactors. The most important of them are hydroxycinnamic acids (Chapter 2.3.1), flavonoids and, in particular, derivatives of their small flavonol and flavone subgroups. Wines from grapes low in cofactors are not able to form much copigmentation. This explains why poorly colored wines sometimes result from darkly colored berries of cultivars such as Pinot noir and Sangiovese despite a carefully conducted extraction process during maceration.

2.2.5. Polymeric pigments

Anthocyanins as extracted from the grapes are quite unstable in the absence of other phenolics, as can be shown by their fast fading in water at wine pH within a few days or weeks. In real wines, they interact with other phenols to initially produce copigments and later stable polymeric pigments.

Polymeric pigments are not originally present in the grapes, but start to be formed as soon as grapes are crushed, progressing during maceration, and throughout storage and aging with increasing contribution to wine color over time. They are a heterogeneous group of reaction products primarily resulting from reactions between anthocyanins, monomeric flavan-3-ols, and tannins, but aldehydes and acids can also participate in these reactions. Briefly stated, the anthocyanins extracted from the fruit are progressively and irreversibly displaced by more stable polymeric pigments. These pigments can account for as much as half of wine color intensity after the first year and approach 100 % in aged wines. They undergo continuous changes during wine aging.

Although anthocyanins can readily react with flavan-3-ols and tannins via direct condensation comparable to the pathways of tannin polymerization outlined in chapter 2.2.3, these reactions are often mediated by acetaldehyde or other aldehydes forming an alkyl[12] bridge between two flavonoid subunits. Aldehydes are formed either by yeast metabolism during alcoholic fermentation or from metal-catalyzed oxidation of alcohols, especially of ethanol to acetaldehyde, through the Fenton reaction (Chapter 7.2).

In addition to aldehydes, other compounds including phenolic acids (Chapter 2.3.1) or fermentation by-products such as pyruvic acid and diacetyl can also react with anthocyanins to form colored structures named pyroanthocyanins and characterized by a new pyran ring[13] at the C4 and C5 positions. All these reactions culminating in the formation of oligomeric and polymeric pigments account for the major part of flavor and color evolution during red wine aging; they can even be considered the driving

[12] An alkyl is a group of an organic molecule that contains only carbon and hydrogen atoms, which are arranged in a chain. The general formula of alkyls is C_nH_{2n-1}. The smallest alkyl is methyl (CH_3), followed by ethyl (CH_2-CH_3), propyl (CH_3-CH_2-CH_2), etc.

[13] A pyran ring is a six-membered heterocyclic ring consisting of five carbon atoms, one oxygen atom, and two double bonds, leading to the general formula C_5H_6O.

force of the maturation process. Possible reaction pathways and products are summarized in Casassa 2017.

A large range of factors such as wine storage conditions, wine matrix composition, and enological interventions affect the rate and course of this continuous shift from monomeric and copigmented anthocyanins towards polymeric pigments. However, this process cannot be considered as a one-way reaction because bond-making mechanisms are accompanied by bond-breaking ones. N, bond-making reactions appear to dominate at least during the first year of aging when molecular weights and degree of polymerization slightly increase.

All these reactions contribute to the dynamic nature and heterogeneity of polymeric pigments and tannins during wine aging. There is still no substantial information about their formation, structure, and the molecular aggregates they produce in a maturing wine. At best, there is evidence that the reactions involved are essentially catalyzed by low pH and temperature, but passive and active oxygen uptake as well as SO_2 status have a tremendous impact on their final outcome. The balance between oxygen supply and the levels of reducing agents such as SO_2 is the most critical part in red wine aging and one of the key issues of this book.

In comparison with intact anthocyanins, polymeric pigments display some fundamental differences:

- They are largely resistant to bleaching by SO_2.
- They are more resilient to pH changes.
- They have a lower color intensity, expressed as the molar absorbance coefficient[14], than the intact anthocyanins they stem from.

In the very long term, the lower color intensity of polymeric pigments causes bottled red wines to become less intense in color as polymerization progresses during extended bottle aging (McRae et al. 2012). This behavior does not conflict with the color increase observed when polymerization progresses in young red wines. Upon the usual SO_2 additions, the color of such wines is much less bleached than the color of young wines derived primarily from monomeric anthocyanins. In other words, the more anthocyanins have already been converted into polymeric pigments, the less color losses occur upon SO_2 additions. In real world winemaking, the resistance of polymeric pigments to SO_2 bleaching is more decisive for final wine color than their lower color intensity.

The formation of polymeric pigments is believed to increase the solubility of tannins as the glucose moiety of the anthocyanin and the electric charge of the flavylium cation decrease the precipitability of the resulting polymeric pigment. In white wines with added tannins, the tannin solubility was shown to increase in the presence of simultaneously added anthocyanins (Singleton and Trousdale 1992).

---

[14] The molar absorbance coefficient is a measure of how strongly a chemical substance at a concentration of 1 mol/L absorbs light (or exhibits color) at a particular wavelength. Values refer to 10 mm path length.

Polymeric pigments generally show lower astringency than colorless, intact tannins with a comparable molecular weight and degree of polymerization. This is explained by the incorporation of the anthocyanin moieties with their glycoside portion that reduces the interaction of the polymer with salivary proteins (Chapter 5). Further modification of the molecular structure by introducing an acetaldehyde bridge lowered astringency even more but also increased bitterness (Vidal et al. 2004 b).

2.3. Nonflavonoid phenols

2.3.1. Phenolic acids

Compared to the flavonoid phenols in red wine, the amount of grape-derived nonflavonoid phenols is much smaller and limited to the phenolic acids, also known as 'acid phenols'. They have only one benzene ring and comprise essentially colorless derivatives of benzoic and hydroxycinnamic acid, also referred to as benzoates and hydroxycinnamates. As shown in figure 7, they differ in the degree of hydroxylation and methoxylation of their benzene ring.

Phenolic acids occur already dissolved in the pulp juice and are found in both red and white wines at concentrations of 100 – 200 mg/L without any extraction being necessary for their transfer into the liquid phase of must. Therefore, the vinification method does not exert a major impact on their final concentrations. At these concentrations, they do not contribute to sourness, bitterness, astringency or any other mouthfeel attribute of red wines to a significant extent. Hence, they are not considered as being a part of what is commonly called the red wine tannin.

Figure 7: Molecular structures of benzoic and cinnamic acid derivatives in wine.

Structure	R_1	R_2	R_3	Compound
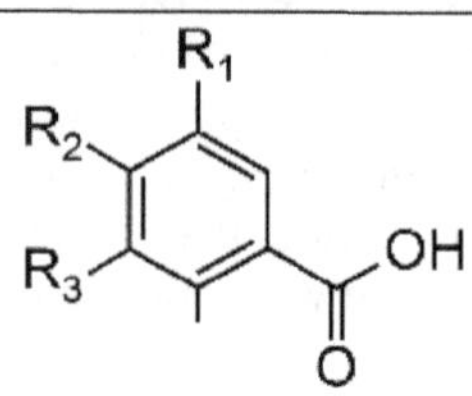 Benzoic acid derivatives	-H	-OH	-H	p-hydroxybenzoic acid
	-OH	-OH	-H	protocatechuic acid
	$-OCH_3$	-OH	-H	vanillic acid
	$-OCH_3$	-OH	$-OCH_3$	syringic acid
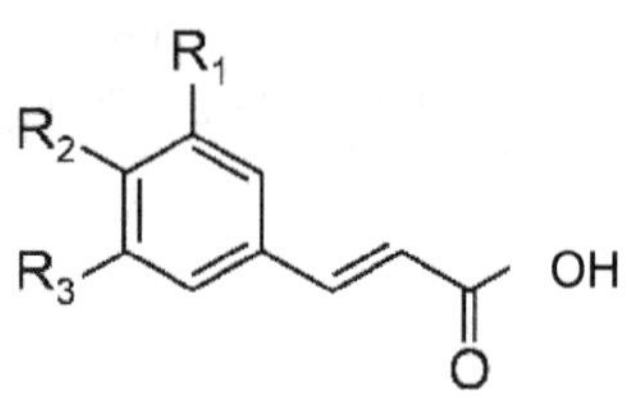 Cinnamic acid derivatives	-H	-OH	-H	p-coumaric acid
	-OH	-OH	-H	caffeic acid
	$-OCH_3$	-OH	-H	ferulic acid
	$-OCH_3$	-OH	$-OCH_3$	sinapic acid

2.3.2. Hydrolyzable tannins

Hydrolyzable tannins are extracted from diverse wood (oak, chestnut, etc.) and fruit (gall nuts) species but do not occur in grapes or wines by nature. They can easily increase natural nonflavonoid levels by some 50 to 200 mg/L when wine is aged in new barrels, supplied with oak chips or any alternative wood products, or enological tannins added as legal flavoring additives. Oak wood in particular provides more hydrolyzable tannins than any other wood, in particular ellagitannins. Although they belong to the family of nonflavonoid phenols, they display mouthfeel properties comparable to those of grape-derived tannins based on flavonoid subunits (Chapter 2.2.3), but they are structurally quite different from them.

Figure 8: Molecular structures of gallic acid and ellagic acid.

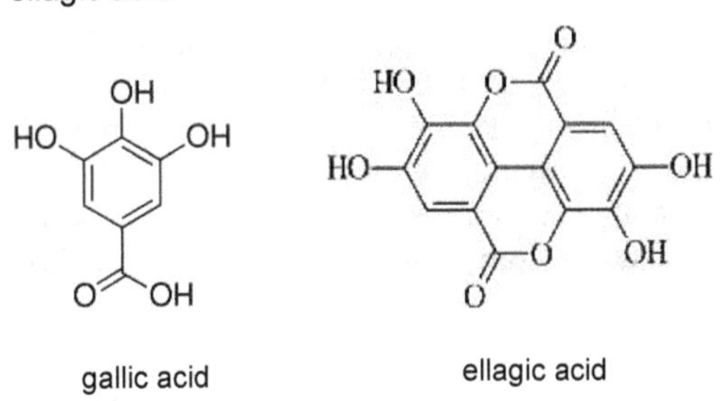

Hydrolyzable tannins consist of complex polymers of ellagic acid and gallic acid attached to a sugar molecule, usually glucose. Molecular structures of gallic acid and ellagic acid are shown in figure 8. Ellagic acid consists of gallic acid groups bonded to dimers.

Depending on whether hydrolyzable tannins are made up of gallic acid or of ellagic acid, one distinguishes between gallotannins and ellagitannins. However, in everyday enological language, both are frequently referred to as ellagitannins in a simplifying way. Therefore, this term is also used sometimes in the framework of this book.

The real ellagitannins differ from gallotannins in that their galloyl groups are linked through C-C bonds, whereas the galloyl groups in gallotannins are linked by depside bonds. Figure 9 depicts the molecular structures of the two most important examples of true ellagitannins, vescalagin and castalagin. Its only purpose is to show how different these molecules are in comparison with the flavonoid phenols depicted in figures 2 and 3. As this book is primarily aimed at winemakers and enologists, it is not necessary to go into more chemical details of these molecules. Their technical properties are more important.

Figure 9: Molecular structures of two ellagitannins: vescalagin and castalagin.

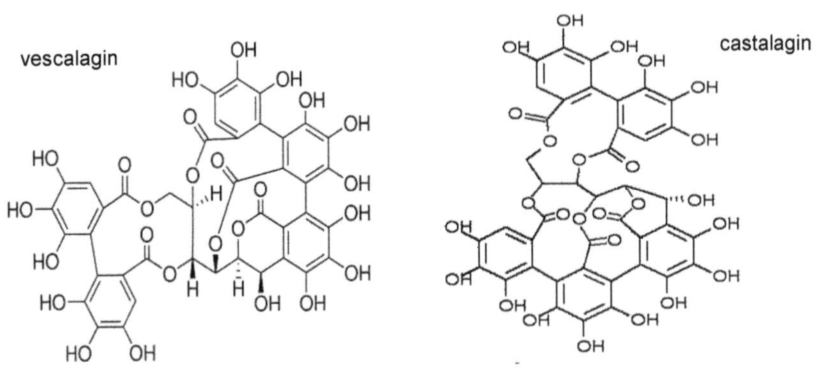

Hydrolyzable tannins display the following characteristics:

- They are odorless.
- They disintegrate in ellagic or gallic acid by acidic hydrolysis, which explains their denomination as hydrolyzable.
- They can also break down by oxidation.
- By reference to their molecular weight, they display more phenolic hydroxyl groups on their molecular skeleton than flavonoid tannins. As a result, they react more intensively with oxygen than flavonoid phenols and also tend to be more astringent and bitter, based on the same concentration by weight.
- Disintegration by acidic hydrolysis or oxidation yields gallic acid and glucose, or ellagic acid and glucose. It results in a decrease of their astringency.
- Due to their chemical heterogeneity and low concentrations, they are difficult to detect and even more difficult to measure in wine than flavonoid tannins. In particular, the currently employed total phenol or tannin measurements in red wine (Chapters 3.1 and 3.4) do not give any information about the presence of ellagitannins. However, it is possible to characterize and differentiate them in chemical terms and, thus, to identify their botanical origin (Chauvet et al. 1992, Versari et al. 2013).

Hydrolyzable tannins are widely available as commercial tannins used as additives in the wine industry since they are less expensive than grape-derived flavonoid tannins. However, commercial tannins are diverse. They can be ellagitannins, gallotannins, grape-derived flavonoid tannins obtained from winemaking residues, or a mix of several of them. They display variable degrees of purity and variable average molecule weight. According to legal requirements, they must be odorless and not impact wine aroma perceived by smell, albeit these requirements have been relaxed in more recent times.

3. Measurement of total phenols, anthocyanins, and tannin

In contrast to standard wine analyses including measurements of alcohol, pH, total acidity, residual sugar, and SO_2, analyses of any phenolics are rarely performed, although most winemakers are aware of the paramount importance of tannins and related compounds to the mouthfeel and overall quality of red wines. The reason for this is the extraordinary chemical complexity of these compounds, the difficulty in interpreting their concentration data, and the technical requirements of many of the available analytical procedures. This chapter describes some basic methods appropriate to any winery lab equipped with a spectrophotometer. They include the measurement of total phenols, anthocyanins, and true tannins. These measurements are useful to determining the moment of devatting after maceration, assessing the phenolic strength of the wines, adjusting their oxygen supply, and estimating astringency and other mouthfeel properties elicited by tannins. Additionally, reference data are given to facilitate the interpretation of the results.

3.1. Measurement of total phenols

3.1.1. The significance of total phenols in red wines

The particular techniques of red winemaking aim at extracting anthocyanins and colorless phenols from the solid parts of the grape cluster and accumulating them in the liquid phase. Together, both groups of substances constitute the total phenols of red wines and are the starting material of their future tannin or, more precisely, their tannin-anthocyanin complex. They are the key to understanding the behavior of red wines in chemical and sensory terms and differentiate them from white wines.

The nature and extent of phenolic extraction depends on the grape's size and composition, and the time and conditions allowed for extraction. Thus, the making of red wines may involve limited extraction, fermentation to dryness on the skins, or prolonged time on skins even after fermentation has ceased. As a consequence, there is no other category of organic composition of red wines that shows anything even approaching such a high order of variability as do phenols.

As to their quantities, total phenols range from as low as 100 mg/L in white wines – these being predominantly nonflavonoid phenols – to more than 4,000 mg/L in red wines. There are also large qualitative differences between these two major sets, as the extractives in red wines consist mainly of reactive flavonoid structures. Figure 10 schematically illustrates how the amounts of total phenols differ between white, rosé, and red wines, and how this quantitative variability is essentially due to varying amounts of flavonoids, while nonflavonoids do not vary widely and can be assumed constant regardless of the kind of wine.

Figure 10: Total phenol levels in white, rosé, and red wines and their composition of nonflavonoid and flavonoid phenols.

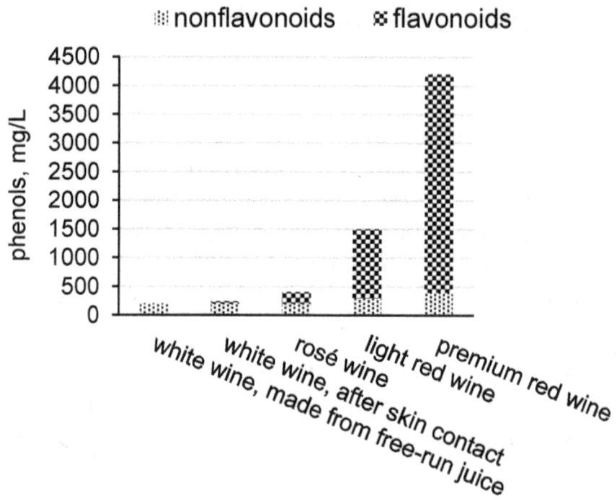

Although most winemakers are well aware of the central role that tannins and related compounds play in mouthfeel, color, and general quality of red wines, they generally do not measure them and have no objective specifications for them. One of the reasons for this is the extraordinary complexity of all these compounds and the lack of experience in interpreting their concentration data. Whilst common analytical parameters such as alcohol, pH, titratable acidity, or residual sugars are clearly defined and easily accessible in routine quality control, tannins and similar compounds are a much more heterogeneous class of substances. Their analytical determination requires thorough understanding of what exactly is measured and how the data obtained are to be interpreted.

As all molecules involved in the tannin-anthocyanin complex are phenols of plant origin, assessing their amounts requires the use of methods of phenol analysis. In contrast to some decades ago, there is now a veritable surfeit of analytical procedures targeting these compounds (Chapter 3.4). Thus, the difficulty lies in determining which of them are relevant and which are much less so, and in evaluating the relevant information so that something useful might result.

It is unquestionable that most of these analytical methods are out of reach of current quality control in most industrial settings, and it is even doubtful whether most of them are of any practical relevance for decision-making in the winemaking process. Within the scope of this book, only certain basic methods are presented as suitable to be run in a commercial winery, where a spectrophotometer is probably the most exotic equipment. They have proven to be useful in current process and quality control and can be implemented in any lab equipped with such a spectrophotometer.

The first step of this approach leads to the determination of total phenols. It is the most elementary parameter in red winemaking. It represents the sum of all single phenolics regardless of their origin in grapes or oak and without differentiating between monomeric flavan-3-ols, tannins, anthocyanins, or polymeric pigments.

One might argue that this kind of measurement greatly simplifies a complex matter as it also includes the fairly meaningless fraction of phenolic acids (Chapter 2.3.1) instead of being specific to the flavonoid extractives. Indeed, it lacks specificity and precision for quantifying the low amounts of flavonoids in white wine. On the other hand, it provides a useful indication of flavonoid levels in red wines, where these molecules represent the overwhelming majority of total phenols, as shown in figure 10. This is to say that in red wines, total phenols correlate quite well with the sum of all flavonoid phenols.

For the aforementioned reasons, measuring total phenols gives a first insight into what is considered red wine's character, its phenolic strength, and its very essence. It is the most critical parameter of red wine composition in wines of any age, variety, and origin. Despite its shortcomings, it is significantly better than no phenol measurement at all. Furthermore, it is easy to execute when some pitfalls are avoided. Therefore, it is presented in detail.

At a first glance, analytical details might appear misplaced in a book addressed to winemakers and practical enologists, but it is crucial to understand how the data the measurement provides are generated and what they mean. There are two reasons for that: Firstly, these data are used throughout this book as key metrics of red wine making collected under real-world conditions. Secondly, the analytical details behind them provide a sound basis for reproducing them in one's own operation.

Strictly speaking, there are even two ways to measure total phenols. They use different approaches but show similar advantages, precision, and drawbacks.

3.1.2. The Folin-Ciocalteu assay for measuring total phenols

The Folin-Ciocalteu (FC) assay is a colorimetric method using the eponymous reagent. It is the most widely performed assay for measuring total phenols. It is based on a chemical reduction of the reagent, which is a mixture of tungsten and molybdenum oxides. In turn, phenols undergo oxidation, which is facilitated in alkaline medium adjusted by addition of sodium carbonate. The products of the metal oxide reduction have a blue color that exhibits a broad light absorption with a maximum around 765 nm (Singleton and Rossi 1965). The intensity of light absorption measured by spectrophotometry is proportional to the concentration of phenols, which are usually expressed as mg/L gallic acid equivalents. OIV reglementations differ as they stipulate to report only the absorption intensity without converting it into concentration units. This is the general way of reporting results in France.

This method was first described by Singleton and Rossi (1965) and has been largely investigated since then (Montreau 1972, Moutounet 1981, Blouin et al. 1972, Slinkard and Singleton 1977, Singleton et al. 1999, Somers and Ziemelis 1980, Magalhães et al. 2010, Danilewicz 2015, Chen et al. 2015, Pereira et al. 2018). There are many variations of it, which make the construction of a calibration curve for the measurement conditions and the particular spectrophotometer used indispensable. Considering its importance, practical details on its implementation and background information are useful. The following implementation protocols have proven useful for dry red wines:

Original method

Figure 11: Calibration curve with gallic acid for total phenols with the Folin-Ciocalteu method at 765 nm over two hours.

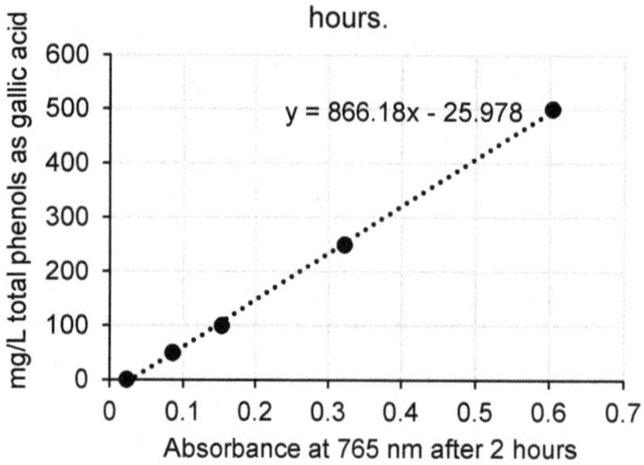

A) Materials:

- Gallic acid
- Folin-Ciocalteu (FC) reagent, stored in the dark and discarded if it becomes visibly green[15]
- Sodium carbonate solution 20 % w/v[16]
- 100 mL volumetric flasks
- Spectrophotometer set to 765 nm, with 10-mm standard plastic or glass cuvettes.

B) Construction of the calibration curve:

- Dissolve 500 mg gallic acid in 10 mL ethanol and then dilute to 100 mL with water. The final gallic acid concentration will be 1,000 mg/L. Dilute this mother solution with water to create standards with 50, 100, 250 and 500 mg/L concentrations, respectively. Store up to two weeks at 5° C.
- Of each standard, place 1 mL in a 100 mL volumetric flask.
- Add ~70 mL water, followed by 5 mL of FC reagent. Swirl to mix and incubate 1 to 8 minutes at ambient temperature.
- Add 15 mL sodium carbonate solution 20 %.
- Add water to make up 100 mL, mix, and incubate 1 hour at room temperature. Set timer!
- Transfer ~2 mL into a 10-mm plastic or glass cuvette and measure the absorbance at 750 nm in a spectrophotometer with respect to a blank prepared with distilled water in place of standard solution.
- Report absorbance data against concentration data in an EXCEL sheet and construct the calibration curve. It will look more or less like the curve given in Figure 11, depending on the spectrophotometer used and reagent purity. The corresponding curve equation allows easy conversion of absorption data into total phenol concentration data. For that purpose, take y as concentration and x as the absorbance reading. Last but not least, bear in mind that the precision of your data depends on the accuracy of your calibration curve!

[15] The FC reagent is cumbersome to prepare correctly but can be purchased from lab suppliers.

[16] Dissolve 200 gram sodium carbonate (Na_2CO_3) in ca. 800 mL warm (40-50° C) warm water, mix periodically, and complete the volume to 1,000 mL after cooling down to room temperature.

- Care should be taken to run the calibration standards at the same temperature (± 2° C) as used for the measurements on wine.

C) Sample preparation:

- Clarify the sample if necessary. For clarification, centrifugation is preferred to filtration as the latter might cause some absorption of phenolic material by the filter medium.
- Prepare a 10 % (v/v) dilution of the red wine sample in water.

D) Measurements:

- Place 1 mL of the diluted sample in a 100 mL volumetric flask or graduated cylinder.
- Add ~70 mL water, followed by 5 mL of FC reagent. Swirl to mix and incubate 1 to 8 minutes at ambient temperature.
- Add 15 mL sodium carbonate solution 20 %.
- Add water to make up 100 mL, mix, and incubate 1 hour at room temperature. Set timer![17]
- Transfer ~2 mL into a 10-mm plastic or glass cuvette and measure the absorbance at 750 nm in a spectrophotometer against water as a blank.[18]
- Use the calibration curve to transform the absorbance readings into concentration units. Be sure to multiply by the dilution factor of 10 (or any other factor if used). Report values in gallic acid equivalents (GAE) using units of mg/L.

Modifications of the original method

With the increasing availability of modern liquid-dispensing equipment, the original method has been modified in several ways in order to better satisfy particular sample matrix requirements, to simplify it, or to make it less time-consuming. To some degree, such modifications are admissible without loss of precision as long as the calibration standards are run under the same conditions as the samples. They include:

a) Downscaling of sample and reagent volumes

The original method is a macroscale one, which produces a lot of hazardous waste. This can be overcome by downscaling reagent and samples volumes to an extent as to perform the reaction in standard 10-mm cuvettes and even to automatize it (Slinkard and Singleton 1977). For the manual procedure usually performed under winery lab

[17] It is of crucial importance to maintain the order of the additions to ensure that the reaction takes place under defined alkaline conditions.

[18] There are some recommendations to use a sample blank without FC reagent to take account of the background interference caused by the wine. As this interference increases readings by only 2 % on average and is, hence, within the relative standard deviation of the method, this step is usually neglected.

conditions, the downscaling by a factor of 10 has been proven feasible. Thus, the sample and reagent volumes of the original method become

- 0.1 mL diluted sample (or calibration standard)
- 7.9 mL water
- 0.5 mL FC reagent
- 1.5 mL sodium carbonate 20 %

= 10.0 mL total volume.

This volume reduction allows to perform the reaction in standard test tubes before transferring the liquid into cuvettes for absorbance measurements. Single-use test tubes should be used and discarded after use since their cleaning for subsequent measurements is often unsatisfactory. Good precision of results requires scrupulously clean glass material.

b) Shortening color development time

Absorbance readings can also be taken after a shorter color development, which allows for less sample dilution or lower sample volume. This procedure is admissible, but it requires some precautions with regard to the timing of the absorbance readings.

The FC method relies on reaction kinetics and not on stoichiometric conversion. Therefore, it exhibits an asymptotic behavior never reaching a defined end point (Figure 12). It is slow and not complete when the reading is taken. Furthermore, the colored product is unstable. Thus, the precise moment of the colorimetric reading is most critical in this measurement.

Figure 12: Folin-Ciocalteu kinetics: Time course of the absorbance at 765 nm (A 765) in two red wines.

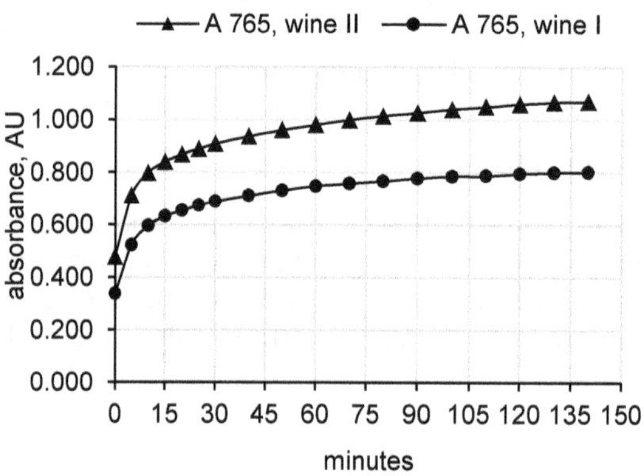

Temperature affects the reaction rate and the degradation of the colored component produced. At higher temperatures, the color development can be accelerated, but the timing of each reading becomes even more critical because the degradation also becomes faster. Hence, analyses that run slightly too long or too short will yield poor results (Singleton et al. 1999) under conditions of manual implementation of the method as those found in almost all winery laboratories. The binomial time/temperature is of crucial importance. As long as the method is not automatized (Slinkard and Singleton 1977), heat should not be used to accelerate the reaction. When FC analyses are run at room temperatures between 18 and 22° C, the temperature effect is negligible.

When fast results are required to make decisions in the course of the winemaking process, there is indeed some interest in shortening the original color development time of two hours at room temperature. This is possible provided that the calibration curve is constructed under comparable conditions. Indeed, OIV specifications stipulate a reaction time of 30 minutes. However, timing of the reading becomes more critical.

One possibility of shortening color development time is to reduce it to 15 minutes. As a consequence, absorption readings will be lower as they would be after two hours. Hence, sample dilution becomes superfluous, and sample and reagent volumes can be as follows:

- 0.05 mL undiluted sample (or calibration standard)
- 7.95 mL water
- 0.50 mL FC reagent
- 1.50 mL sodium carbonate 20 %
- 10.00 mL total volume.

As can be seen in figure 12, color development is the fastest during the first 30 minutes and flattens thereafter. The shorter the reaction time and the earlier the absorption reading takes place, the greater the error caused by imprecise timing. Hence, precise timing of the absorbance reading becomes more critical. Furthermore, there are limits to the number of samples than can be handled at once because of the need to exactly time absorbance readings after sodium carbonate addition and mixing. More than 10 samples should not be run within one set.

When absorbance readings exceed 1.0 AU, the photometric precision decreases, and the measurement should be repeated with less sample volume or after appropriate sample dilution. It is not admissible to dilute the mix of sample plus reagents with water before the measurements takes place.

c) Measuring at different wavelengths

Many method protocols published in the technical literature for the FC assay recommend the measurement at wavelengths other than 765 nm, for example at 760, 750, or even 720 nm. Figure 13 shows that the color produced indeed exhibits rather a very broad maximum absorption between 700 and 800 nm than at a clearly defined wavelength. Hence, it is admissible to run measurements at any wavelength within that range without affecting the accuracy of the results, provided that the calibration curve is constructed at the wavelengths chosen. On the

Figure 13: Folin-Ciocalteu kinetics: Impact of wavelength on absorption in a red wine after two hours at 20° C.

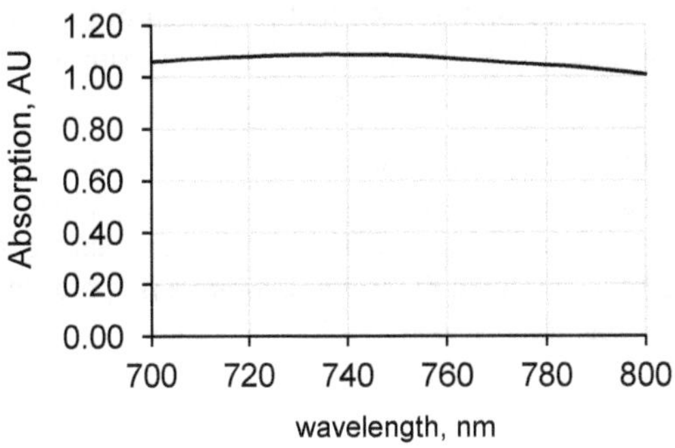

other hand, there is no real advantage in doing so. OIV rules stipulate measurements at 750 nm.

d) Using calibration standards other than gallic acid

Since the FC assay for total phenols reveals the presence of many different compounds in one result, the only practical standard is a single compound. Gallic acid has been established and widely accepted as the standard for wine. There is no reason to change that because the use of a single standard makes it easier to compare data. Furthermore, gallic acid has the practical advantage of being relatively inexpensive in pure form and stable in dry form.

In principle any other phenol could also be used as a standard. Thus, catechin is sometimes used because it facilitates comparing total phenol data with data obtained by other methods measuring specifically catechins or flavan-3-ols (Chapter 3.4) and requiring catechin as a standard. Thus, difference values can be calculated and balance sheets prepared.

Figure 14: Folin-Ciocalteu kinetics: Color yield (absorption units) of the FC reagent with catechin as compared to gallic acid at 765 nm after two hours.

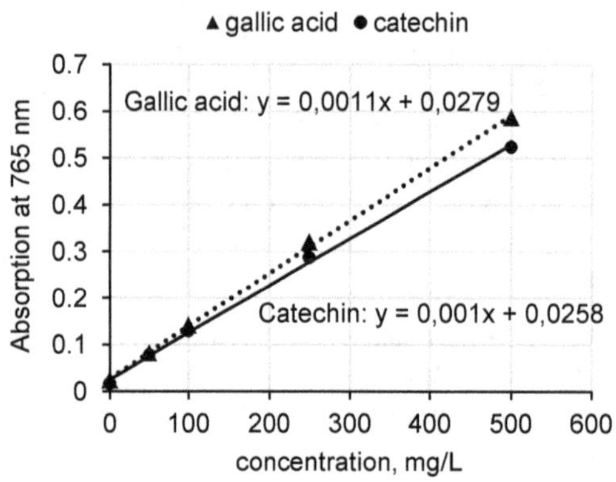

Fortunately, the FC procedure has a fairly equivalent response to the different phenolic compounds in wine. The mass response factor of most of them relative to gallic acid ranges from 0.9 to 1.1 (Singleton and Rossi 1965). This means that under comparable conditions and for a comparable concentration of for example 100 mg/L, they produce between 90 and 110 % of the color yield and absorbance readings of gallic acid. Figure 14 shows that difference for catechin. It yields 10 % less color than gallic acid. Hence, total phenol concentrations expressed as gallic acid equivalents are 10 % lower than those expressed as catechin equivalents.

e) Interferences

The FC assay suffers from several interferences, which fortunately do not apply when it is performed on dry red wines. As it is based on the chemical reduction of the reagent, oxidizable compounds other than phenolics increase results. Interfering compounds in wine can be sulfur dioxide, ascorbic acid, and residual sugar.

Free and bound sulfur dioxide only cause a significant interference in white wines, where they occur at relatively high levels in the presence of low levels of total phenols (Figure 10). This effect makes the method unusable in white wines (Somers and Ziemelis 1980) when sulfur dioxide is not eliminated by previous sample preparation. It is negligible in red wines with total phenol levels typically more than ten times higher.

The interference of ascorbic acid is comparable with that of sulfur dioxide, but it does not apply to red wines, which are usually not supplied with ascorbic acid.

Sugars like fructose and glucose exert a synergistic effect in that the absorbance of a mixture of phenols plus sugar exceeds the sum of the absorbance of their individual solutions. Moreover, it increases the interference by sulfur dioxide (Moutounet 1981, Magalhães et al. 2010). In dry red wines and those with less than 20 g/L sugars, this interference is negligible, but it makes the FC assay unsuitable for sweet wines and musts if no standards with the same sugar content are run. Correction factors for fructose and glucose have been proposed, but they lack reliability because of the chemical complexity of that kind of interference.

3.1.3. Spectral measurement of the UV total phenol index

All phenols absorb UV light around 280 nm, which is the wavelength where they show maximum light absorption. This property can be used to assess total phenols by absorption measurement at this wavelength. Unlike visible color, the UV absorbance is not affected by pH, and the Beer-Lambert[19] laws regarding concentration and light path length are strictly obeyed. Because of the intensity of the absorbance, it is necessary to dilute the limpid samples to permit measurement in standard 10-mm cuvettes.

However, the various phenolic compounds exhibit different absorptivity. Therefore, the results cannot be related to any specific standard but are reported directly in absorbance units (AU). The UV total phenol index, A 280, is the absorbance value for undiluted wine in 10-mm path length at 280 nm. It is obtained by multiplying the actual reading by the dilution factor.

The advantage of this method is its extreme simplicity and speed, whilst its basic drawback is the need of a spectrophotometer equipped with a UV lamp and quartz optical cells that most winery labs do not have at their disposal.

A) Materials

- Spectrophotometer, set to 280 nm
- Quartz or polycarbonate cuvettes with 10 mm path length, transparent at 280 nm[20]
- Volumetric flasks of 100 mL
- Pipette of 1 mL

B) Measurement

- Clarify the sample if necessary. For clarification, centrifugation is preferred to filtration as the latter might cause some absorption of phenolics by the filter medium.

[19] The Beer-Lambert law relates the attenuation of light to the properties of the material through which the light travels. When it applies, there is a linear relationship between absorbance and the concentration of the absorbing material in the sample measured.

[20] Current silicate and plastic cuvettes are not suitable for this purpose as they absorb in the UV spectrum.

- Prepare a 1 % (v/v) dilution of the sample in water and mix.
- Transfer sample into the cuvette and read its absorbance at 280 nm against water as a blank.
- Correct the absorbance to the original concentration, multiplying it by 100. Subtract 4 AU to get the final result

Interferences

Wines contain nonphenolic substances contributing with ~4 absorption units on average to the readings. This nonphenolic background should be subtracted to obtain the true absorbance due to phenols (Somers and Ziemelis 1985). However, many if not most laboratories do not respect this requirement nor specify in their result reports whether they do so or not. In case of doubt, one should ask.

Besides that, this method is only subject to interference by sorbate, which has a strong absorbance at 280 nm resulting in exceptionally high readings. Since there is no possibility to reliably eliminate this interference by sample treatment or correct for it, wines containing sorbate cannot be assessed by this method. In practice, absorbance readings that exceed the expected ones by 30 to 40 AU are a clue to the presence of sorbate. However, sorbate is only added to wines with residual sugar and can easily be anticipated in known wines.

No measurement modifications

There are no further method alterations. Figure 15 displays the absorption spectrum of three red wines in the UV range. Depending on the individual wine, the absorption maximum lies between 277 and 283 nm with 280 nm as an acceptable average. The use of another wavelength 1 or 2 nm distant from 280 nm does not greatly alter results. This helps to make minor wavelength differences between spectrophotometers meaningless.

Figure 15: Impact of the wavelength on the absorption of three red wines in the UV area.

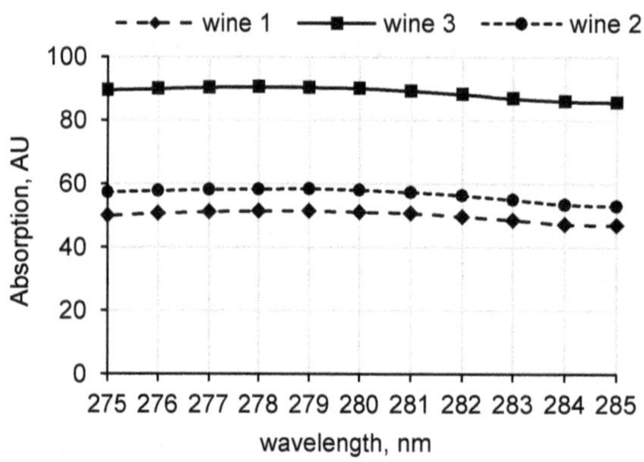

Correlation with total phenols measured by the FC method

Measuring total phenols as their absorption at 280 nm is quite a bit simpler and faster than the FC method. However, many people in the wine industry are used to work with concentration units, whilst the measurement of the absorbance at 280 nm just provides a dimensionless result as long as absorption units are not considered a measurement unit. Thus, there is some interest in understanding to what extent A 280 is related to total phenol concentrations as measured by the FC assay.

Figure 16: Correlation between total phenols measured as gallic acid equivalents (GAE) using the FC method and the UV index at A 280 nm.
UV index corrrected by 4 AU contribution of nonphenolic substances.

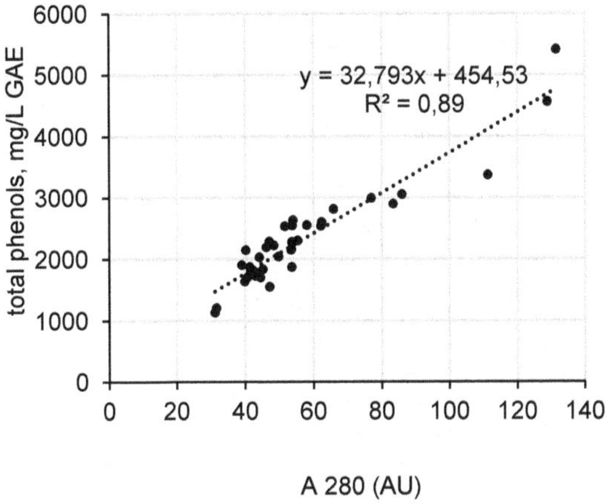

Figure 16 shows the correlation between these two parameters based on a set of 37 randomly chosen red wines samples from all continents. With $R^2 = 0.89$, it is a satisfactory correlation. It means that A 280 explains 89 % of total phenols as measured with the FC method in this range of wines. Similar measurements on more uniform samples are able to yield better correlations with R^2 around 0.95. This strong correlation makes the total phenol content measured by these two methods comparable (Aleixandre-Tudo et al. 2017).

Prospects of further development

Recently, another UV absorption maximum at 230 nm has been identified and shown to better correlate with astringency assessed by tasting. This wavelength was selected because of its location within the short spectral region for which tannin has at the same time high and close together lying specific absorbance values when compared to those of other phenolics. The absorbance of wines at 230 nm is mainly due to flavan-3-ols because of their higher concentration and specific absorbance value, though other phenolics also contribute (Boulet et al. 2017). However, after 50 years of use of the 280 nm measurement, the 230 nm measurement has not yet found practical application in the wine industry.

3.1.4. Interpretation of results

In conclusion, measurement of total phenols as the UV index at 280 nm can be a good substitute for the more tedious FC assay in red wines. In average red wines, one absorption unit at 280 nm equals approximately 40 mg/L GAE of total phenols measured with the FC assay and expressed as gallic acid equivalents. However, there is no exactly defined conversion factor as it would to some degree depend on the set of wines used to calculate it. At best, it allows for a rough ranking of red wines according to their phenolic strength and its impact on general quality. This ranking is given in table 1.

Table 1: Ranking of red wines according to their phenolic strength.		
Total phenols measured as mg/L GAE (FC assay)	Total phenols measured as A 280	Ranking
< 1,500 mg/L	< 30 AU	Ordinary low-phenol red wines of dubious quality and not suitable for barrel aging.
1,500 – 2,000 mg/L	30 – 45 AU	Low-phenol red wines of medium quality requiring care with barrel aging.
2,000 – 3,000 mg/L	45 – 80 AU	Common red wines suitable for or even requiring barrel aging.
3,000 – 4,500 mg/L	80 – 110 AU	Rarely found, excellent high-phenol red wines with high aging potential and requiring several years of maturation.
> 4,500 mg/L	> 110 AU	Overextracted wines requiring especial treatment in most cases.

Differences are due to variable amounts of extractable phenols in grapes based on variety and growing conditions. Moderate or cool climate growing areas tend to yield lower total phenol levels. Additionally, vinification techniques exert a strong effect on the extent of phenol extraction from fruit, with longer skin contact times and higher temperatures increasing extraction (Chapter 4.1.4 and 4.1.9).

During the first year of storage, there is a loss of 15 to 20 % of total phenols as measured right after pressing. These losses are due to polymerization reactions, to a decrease of the UV absorption ability of anthocyanins, to absorption of tannin by suspended pulp material and yeasts, and precipitation with potassium bitartrate. Once the new wine has been stabilized against precipitation of phenolic sediments, which naturally occurs during the first winter after harvest, total phenols remain essentially constant regardless of how they are measured. In contrast to color, they are a fairly stable wine feature.

It should be noted that the results obtained by both methods decrease by about 15-20 % in the first year post fermentation. This decrease can be attributed to a decrease of the UV absorption ability of anthocyanins and to a lesser extent to a decrease of tannin concentration through precipitation.

Uptake of phenolics in the form of wood tannins is marginal in relation to the total phenol content and usually much less than 5 % of the total. The significant extract in sensory terms, as for new oak, consists of volatile components in trace amounts adding to complexity of the wine aroma.

For those who are not used to interpret total phenol data, it is easier to grasp the information they provide by analogy with another analysis value everybody in the wine business is familiar with, for example total acidity. Different wines with identical total acidity can display variable sourness perceived on the palate because they contain variable proportions of individual acids with variable strength. This picture is even more differentiated by variable pH values, which reflect variable amounts of potassium and other mineral cations able to mask sourness to a certain extent (Schneider and Troxell 2018). Moreover, alcohol and residual sugar mask sourness and relativize the information that total acidity measurements provide. Despite these shortcomings, measuring total acidity is common practice to classify wines and to complement sensory evaluation. When, for example, such a measurement yields only 4.5 g/L total acidity, the wine is considered too flat and requiring some acidification.

What does that mean for total phenols? Bottling a young red wine with some 4,000 mg/L total phenols is comparable to expecting the consumer to like a wine with 9 g/L total acidity. Only after several years of maturation in conjunction with adequate amounts of oxygen incorporation, will the tannin of this red wine have changed in a way that it becomes a quality feature making the wine drinkable. If there is not such a possibility of aging, one would consider a fining to lower total phenols and astringency (Chapter 12.1). One might also consider shortening skin contact time and extracting less tannins (Chapter 4.1) in subsequent years. On the other hand, when total phenols are only some 1,500 mg/L, red wine quality will be probably be questioned if not rejected, although color intensity might be satisfactory. Blending with a high-phenol wine would be a solution.

Correlation with perceived astringency

The largest phenol fraction comprised in the total phenol content is tannin, which is the basic driver of perceived astringency in red wines. Thus, some correlations (r^2)[21] between total phenols and astringency must be expected and have actually been found in different studies. These include:

- $r^2 = 0.40$ in a set of 40 red wines from Washington State when total phenols were measured as A 280 (Kennedy et al. 2006).
- $r^2 = 0.59$ in European wines when total phenols were measured with the FC reagent (Schneider 1998).
- $r^2 = 0.74$ for astringent aftertaste in 35 wines from British Columbia when total phenols were measured with the FC reagent. However, a predictive one-way model based on this correlation yielded only $r^2 = 0.42$ in another set of wines (Cliff et al. 2002).

From a sensory perspective, the information about astringency provided by these correlations is not bad, taking into account the easy execution and lack of selectivity of total phenol measurements on one hand, and the complexity of the astringent perception on the other hand. Chapter 5 explains how this perception is modulated by matrix variables and other wine constituents than phenols. Nevertheless, astringency, type, and quality of red wine are primarily a question of total phenol content. As long as

---

[21] r^2, also known as R-squared or R^2, is the coefficient of determination and explains how much variability of one factor is caused by its relationship with another factor. It is represented as a value between 0.0 and 1.0, where a value of 1.0 represents the best possible correlation between both factors, while a value of 0.0 indicates that there is no correlation at all. In this context, a value of 0.50 suggests that 50 % of the astringency can be predicted by the total phenol content.

there is no awareness of the significance of this basic parameter, the elaboration of red wine style and taste profiles will depend on empiricism and chance.

Premium wines have higher than average phenol levels and require a longer maturation phase during which the wine quality is maintained or improved. There is a sort of continuum from low to high phenol levels, wine quality, and marketing level. A positive correlation between market value grades allocated by winemaker panels, total phenols, and tannins has been shown for Australian conditions (Mercurio et al. 2010), and this might apply for most wine types around the world. However, quality grades are vague terms. The fact that a red wine of UV index around 45 can have many times the commercial value of another having only 30 AU demonstrates the essentially non-linear structure of total phenols in relation to the sensory effect.

This non-linear relationship is partially due to the fact that tannin composition and color play an additional role. Thus, the concentration of skin tannins and their proportion in total tannins (Kassara and Kennedy 2011), as well as anthocyanins and color intensity (Chapter 5.3.2) have been shown to positively correlate with grade allocation. Total phenol measurements do not take into account these compositional differences. In particular, they include the highly variable amounts of anthocyanins that, as such, are not tannic. Therefore, measurement of anthocyanin levels would be a logical next step.

3.2. Measurement of total anthocyanins

3.2.1. Principles, objectives, and limitations

There is some interest in measuring anthocyanins, including their co-pigmented (Chapter 2.2.4) forms in young red wines. In these wines, they account for 5 to 50 % of the total phenol content, depending on variety and vinification method. Their quantitative relation to colorless tannins in this early stage strongly impacts the sensory outcome of maturation, aging, and all enological interventions involved in that process, in particular micro-oxygenation and any other form of oxygen supply (Chapter 8). This leads to the so-called tannin-anthocyanin ratio, which is an index than can also be obtained in a simplified way be calculating the total phenol-anthocyanin ratio.

Red color intensity as perceived with the human eye or measured at its maximum absorbance at 520 nm gives no information about the anthocyanin content. This lack of correlation is due to the interference of a number of variables such as co-pigmentation, polymeric pigments, pH, and free sulfur dioxide. In young red wines, color intensity is temporarily increased by co-pigmentation, whilst its increase during aging is explained by losses of sulfur dioxide and by the formation of polymeric pigments associated with the decrease of monomeric anthocyanins. After four years, there are no more anthocyanins in measurable amounts.

The most correct way of quantifying total anthocyanins would be measuring the amounts of the individual ones by HPLC[22] and adding them up to a total. This approach is used for cultivar authentication, but it is too cumbersome for routine analysis. Instead, spectrophotometric methods are used. They rely on two principles (Ribéreau-Gayon and Sudraud 1965):

- The first method, the pH shift method, is based on the property of anthocyanins in an acidic medium to occur as two forms, the colored flavylium ion and the uncolored forms (Chapter 2.2.4). Both forms are in a dynamic equilibrium, whose position depends on pH. Thus, their color change between two pH levels is proportional to their concentration. The pH levels used are pH 3.5 obtained by addition of a buffer solution, and pH 1.0 generated by addition of hydrochloric acid. At the latter pH, all anthocyanins are present in their colored form. Subsequently, absorbances are measured at the visible maximum at 520 nm, and the absorbance difference is used to calculate the anthocyanin concentration from the molar absorptivity of the predominant anthocyanin present. In the meantime, this method has been largely abandoned because it is affected by various interferences, which can only be overcome by a more recent analysis protocol (Giusti and Wrolstad 2001) that is hardly suitable for routine measurements in a winery setting.
- The second method relies on the property of anthocyanins to be completely bleached and converted into an uncolored adduct with bisulfite (HSO_3^-) when SO_2 is added in excess (Chapter 2.2.4). One aliquot of the acidified sample is supplied with a concentrated SO_2 solution, and another aliquot with the equivalent amount of water. The absorption difference between both aliquots is measured at 520 nm and used to calculate the anthocyanin concentration as described above.

There are two basic problems with these direct spectrophotometric measurements based on color shift:

- A variable interference by red polymeric pigments reacting similarly to anthocyanins upon acidification and addition of SO_2, though to a lesser extent. Thus, results tend to be higher than those obtained by HPLC (Bakker et al. 1986, Rivas-Gonzalo et al. 1992).
- The choice of the reference compound used to express the total anthocyanin content. The various anthocyanins present in grapes and wines exhibit quite different molar absorptivities, which unavoidably lead to different results expressed in mg/L. As an example, the original method (Ribéreau-Gayon and Stronestreet 1965) used an unspecified anthocyanin mix extracted from grapes as a reference.

Hence, direct spectrophotometry results must be viewed as approximate rather than absolute values. They only provide realistic information during the first year after harvest. This is the period when their use has the most practical relevance. Thereafter, their accuracy is compromised because the increasing polymeric pigments feign higher levels of total anthocyanins than are really present (Tamborra 1993).

[22] High pressure liquid chromatography.

3.2.2. The sulfite bleaching method for measuring total anthocyanins

For the reasons discussed in the previous chapter, there is not yet any official method for measuring total anthocyanins. However, in order to meet practical needs, the spectral method of Somers and Evans (1977) has become the most widely used for that purpose. It is an improved version of the original SO_2 bleaching method and is based on two assumptions:

- SO_2 bleaches only monomeric anthocyanins.
- Upon acidification, anthocyanins are converted into their colored form, but the color of polymeric pigments also increases by a factor of 1.67.

In summary, this assay determines in a 10-mm path length cuvette the absorbance at 520 nm of the highly diluted wine after acidification to pH<1 with hydrochloric acid. At low pH all anthocyanins are in the colored flavylium form. Hence, this reading corresponds to anthocyanins plus polymeric red pigments. In a second sample aliquot, the residual absorbance after addition of excess SO_2 is measured to account for polymeric pigments not reactive to SO_2. Total anthocyanins are calculated by comparing the wine aliquot adjusted to pH<1 to the aliquot that has been decolorized with SO_2. Measures can be performed in a 1-, 2-, or 10-mm path length cuvette, and all absorbance measurements are converted to a 10-mm path length equivalent.

In order to facilitate the implementation of the sulfite bleaching method according to Somers and Evans (1977), it is described step for step as follows:

Sample aliquot I, A_{520}^{HCl} :

- Add 100 µL wine to 10.0 mL 1n-HCl[23] and mix.
- After 3 to 4 hours, measure absorption at 520 nm against water as a blank in a 10-mm cuvette.
- Multiply the absorbance by 101 to correct for the dilution. Result = A_{520}^{HCl}

Sample aliquot II, A_{520}^{SO2} :

- Add 0.15 mL 20 % (m/v) sodium disulfite[24] to 10.0 mL wine and mix.
- After one minute or more, measure absorption at 520 nm against water as a blank in a 1-mm cuvette. Convert absorbance reading to absorbance A 520 at 10 mm path length by multiplying by 10.
- Multiply by 1.015 to correct for the dilution. Result = E_{520}^{SO2}

Calculation:

$20 \times (A_{520}^{HCl} - 1.67 \times A_{520}^{SO2})$ = mg/L total anthocyanins (as malvidin-3-glucoside)

[23] A 1n-solution of HCl has 3.6 %.

[24] To prepare a 20 % (m/v) sodium disulfite solution, dissolve 200 grams of sodium disulfite in water and make up the volume to 1,000 mL. This solution has limited shelf life and should be prepared freshly every month.

In this formula, the factor 1.67 corrects for the color increase of polymeric pigments upon acidification, whilst the multiplication by 20 is based on the absorption of 500 AU for 1 % malvidin-3-glucoside used as a standard.

Compared to older methods, this assay provides the following advances and benefits:

- It uses malvidin-3-glucoside as an exactly defined anthocyanin reference to express results in mg/L.
- It uses a higher dilution factor to better include co-pigmented anthocyanins in young wines, which only disintegrate into single, measurable anthocyanins in dilutions greater than 50.
- It introduces a correction factor to account for the color increase of polymeric pigments upon acidification.
- As a consequence of the preceding advances, results obtained are 30 to 40 % lower than those obtained by the traditional method and closer to those provided by HPLC.

This assay is part of a more elaborated spectral measurement system that can be used to determine other details, such as the percentage of anthocyanins that are ionized as colored flavylium ions in the wine, the fraction bound with SO_2, the proportion of polymeric anthocyanins, and features of chemical age describing the extent to which monomeric anthocyanins have been replaced by polymeric anthocyanins during aging. These determinations have been proven to be extremely useful in research, but less suitable for routine quality control. Refer to the original work (Somers and Evans 1977) and its basic principles (Somers 1998) for details.

3.2.3. The total phenol-anthocyanin ratio

As outlined in section 3.1.1, the total phenol content is a superordinate term, which includes anthocyanins as one of its various subgroups. Thus, by dividing total phenols by total anthocyanins, one obtains a valuable information about the proportion of anthocyanins contained in the total phenol content. In some way, this is a simplified index of the tannin-anthocyanin ratio. It completes the information provided by the total phenol content and is of particular importance when issues such as micro-oxygenation (Chapter 8.3) and barrel aging (Chapter 10) are discussed.

However, there are better means of calculating a tannin-anthocyanin ratio, for example calculating directly the quotient of flavan-3-ols and anthocyanins (Chapters 2.2.2. and 2.2.3) (Durner et al. 2015). Inherently, the result of such indices always depends on the analytical methods used and on what is really measured. In the absence of any binding definition of the tannin-anthocyanin ratio, it is indispensable to carefully specify the analytical methods from whose results that ratio is derived.

It goes without saying that subtractions or additions of phenolic subgroups are not allowed since different analytical approaches and different reference substances are used.

3.3. Color intensity and color hue

Visible color of red wines is composed of three components:

- Red color, which is in young red wines primarily caused by monomeric anthocyanins and in old wines mainly by polymeric pigments, i.e., tannin-anthocyanin associations. It is measured as absorbance at 520 nm (A 520).
- Brown color caused by tannins containing no anthocyanin moiety and measured as absorbance at 420 nm (A 420).
- Purple color, which is elicited in some wines by elevated concentrations of certain anthocyanins. It as measured as absorbance at 620 nm (A 620).

Wine color intensity (I) is presented by the sum of A 420 + A 520 or by the sum of A 420 + A 520 + A 620. The results are often reported as % brown, % red, and % purple providing thus a more complete interpretation of wine color properties.

The hue or tone (T) of red wines is calculated as the ratio of A 420 and A 520. It reports the browning of wine and is used to monitor wine aging because of the shift from red to brick red that occurs when monomeric anthocyanins polymerize with tannins during aging.

Measurements

The absorbances of the undiluted wine is measured at the respective wavelengths in cuvettes of an appropriate path length, usually 1 mm. Subsequently the absorbance readings are converted to their 10-mm path length equivalent. Sample dilutions are not admitted because the color decrease is not linear to the dilution factor.

Calculations

Color intensity I = A 420 + A 520

Color intensity I = A 420 + A 520 + A 620.

As both versions are in use, it must be specified how color intensity was has been measured. When A 620 is included, it does usually not exceed 10 % of the total. Most red wines display a color intensity comprised between 5 and 10.

Hue or tone T = A 420 : A 520

When T exceeds 1, a visible browning can be observed.

A more sophisticated measure of wine color is the so-called CIElab color space obtained from the information contained in the whole visible spectra of wines. Three color components result from the integration of the visible absorption features, and three chromatic coordinates X, Y, and Z determine the color characteristics[25]. The color of wine is thus described by the intensity of the color (chromaticism), the luminosity of the wine, and the color itself based on red, yellow, green, and blue compo-

[25] Technical details can be taken from http://www.oiv.int/public/medias/2478/oiv-ma-as2-11.pdf

nents (tonality). This method aims to better simulate the perception that human observers have for the color properties. In a winemaking context, it does not provide any information that could be helpful in enological decision-making.

Significance and limitations

There is no linear relationship between color measurements and the levels of anthocyanins or tannins because they are impacted by matrix factors such as pH, SO_2, and co-pigmentation in young wines. As an example, increasing free SO_2 by only 10 to 20 mg/L in young red wines of the first year can decrease A 520 and color intensity substantially, causing color hue shifting towards brown.

Because of these relationships and interfering factors, color measurements can at best describe numerically the visual appearance of wines at a given moment and for a given level of free SO_2. Thus, they are useful in the context of monitoring a single wine as it ages, but their utility for comparing multiple wines is very limited. They are unable to provide information about how different phenolic components will change during wine processing and aging, making it difficult to interpret the results. Ultimately, they are useless for enological decision making. Only their simplicity makes them an attractive means to provide an idea of the instantaneous color of wines.

3.4. Fractionation of total phenols, tannin assays, and their validity

3.4.1. Principles, objectives, and limitations

As outlined in section 3.1., total phenol measurements provide a basic information about the phenolic strength and general quality of red wines. However, they lack selectivity for tannins. Hence, they can only provide a rough estimate of the astringency that tannins elicit.

Given the expense of conducting controlled sensory evaluation of astringency (Chapter 5.2), one of the most important challenges for practitioners in the wine industry is the selection of an analytical method that correlates well with perceived astringency. Additionally, such a method must be reproducible and inexpensive, and require minimal analytical equipment and skills for its implementation. Therefore, this chapter about phenol analysis would not be complete without touching upon more advanced measurements than just total phenols and anthocyanins. Indeed, some readers might look for a better understanding of what they are tasting or doing in their winery. Thus, the following gives an overview of what can be done without going into details. Background information can be found in the references cited.

Considering the complexity of wine phenolics, there is no single best way to characterize the phenolic composition of red wines. The broad diversity of phenolics with regard to molecular weight and other properties limits the utility of a single measurement. Choosing an appropriate phenolic assay depends on what information is required. Sometimes a combination of assays is the best approach to properly characterize the phenolic composition of a sample. Results of the various methods correlate between themselves to a certain extent, but attention should be given to what they really measure and how results are to be interpreted (de Beer et al. 2004). As red wine tannins are extremely heterogeneous (Chapter 2.2), reported values vary considerably depending on the analytical method.

While there are numerous spectrophotometric methodologies for measuring phenolics, many of them have been neglected in favor of modern liquid chromatographic (HPLC) methods. Nevertheless, these methodologies remain valid and readily accessible to winery laboratories equipped with a few pieces of basic equipment and a spectrophotometer. Their major advantage is that they are able to quantify phenolic fractions according to their technical or sensory importance. These fractions comprise phenols with similar characteristics and report them as a cumulative sum. Astringency is one of the most important of these characteristics.

Fractionation is the basis of most modern methods of spectrophotometric phenol analysis. It is a separation process in which the bewildering mix of wine phenols is broken down into a manageable number of smaller fractions, which consist of phenolics that show comparable properties. In contrast, chromatographic methods allow for measuring an unmanageable multiplicity of individual compounds frequently without any practical interest. One cannot see the wood for the trees.

In section 3.1, it has been shown how total phenols can be quantified. It is a generic term that gives information about the phenolic strength, which is the most important sensory feature of red wines. In the broadest sense, it also gives a clue about the color intensity and astringency to expect. Astringency is the key property of tannins and plays a crucial role in the gustative characteristics of red wines, but it is cumbersome to exactly measure its intensity by sensory means. Hence, most fractionation methods of total phenols aim at quantifying phenolic subgroups eliciting it and thus support associated sensory analysis. In the broadest sense, a more detailed analysis of the phenolic make-up of red wines pursues the following objectives:

- Identifying phenolic fractions that are relevant to specific mouthfeel, taste, and color properties and improving understanding of them.
- Aquiring more knowledge to establish and meet present and future phenolic specification with importance to red wine quality and style.
- Increasing the capacity to meet wine specifications in order to match consumer expectations.
- Supporting decision-making.

Figure 17 depicts schematically how one can break down total phenols in increasingly smaller subgroups or fractions. Each fraction exhibits different sensory, chemical, and chromatic properties. The relative proportions of the fractions change during storage and aging. The driving forces of these changes are reactions of polymerization, acid-catalyzed cleavage of polymers, and oxidation. In this process, the levels of monomeric anthocyanins decrease as they polymerize with colorless phenolics to the tannin of matured red wines.

Figure 17: Fractionation of total phenols in red wines.

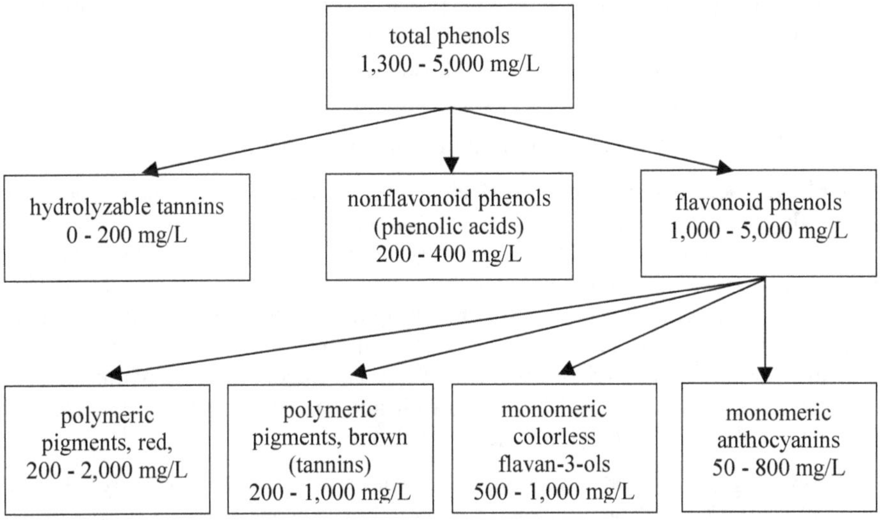

The methods used to fractionate total phenols comprise precipitation steps, specific color reactions suitable for spectrophotometric evaluation of particular phenolic subgroups, or a combination of both. There is a large number of review reports evaluating their usefulness and explanatory power (Blouin et al. 2000, de Beer et al. 2004, Herderich and Smith 2005, Harbertson and Spayd 2006, Kennedy et al. 2006, Lorrain et al. 2013). Only a few of them have been occasionally used in current quality control. This is no surprise given that many small wineries do not even attach any importance to the more important total phenol content.

Whilst phenolic acids (Chapter 2.3.1) and wood-derived hydrolyzable tannins (Chapter 2.3.2) occur only in relatively small amounts with secondary importance for perceived astringency and other mouthfeel sensations, flavonoid phenols extracted from the grape solids constitute the phenolic subgroup that understandably arouses most interest. Hence, most fractionation methods focus upon flavonoid phenols and their subfractions. Their essential purpose is to better predict astringency than the total phenol content does.

3.4.2. Fractionation of total phenols by colorimetric assays

As outlined in chapter 3.1, measuring total phenols is the very first step for approaching the phenolic make-up of red wines and can be done in any laboratory equipped with a spectrophotometer. However, it is only a sum parameter including all phenols regardless of their sensory properties. Therefore, one cannot expect a good correlation between total phenol content and perceived astringency. In one study comprising 18 red wines of three cultivars produced under comparable growing conditions, it yielded a coefficient of determination of $r^2 = 0.59$, meaning that it explained only 59 % of astringency on the palate (Schneider 1998). Hence, there have been early attempts to find better colorimetric methods for specifically determining its subgroups with sensory importance, in particular its flavonoid fraction containing the grape-derived tannins but without including tasteless anthocyanins and phenolic acids.

Differentiation of total, flavonoid, and nonflavonoid phenols

Under highly acidic conditions, formaldehyde can be used to cross-link and precipitate all flavonoid compounds. The precipitation reaction takes 24 hours and requires subsequent centrifugation or filtration. The flavonoid content can be determined by measuring the remaining nonflavonoids and subtracting them from the total phenols. Both phenol measurements can be performed by use of the Folin-Ciocalteu reagent (Kramling and Singleton 1969) yielding concentration units expressed as gallic acid, or by just measuring the UV index (Glories 1974). The major drawback of this method is the use of formaldehyde, which is a known health hazard and suggests the use of a fume hood not normally found in winery laboratories.

Furthermore, grape-derived nonflavonoids are known to not vary appreciably between wines (Chapter 2.3.1). Thus, flavonoid phenols in red wines correlate in some way with total phenols. Hence, for predicting astringency, measuring flavonoid phenols by this approach does not provide more information than the total phenol content. With $r^2 = 0.59$, the coefficient of determination was identical (Schneider 1998).

Measurement of monomeric and oligomeric flavonoids

Cyclic aldehydes such as 4-dimethylaminocinnamaldehyde (DAC) react with flavonoids at their C8 position in a highly acid medium, yielding a colored condensation product that enables their measurement at 640 nm (Zironi et al. 1992). Since nonphenolic wine constituents do not interfere, this measurement is fast and easy to execute, but results require careful interpretation.

When the C8 position of flavonoids is involved in polymerization reactions (Chapter 2.2.3), the reaction with DAC does not take place. Therefore, results reflect essentially monomeric and oligomeric flavonoids, depend on the degree of tannin polymerization, and decrease to the extent polymerization further proceeds during aging. Thus, they can be used to estimate the degree of polymerization of tannin in conjunction with other tannin measurements (Nagel et Glories 1991), but they are not suitable to quantify the tannin content. Moreover, they poorly correlate with perceived astringency and are not suitable to predict this sensory parameter in red wines.

Measurement of tannins by acid hydrolysis

In this colorimetric assay, brown tannins are depolymerized to yield individual subunits, which are then oxidized to colored anthocyanins. The reaction takes place under acidic conditions when the wine is heated under reflux in boiling water after dilution with hydrochloric acid (Ribéreau-Gayon and Stonestreet 1966) or, better, with a mix of hydrochloric, butanol (5:95 v/v), and iron sulfate (Porter et al. 1985). The red color produced is measured at 550 nm using the unheated sample as a blank and converted into concentration units obtained from a standard.

This method is selective to flavan-3-ols and polymeric tannins in that it does not detect hydrolyzable tannins. However, it sometimes overestimates tannin concentrations with higher values reported for tannins than those for total phenols. Moreover, it is also common to observe an increase of the tannin content during wine aging. Obviously, the kinetics of color formation is affected by several variables and side reactions, which lead to incorrect concentration values. Therefore, it poorly correlates with astringency (Cáceres-Mella et al. 2013). Ultimately, the major drawback of this method is that it is used in many modifications giving different results (Vivas et al. 2003 a), and that it has been rarely implemented in wine quality control so that there is a lack of practical experience in interpreting the results.

More recently, all these colorimetric methods of determining phenolic fractions have lost their importance for red wines as they do not reliably reflect the tannins responsible for red wine astringency. However, they continue having some importance for white wines, whose tannin levels are far too low to be detected by methods that are more suitable for quantifying tannins in red wines.

3.4.3. Measurement of tannin by precipitation assays

Flavonoid phenols require a certain degree of polymerization to act as tannins and elicit astringency. However, the colorimetric methods described in the previous chapter lack selectivity to quantify these flavonoids acting as true tannins and the astringency they elicit. To overcome this problem, they require an additional precipitation step. Thus, they become precipitation assays.

One of the main benefits of precipitation-based methods is that a theoretical positive correlation with astringency is foreseen. This hypothesis is based on the assumption that these methods simulate the phenomena that naturally occur in wine when it interacts with the salivary proteins of the oral cavity. An insoluble macromolecular complex is then formed that precipitates from solution causing the drying and puckering sensation known as astringency (Chapter 5.2).

Consequently, all precipitation assays are based on the precipitation of tannins by proteins, and sometimes by other reagents, that result in an insoluble complex which precipitates and is removed by centrifugation. They are subtractive measures requiring the preparation of a control and a treatment sample. The control sample allows measurement of all phenolics present in the sample, whereas the treatment sample allows measurement of the phenolics remaining in the supernatant solution after the tannin has been removed by precipitation.

The phenolic content is monitored either by measuring the absorbance at 280 nm (Chapter 3.1.3), which can be used as an arbitrary value or converted into concentration units of monomer equivalents like catechin or gallic acid, or by chemometrical methods. By subtracting the phenol content of the treatment sample from the phenol content of the control sample, the tannin of the sample can be determined. The basic difference between the various precipitation assays is the precipitation agent they use, thus leading to results which are not directly comparable from one assay to another.

Cinchonine

A precipitation step with cinchonine[26] sulfate yields 'tannin' flavonoids plus hydrolyzable tannins in the precipitate, and 'nontannin' phenols in the supernatant. Nontannin phenols comprise phenolic acids plus nontannin flavonoids. Combining cinchonine sulfate with formaldehyde allows measurement of phenolic acids, hydrolyzable tannins, flavonoid tannins, and total flavonoids (Peri and Pompei 1971). This represents an improvement of the formaldehyde method (Chapter 3.4.2), which does not permit differentiation between nontannin and tannin flavonoids.

The drawback of this method is that it defines tannins by their ability to precipitate with cinchonine, one of the various precipitation agents used for tannin precipitation and phenol fractionation. This would mean that all phenols not precipitated by cinchonine are non-tannins, a clearly arbitrary definition. As will be shown in chapter 5.2, tannins are defined by their ability to precipitate human salivary proteins and, thereby, elicit their typical astringency. It is not known to what extent cinchonine is comparable to salivary proteins, but it is reasonable to expect some similarity. The same applies when cinchonine is replaced by PVP/trichloroacetic acid (Burckhardt 1976). The correlation between tannins measured by these methods and perceived astringency was $r^2 = 0.60$ (Schneider 1998).

Proteins

Tannins based on a flavonoid structure as well as hydrolyzable tannins can interact with proteins, forming insoluble complexes that precipitate. This is the basic and most important property of all tannins. The extent of this reaction depends on the chemical characteristics of the tannin and the experimental conditions used.

Since the mechanistic basis of astringency is the interaction of tannins with salivary proteins resulting in insoluble complexes, protein precipitation assays are considered the most reliable techniques for astringency assessments. Their operating principle is similar to the human physiological response to tannins in the oral cavity. They are supposed to more reliably reflect the interaction of saliva proteins with tannins than any other methods of phenol or tannin measurement.

All these methods are based on measuring the difference of phenol concentration before and after the precipitation step with the respective protein. They circumvent the lack of specificity of colorimetric assays (Chapter 3.4.2) and are not affected by interferences from other wine constituents such as sugars and SO_2 (Harbertson and Spayd 2006). On the basis of these advantages many protein precipitation methods have been developed to quantify tannins. Some of them can be implemented in a winery laboratory setting and these are presented in the following paragraphs. Details can be taken from the literature cited.

Gelatin: One of the first protein precipitation assays used on red wines is based on gelatin. After a precipitation step taking three days, tannins are measured in the treated and untreated sample using acid hydrolysis (Chapter 3.4.2). The results obtained report the total tannins and the tannins that do not react with gelatin, respectively. The difference yields the gelatin-reactive tannins. They are expressed as a percentage of

[26] Cinchonine is a natural alkaloid frequently used as a catalyst in organic chemistry.

total tannins. This percentage is known as the gelatin index. Its values vary from 25 to 80. A gelatin index of 50 sometimes given as standard may characterize an astringent wine if total tannins are lower than 2.5 g/L, but another wine with 3 g/L may seem well balanced. Values above 60 indicate the presence of highly reactive tannins responsible for a rough astringency (Glories 1984).

In one study on wines from the Bordeaux area, tannins measured as gelatin index correlated incredibly well with perceived astringency with an incredibly high r^2 = 0.935 (Blouin et al. 2000). The basic problem with this method is that the precipitation step takes a very long time. Furthermore, results depend of the kind of gelatin used. Gelatin is known as a well performing precipitant of tannins, which explains its widespread use as a fining agent to remove excessive astringency from red wines (Chapter 12.1.3). However, there are many gelatins on the market with a heterogeneous composition. As a consequence of their variable specification (Versari et al. 1999, Maury et al. 2001), gelatin has largely lost its importance for analytical purposes.

Ovalbumin: Instead of gelatin, ovalbumin, a purified chicken egg albumin, can also be used to precipitate tannin. In this assay, total phenols are measured as UV index (Chapter 3.1.3) before and after the precipitation step. After the initial measurement of the untreated sample, 12 different amounts of diluted ovalbumin are added to fixed amounts of wine and a tannic acid standard. After 10 minutes of incubation, residual phenols are measured in the clarified samples. Correlation with perceived astringency was reported as $r^2 = 0.77$ (Llaudy et al. 2004).

Running various samples of each wine is necessary because the UV index used to monitor phenols suffers from interference by residual proteins. This would also apply if the Folin-Ciocalteu assay was used for this purpose. Thus, the amount of protein in the supernatant must be minimized to optimize reproducibility, and the absolute amount of precipitation must be found for each sample at the expense of efficiency. These requirements make this tannin assay rather time-consuming.

Bovine serum albumin (BSA) is another protein that was initially used to measure tannin in plant extracts. The tannin precipitate is redissolved in alkaline solution, where its absorption is measured at 510 nm after addition of ferric chloride (Hagerman and Butler 1978). After adaptation of this method to wine (Harbertson et al 2002), results correlated with $r^2 = 0.82$ with astringency in a set of 40 wines from Washington State (Kennedy et al. 2006), and with $r^2 = 0.887$ in a hundred wines from the Bordeaux area, where they were also strongly correlated with the gelatin index (Blouin et al. 2000). When this assay was applied to 1,325 red wines from different countries and varieties, a mean tannin concentration of 544 mg/L catechin equivalents (CE) within a range from 30 to 1895 mg/L CE (Harbertson et al. 2008) was measured.

This so-called BSA method has found relatively widespread use in more recent times, though its implementation might appear too cumbersome for most winery laboratories. Furthermore, precision and reliability of this assay have been seriously questioned after it had shown a coefficient of variation of ±27 % and an intralaboratory validity recovery of only 55-63 %. Results differed over 2-fold between different laboratories and between analysts within the same company (Brooks et al. 2008). One of the reasons is that accuracy of this measurement is based on obtaining the appropriate wine dilution because concentrated or very diluted samples tend to underestimate the tannin content. Thus, accuracy is compromised by a first-order error related to the method and second-order errors derived from its execution.

A modification of this method (Harbertson et al. 2003) includes a bleaching step with bisulfite. The reason for this adaptation was to also analyze monomeric anthocyanins, small polymeric pigments that do not precipitate with BSA, and large polymeric pigments that do precipitate with BSA. If these analyses are not required, the process can be considerably shortened in order to only obtain protein-reactive tannin expressed in catechin equivalents.

Methyl cellulose

Methyl cellulose (MC) is not a protein but a polysaccharide polymer able to precipitate tannins in the presence of ammonium sulfate similarly to proteins. In a nutshell, the sample is mixed with 0.04 % MC solution and saturated ammonium sulfate solution, allowed to react for 10 minutes, and clarified by centrifugation. Tannin values are calculated from the difference between the absorbance values at 280 nm (Chapter 3.1.3) of the sample without the MC addition and the sample with the MC added and precipitated. The absorbance units can be converted into epicatechin or catechin equivalents.

This subtractive method has been shown to measure all grape-derived tannins and not suffer interference from other A 280-absorbing phenolics such as phenolic acids, anthocyanins, and catechins. However, its response to commercial tannins is variable due to their structural diversity and needs clarification (Sarneckis et al. 2006).

In a further study (Mercurio and Smith 2008), this assay was compared with the protein (BSA)-precipitation assay. Although both methods showed with $r^2 = 0.80$ a good correlation between tannin levels in 41 red wines, the MC method yielded approximately 3-fold higher tannin concentrations than the BSA method. This difference is explained by the different detection methods used because the amount of tannin precipitated in both assays is very similar.

Correlation with astringency was $r^2 = 0.83$ for the MC assay and $r^2 = 0.90$ for the BSA assay in this set of wines (Mercurio and Smith 2008). However, it was lower in another study (Cáceres-Mella et al. 2013) with $r^2 = 0.59$ for the MC assay and $r^2 = 0.47$ for the BSA assay. Both assays were considered to predict wine astringency with a reasonable confidence level that depends on wines and tasters used in each study, but the MC test is easier and faster to execute.

Which assay to choose?

After reading all the preceding details concerning tannin measurements, any wine industry practitioner will be frustrated and wondering which assay to choose for their own purpose. It has become evident that results might reflect the analytical method as much as they reveal the sample composition. Hence, it is difficult to compare data that have been obtained by different methods. In contrast to routine parameters such as alcohol, pH, total acidity, or residual sugar, tannin measurements lack official methods, which require validation by regulatory organizations or independent organizations.

As we are moving in a field of on-going research, the question of what kind of tannin assay to choose cannot be clearly answered, though there is a trend towards precipitation assays using either methyl cellulose or bovine serum albumin. Ultimately, the choice will depend on one's individual needs and the available laboratory equipment. On the other hand, there is general agreement on the requirements such an assay must fulfill: They can be summarized as easy and fast execution, limited demands on lab equipment, ruggedness, selectivity, accuracy, precision, and repeatability. It seems that the methyl cellulose tannin precipitation assay complies more with these requirements that most other assays. Moreover, it has been established as an industry standard in many wine growing countries. Therefore, its practical implementation is outlined in the following section.

3.4.4. Implementation of the methyl cellulose precipitation assay

A) Reagents

Saturated ammonium sulfate solution:

Fill a 500-mL screwcap flask with 300 mL distilled or deionized water and add ammonium sulfate crystals in excess while stirring until they will no longer dissolve. Keep adding until approximately 1.5 mL of ammonium sulfate crystals remain visible on the bottom of the flask. The solution is stable for at least six months.

0.04 % Methyl cellulose (MC) solution:

Heat 300 mL of deionized water to 80° C and separately cool another aliquot of approximately 700 mL to 0-5° C. Weigh 400 mg of methyl cellulose (Sigma M-0387). Transfer the 300 mL of hot deionized water into a 1 L-volumetric flask and add small portions of the MC powder whilst stirring quickly on a magnetic stirrer at a speed such that a large vortex forms. Use a pipette to rinse hot water down the inner side of the volumetric flask to remove any MC stuck to the neck. Add the remaining 700 mL of cold water to lower the temperature and keep on stirring for half an hour immersed in ice water (0-5° C). Remove from ice water and continue stirring until the solution becomes clear. After the solution has reached room temperature, adjust the final volume to 1 liter with deionized water. This solution is stable for approximately one month at room temperature. Discard it when MC falls out making it turbid.

B) Construction of the calibration curve

- Dissolve 100 mg (+)-catechin[27] in 10 mL ethanol and then dilute to 100 mL with water. The final catechin concentration will be 1,000 mg/L. Dilute this mother solution with water to create standards with 10, 20, 50, and 100 mg/L concentrations,

[27] In the original work (Sarneckis et al. 2006) epicatechin was used, but its replacement by (+)-catechin has only minimal influence on slope and intercept of the calibration curve (Mercurio and Smith 2008). Catechin is more easily available and much less expensive than epicatechin, but attention should be paid to whether it is pure catechin or catechin hydrate. In the latter case, the amounts of catechin have to be multiplied by 1.062 to correct for the hydration water.

respectively. Catechin standards are unstable and should be freshly prepared before use.

- Measure the absorbance of the catechin standards in a 10-mm quartz cuvette against water as a blank using a spectrophotometer set at 280 nm.
- Record absorbance data against concentration data in an EXCEL sheet and construct the calibration curve. The corresponding curve equation obtained will allow easy transformation of absorption data readings into catechin concentration data. Figure 18 shows such a calibration curve as an example. Note that the curve will need to be prepared on the spectrophotometer that is to be used for the assay!

Figure 18: Absorbance of catechin at 280 nm.

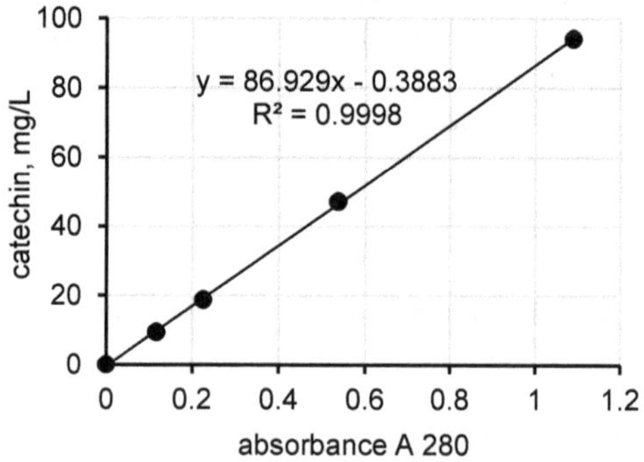

C) Procedure

Prepare the treatment sample adding into a test tube

- 0.25 mL sample
- 3.00 mL methyl cellulose solution, shaking and leaving to stand for 2-3 minutes,
- 2.00 mL ammonium sulfate solution
- <u>4.75 mL</u> water, shaking.

=10.0 mL

Prepare the control sample adding into another test tube

- 0.25 mL sample
- 2.00 mL ammonium sulfate solution
- <u>7.75 mL</u> water

=10.0 mL

Allow both samples to sit for 10 minutes at room temperature and centrifuge them (~ 5 minutes at 4,000 rpm)

From each sample, pipette clear supernatant into a 10-mm quartz cuvette and read absorbance at 280 nm (A 280).

D) Calculations

The A 280 value of the control sample represents the value for total phenols[28], whereas the A 280 value of the treated sample indicates the value for phenolics remaining in solution after the MC-reactive tannin has precipitated. Thus:

Subtract A 280 of the treatment sample from A 280 of the control sample to obtain A 280 of the tannin precipitated:

A 280 (tannin) = A 280 (control) – A 280 (treatment)

Use A 280 (tannin) to calculate tannin (in mg/L catechin equivalents) from the calibration curve.

Multiply by the dilution factor of 40, corresponding to the dilution having taken place in the test tubes.

Generally, tannin measured in this way will make up approximately half of total phenols. When total phenols are determined with the Folin-Ciocalteu method and expressed likewise as catechin equivalents (Chapter 3.1.2), tannin will amount to between 40 and 80 % of the total phenol content in most red wines. In young wines with high levels of monomeric anthocyanins, this percentage will tend to the lower side of the range.

Comparing tannins with total phenols

The box-and-whisker diagram of figure 19 shows the dispersion of tannin levels measured with the MP assay and total phenols obtained with the Folin-Ciocalteu method in 25 red wines from various cultivars and European countries. In this set of wines, tannins were 61 % of total phenols on average, whilst the correlation r^2 between both was only 0.44. It seems that such correlations are better when the wine matrix is more similar, for example when the wines are from one single cultivar or geographical origin.

Figure 19: Tannins and total phenols in 25 red wines of various European countries and cultivars.

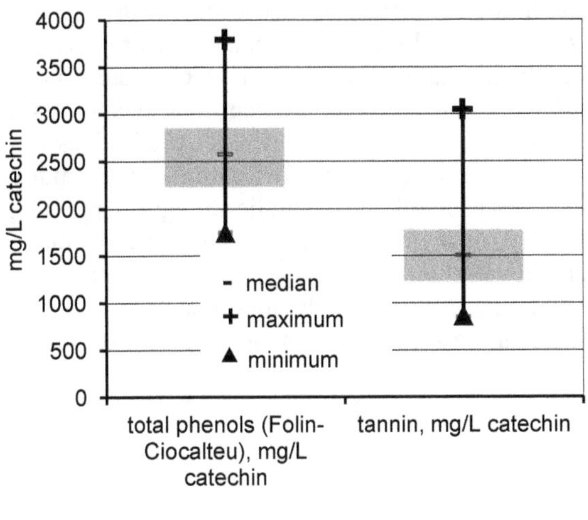

[28] The addition of ammonium sulfate causes a slight increase of the absorbance at 280 nm, which is greatly enlarged after multiplication with the dilution factor. Therefore, this measurement should not replace the standard determination of the UV index at 280 nm as described in chapter 3.1.3.

Limits of the methyl cellulose assay

As will be shown in chapter 5.3, the wine matrix has as much influence on perceived astringency as the amount and quality of tannin. Hence, within a selection of samples, the more similar the constituent wine matrices are with regard to pH, total acidity, mannoproteins, alcohol content, etc., the better the results of tannin measurement will correlate with perceived astringency. Total acidity and pH have a strong impact on perceived astringency. It strongly increases when a wine is acidified by addition of any acid. In contrast, it is lowered when total acidity is reduced by any means of deacidification (Chapters 5.3.4 and 12.1.1). Modifying total acidity by 1 g/L can elicit a change in astringency of as much as 30 %. However, the methyl cellulose assay is not able to reflect relatively strong changes in perceived astringency trigged by changes in pH, total acidity, or by mannoprotein additions.

As is the case with any tannin precipitation assay, the response of the methyl cellulose assay to commercial tannins is also highly variable due to their structural diversity. Many of these preparations are not detected as tannin when they are added in a typical amount of 100 to 200 mg/L to finished wine, though they can cause a considerable increase of perceived astringency (Chapter 9). Similarly, the strong and temporary increase of astringency observed during the first months of aging in new oak barrels and caused by oak-derived ellagitannins is barely reflected in the measurements.

All things considered, tannin measurements only provide a rough classification of the astringency to expect in a large range of samples. They allow one to better order unknown samples as a function of their astringency and facilitate quality assessments in wine competitions. However, they cannot replace the sensory evaluation of astringency when it comes to deciding whether it displays the appropriate intensity for a given wine and marketing segment.

When the expense, complexity, and limited reliability of such a specific tannin measurement are inappropriate, then simple total phenol measurements can replace it to some extent. It is the first step to evaluate what is considered style, character, and the very essence of red wines. Section 3.1.4 reports to what extent total phenol assays have allowed prediction of perceived astringency in various sets of red wines from around the world. However, there is one basic difference between these techniques: Whilst the total phenol content describes primarily the phenolic strength of red wines and their ability to consume, withstand, or require oxygen uptake post fermentation (Chapters 6 and 7), the tannin content better reflects the astringency to expect. Nevertheless, from a sensory perspective, the information the total phenol content provides should not be underestimated. A more specific tannin measurement is meant to complement it when needed.

4. Extraction of phenols in red winemaking

This chapter covers hands-on issues. It deals with the different techniques of red winemaking, putting special emphasis on maceration, also known as skin contact, as the most popular of these techniques. It describes how technical variables such as maceration length, extended maceration, temperature, juice runoff, cold soak, and the various means of cap management impact tannin, color extraction and the quality of the finished wine. The use of pectolytic enzymes, commercial tannin additions, and the choice of the yeast strain are additional measures that are discussed. It is emphasized that the completion of fermentation does not give any information about how much phenolic material has been extracted from the fruit because its extraction kinetics do not follow fermentation kinetics in a linear way. Measuring total phenols is a good means to assess the tannin extracted and to adjust the time point of pressing to the desired wine style. The chapter concludes by also presenting the benefits and drawbacks of extraction techniques other than maceration.

4.1. Maceration

4.1.1. Decisions to be taken upon crushing

The first step in handling harvested grapes is to crush and destem them. The objective of this process is to gently break the berry skin and to remove the stems. There are, however, some exceptions to this rule that require decisions as early as on the crush pad.

Some winemakers add a small fraction of stems to the must in order to extract some extra tannins they contain. Whilst this can be beneficial for a very few wines, typically it gives rise to undesirable harsh tannins and color losses in the vast majority of cases. Besides phenols, sometimes undesirable herbaceous vegetative aroma compounds can also be extracted (Fischer 1997 b). Moreover, the stems make the cap formed during fermentation more compact and its punching down more difficult. Therefore, complete destemming is a standard practice in red wine making.

Other winemakers wish to leave a fraction of whole grape clusters or intact berries in the must. The rationale is that intact berries produce a strong fruity aroma through a pathway called 'carbonic maceration' under conditions of oxygen exclusion (Chapter 4.2.1). These berries tend to display higher sugar levels at devatting with the consequence of elevated sugar levels in the whole wine volume after pressing. This can lead to microbial complications which are difficult to control, in particular during subsequent malolactic fermentation.

The manner of transferring crushed grapes to the fermentor is another hotly debated issue. Whilst most wineries use pumps for that purpose, some wineries opt for gravity flow transfers, which are facilitated by multilevel winery designs or achieved by the use of forklifts. The alleged advantage of gravity flow is less mechanical impact on the fruit. Obviously, it is questionable whether that advantage proves significant in light of the mechanical impact exerted by the various means of cap management applied during subsequent maceration (Chapter 4.1.12).

Pre-fermentation SO_2 additions

The issue of SO_2 additions to the must needs some consideration. Some winemakers do not add any SO_2 to red must prior to fermentation. The rationale is to minimize SO_2 levels in the wine, to facilitate malolactic fermentation, and maybe to achieve more flavor complexity by allowing indigenous yeasts to participate in alcoholic fermentation. This approach works out at low harvest temperatures in cool-climate growing areas, but leaves the must vulnerable to the activity of spoilage causing bacteria and wild yeasts when temperature exceeds some 15° C (59° F). Therefore, most winemakers add 30 to 60 mg/L SO_2 to the must upon crushing for microbial safety. About half of that amount gets lost by oxidation so that the remainder after primary fermentation does not suppress malolactic fermentation.

Regardless of its microbial effect, SO_2 is known to facilitate the extraction of anthocyanins and to protect them against oxidative losses (Joutei and Glories 1994), especially at lower temperatures like those occurring during cold soak. However, to make practical use of that effect, many times higher additions than practically possible would be required. When total SO_2 levels in must were increased from 0 to 75 to 150 mg/L, extraction of anthocyanins also increased, but there was little difference of anthocyanins and total phenols in the finished wines, which only displayed a higher color percentage due to polymeric pigments and slightly more brownness. SO_2 additions have negligible effect on aroma and flavor attributes after 18 months (Bakker et al. 1998).

Ultimately, standard SO_2 additions to must do not exert a noticeable effect on phenol levels and other compositional parameters in the finished wines. This is all the more the case as any SO_2 not lost by oxidation is quickly bound to acetaldehyde produced by yeast metabolism as soon as primary fermentation starts. As a consequence, no more free SO_2 remains to protect against oxidation from that moment on.

pH and acidity corrections

Further must adjustments may comprise sugar additions to achieve the desired alcohol level, or corrections of acidity and pH. Generally a total acidity in the range of 6 to 8 g/L and a pH value of 3.4 to 3.6 is preferred in must. After malolactic fermentation and bitartrate precipitation, wines display less acidity than the musts they have been obtained from. These losses are greater the higher the initial must acidity level.

When fruit is very low in total acidity content (e.g., less than 5 g/L) or high in pH (> 3.7), these numbers should be corrected by tartaric acid additions. However, care must be taken to avoid over-acidification in the pursuit of an excessively low pH for microbial safety. The balance on the palate would be compromised since acidity interacts with the perception of tannin. High acidity levels increase its astringency and lower the perception of body and creaminess (Chapter 5.3.4). Acidity corrections entail far-reaching and complex sensory consequences, which cannot be mastered relying only on analytical numbers (Schneider and Troxell 2018). However, it is important to state that they do not affect phenol extraction to a noticeable extent, though the extraction of anthocyanins slightly increases as the pH decreases.

Most red wines from around the world display a total acidity of 4.3 to 6.0 g/L associated with pH values higher than 3.5 and hence those required for absolute microbial safety. In the production of the internationally renowned premium wines, producers have learnt to cope with high pH values that would routinely be decreased by tartaric acid additions in many other growing areas. It is crucial to understand that acidity in red wines is much less important for building structure and aftertaste than in white wines because tannins and other phenolics take this role to a large extent.

4.1.2. General reflections on phenol extraction

Of all the phenolics in destemmed grape berries, more than 45 % are found in the skins and in the seeds, respectively, whilst only some 5 % occur in the flesh and in the pulp juice, respectively. Each of these berry components contains different classes of phenolic compounds. Phenolic acids and their esters are found primarily in the juice and, to a lesser extent, in the skins. Polymeric tannins and monomeric flavan-3-ols such as catechins are found in the skins and seeds, whilst small amounts of flavonols only occur in the skins. In most varieties, anthocyanins can only be found in the skins, with the exception of the so-called teinturier varieties that also display high levels of these compounds in the pulp.

Since the characteristic phenolics of red wines are found in the skins and seeds, red wines are normally made with skin contact, also called maceration. All red winemaking techniques have been developed to promote the extraction of these compounds during vinification, and most of them are directed at enhancing their extraction within a short period of time. However, the amounts and kinds of phenolics that can be potentially extracted are determined by fruit composition as a result of variety, climate, growing conditions, and viticultural practices.

There are many extraction techniques used for red winemaking, including some less familiar ones such as thermovinification, flash détente, and carbonic maceration, but the most widespread of them is maceration. Indeed, it can even be considered as the standard vinification procedure for red wines. Nevertheless, it is not a clearly defined technique. There are various technical options when implementing maceration, including a wide array of physical and chemical parameters that impact phenol extraction. They determine to what extent extractable phenolics are actually extracted from the fruit. Hence, they play as much a role in determining the wine style as fruit composition does.

In traditional red winemaking, the skin contact period coincides with primary fermentation because the must is pressed as soon as fermentation is finished. This is based on the assumption that the extraction of tannins and color goes together with the depletion of sugars and is finished when primary fermentation is completed. However, in reality, these are two different and independent processes. The completion of fermentation does not give any indication of whether the amount of tannin extracted is satisfactory, too high, or too low for a given style of wine. Inevitably, the result of such a simplified approach is a product of chance.

Many winemakers suppose they are able to evaluate tannin during skin contact by tasting. However, such sensory assessments are dubious due to a strong 'background noise'. Astringency and other mouthfeel perceptions derived from tannin are heavily impacted by suspended yeasts and grape tissue particles, acidity, sugar etc. In particular astringency is a multicausal perception, which is controlled not only by tannin but also by a multitude of matrix variables (Chapter 5). Sample centrifugation facilitates sensory tannin evaluation to a certain extent, but does not standardize the variable matrix composition impact as it only removes suspended solids. This is a typical situation when total phenol (Chapter 3.1.1) or tannin measurements (Chapter 3.4.3) come into play.

One of the central issues of contemporary maceration management is to know how much time it must last under given boundary conditions (temperature and cap treatment intervals) to extract a certain amount of tannins. To answer this question, the fast and simple analytical measurements mentioned above, in particular total phenol measurements as outlined in chapter 3.1, can make a valuable contribution to decision-making in a winery setting.

4.1.3. Varietal differences

Grape varieties differ by their amounts of extractable anthocyanins as well as skin and seed tannins, and these differences even extend to their clones. Data of flavan-3-ols are considered useful to classify varieties. The large variability of grape flavan-3-ol and proanthocyanidin composition provides at least a partial chemical explanation for the large variability of tactile-astringent sensations of red wines (Mattivi et al. 2009), although their extraction conditons and fruit ripeness also play a major role.

Figure 20 shows the extraction kinetics of total phenols and anthocyanins in musts from two fairly different cultivars, Merlot and Pinot noir. In this case, skin contact results in two completely different extraction patterns concerning the amount and the composition of the tannin-anthocyanin complex. Data were collected under normalized conditions on a pilot scale allowing for maintaining a constant temperature at 25° C (77° F). This temperature was chosen because it represents a good average of the temperature most red wine fermentations are conducted at. Musts were homogenized in regular intervals of eight hours.

Figure 20: Extraction kinetics of total phenols and anthocyanins during maceration of two different red grape varieties.

Temperature = 25° C (77° F) constant, pumping over three times per day during active fermentation and once per day after fermentation completed.

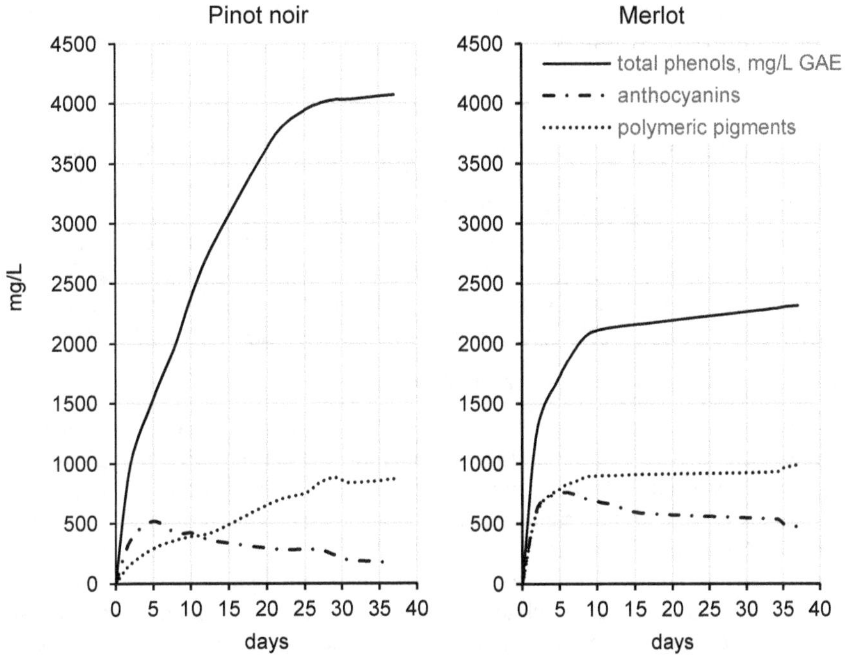

In the Merlot must, the phenolic material occurring in the fruit was almost entirely depleted after 10 days of skin contact. Further maceration did not yield a meaningful increase of total phenols, but caused qualitative alterations of the total phenols. They can be summarized as a decrease of anthocyanins accompanied by a slight increase of polymeric pigments. These reactions are basically identical with those occurring after pressing and during aging. More important, however, is the observation that after extraction of 2,300 total phenols (as GAE)[29], the phenolic potential of the fruit was exhaustively extracted, indicating that it was not sufficient for producing a premium red wine.

In contrast, in the Pinot noir must, extraction proceeded over one month before the extraction curve flattened at approximately 4,000 mg/L total phenols (as GAE). The grapes had higher amounts of extractable phenols, which could be extracted by means of a longer skin contact period. In this case, an exhaustive extraction only makes sense when the fruit has achieved a high level of phenolic ripeness ensuring good sensory quality of the tannin (Chapters 5.3.3), and when the intention is to produce a rich, full-bodied red wine with at least 3,500 mg/L total phenols, aging potential, and scheduled for prolonged barrel aging. Otherwise, one would stop skin contact after 8 days at

[29] GAE = gallic acid equivalents

2,000 mg/L total phenols and press to obtain a fruity red wine ready to be marketed and drunk in a few months.

Merlot and Pinot noir are two opposed varieties, which display a different phenolic structure for a comparable level of total phenols. Such differences are partially explained by variable amounts of extractable anthocyanins. Those reported for Pinot noir in figure 20 are within the low concentration range typical for wines from that lightly colored variety. To some degree, they correlate with the color intensity displayed by the wines after pressing and aging.

In Merlot, anthocyanins and the polymeric pigments resulting therefrom constitute a much larger portion of total phenols than in Pinot noir. They convey a tannin structure that is described as relatively soft. This facilitates harmonization of the wine. In contrast, musts from Pinot noir tend to yield more monomeric flavan-3-ols and low-molecular weight tannins in conjunction with considerably less anthocyanins. This specific phenolic composition of Pinot noir requires much sensitivity with regard to oxygen incorporation during the aging process. The reasons are described in chapter 7.4.

Fast extraction of anthocyanins and color

The only common feature of both varieties was a maximum anthocyanin concentration observed early at the beginning of maceration. This behavior is typical for all Vitis vinifera varieties. Anthocyanins are easily soluble in an aqueous medium and start to dissolve in the liquid right after crushing and largely regardless of the presence of alcohol. In musts of 13 cultivars subject to a constant 25° C and three pump-overs per day during active fermentation, maximum anthocyanin concentration was achieved after 5.5 ± 1.5 days (Schneider 2009 a). After that period, anthocyanins start being converted into polymeric pigments and decreasing, a process that continues to proceed over many years after fermentation until their complete depletion.

Practical hint

In contrast to wine, red must is a heterogeneous mixture of solids and partly fermented juice, thus making it difficult to obtain representative must samples for phenol or any other measurements. Therefore, samples should only be taken immediately after a thorough must homogenization using pumping over, punching down, or any other technique of cap management referred to in chapter 4.1.12. Otherwise data obtained will be far from the average of the fermentation bulk, and extraction curves will become irregular. When working on large volumes, mixing several samples taken at the same point of time but from different heights of the fermentation vessel will help make results more accurate.

4.1.4. Importance of maceration length

After anthocyanin accumulation in must is completed after a few days, any further increase of total phenols is only due to the extraction of colorless phenolics comprising flavan-3-ols (Chapter 2.2.2) and tannins (Chapter 2.2.3). In the course of this process, highly variable levels of total phenols and tannins can be achieved. The difference lies in fruit quality and ripeness. Figure 21 depicts this dissimilarity on the basis of three musts of the same grape variety, Cabernet Sauvignon. The levels of total phenols provided by these musts allow classification of the quality and style of the final wines from a phenolic perspective: Must I fulfills the conditions to yield a full-bodied, rich and dark premium red wine with a high aging potential, must II will not provide more than an average standard red wine, and must III will give a thin, ordinary red wine of dubious quality making the transition to rosé wines. In ripe fruit, tannins show higher concentrations and a higher extractability than in unripe fruit.

Figure 21: Total phenols extraction kinetics at a constant temperature of 25° C (77 F) in three batches of Cabernet Sauvignon grown under different viticultural conditions and harvested at different degrees of ripeness.

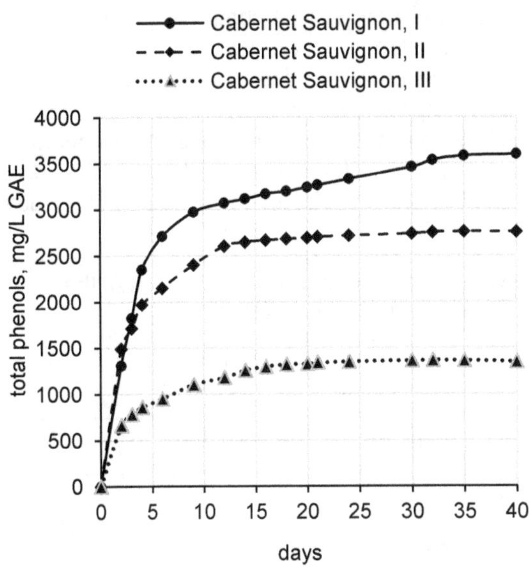

Moreover, figure 21 illustrates some general features of tannin extraction that apply to all grape varieties:

- Depending on cultivar and fruit quality, grapes contain highly variable amounts of extractable phenols in general and tannins in particular. These differences can exceed a factor of 3.

- Thus, when working with unfamiliar cultivars or in unfamiliar growing areas, extractable tannin must be determined by analytical or empirical means.

- Under comparable conditions, tannins dissolve at a highly variable rate in the liquid, whilst anthocyanins dissolve in a predictable period not exceeding one week.

- The extraction of tannin during maceration has no connection with fermentation progress.
- Extending maceration beyond the end of primary fermentation can extract additional amounts of tannin, but should not automatically be allowed to do so.
- Total phenol extraction during maceration comprises three distinct phases: It starts with an initial fast increase, which is due to the rapid dissolution of anthocyanins. It comes to an end when the extraction curve reaches a third flattened section, in which the daily total phenol increase is no longer significant and levels out asymptotically. The second stage in the middle shows a decrease of the total phenols extraction rate and can make, depending on the fruit, a sharp or a fluent transition from the first into the third stage.

- Under comparable maceration conditions, the duration of maceration does not allow prediction of the ultimate tannin level that will be achieved. On the contrary, for a given maceration length, the tannin extracted will depend on fruit quality as affected by cultivar, growing conditions, and phenolic ripeness, and in particular the amount of extractable phenols the fruit contains. Fruit composition tends to have more impact on the phenolic composition than the maceration regime (Zimman et al. 2002).
- Measuring total phenols is a useful way to optimize the timing of pressing.

Most tannins are extracted after 10 days of skin contact, but there are exceptions

Under real-world conditions, container capacities for running macerations are limited. Therefore, there is much interest in knowing on a more general basis how much time is required to extract a satisfactory amount of tannin, and how much tannin is dispensed with when pomace contact is shortened.

The musts featured in figure 21 were part of a larger trial comprising a total of 13 musts from known and less known varieties grown in various European countries. They were submitted to maceration under conditions as described before, i.e. 25° C (77° F) and three pump-overs per day during active fermentation and once per day after completion of fermentation. Table 2 shows, among other data, the maceration length required to achieve the maximum possible level of total phenols (TP_{max}) in these musts. The TP_{max} value was defined as the total phenol concentration at the day after which its increase became insignificant, i.e., less than 15 mg/L per day.

Table 2: The relationship between extractable total phenol content (TP_{max}), the maceration length to achieve it, and press yield after maceration.

cultivar	press yield (% v/v)	TP_{max} (mg/L GAE)	maceration length (days) to achieve TP_{max}	total phenols after 10 days maceration	
				mg/L GAE	% of TP_{max}
Pinot noir I	72.9	4,018	27	2,379	59.2
Pinot noir II	75.2	3,621	42	2,197	60.7
Cabernet Sauvignon I	80.5	3,560	32	2990	84.0
Cabernet Sauvignon II	76.9	1,316	18	1170	88.8
Cabernet Sauvignon III	72.7	2,666	16	2,417	90.6
Sangiovese	72.9	3,381	24	2,696	79.8
Merlot	73.8	2,293	24	2,103	91.7
Zweigelt I	73.5	3,182	25	2,952	92,8
Zweigelt II	75.1	2,685	20	2,511	93.5
Dornfelder I	78.6	1720	14	1,666	96.9
Dornfelder II	82.3	2313	9	2,318	99.8
Regent	80.8	3,599	22	3,319	92.2
Portugieser	87.9	1,423	16	1,294	90.9
average	*77.2*	*2,752*	*22.2*	*2,308*	*83.9*

Results from table 2 can be summarized as follows:

- Maceration length required to completely transfer extractable phenolics (TP_{max}) from fruit into must varied between 9 and 42 days under the conditions described.
- When TP_{max} increases, maceration length also increased ($r^2 = 0.46$).
- The average of total phenols extracted after 10 days of maceration was 83.9 % of TP_{max} and ranging from 1,316 to 4,018 mg/L GAE. This is an important assessment for practical reasons because container capacity for maceration is limited under real-world conditions. Thus, there is a trend to shorten macerations and press early, frequently at the end of fermentation, and dispensing with some tannins left in the fruit in favor of free fermentor capacity.
- Pinot noir was an exception with only 60 % of TP_{max} extracted in 10 days, which amounts to a loss of 40 %. This behavior is in line with the slow extraction kinetics of Pinot noir shown in figure 20 and typical for that variety. Clearly, Pinot noir is a cultivar that requires longer maceration periods for extracting its phenolic material, i.e. its tannin, than most other varieties.
- There was a weak but significant correlation ($r^2 = 0.18$) between berry size expressed as press yield and TP_{max}. At a comparable ripeness, varieties characterized by small berries tend to yield more total phenols than large-berry varieties because they provide less juice to dilute phenols. Taking into account the importance of the juice-skin ratio, varieties with small berries are preferred for the production of premium red wines.

Commercial tannin additions rarely compensate for tannin deficiencies

For producing powerful and vigorous red wines characterized by relatively high total phenol levels, the potential of extractable phenols the fruit displays is at least as important as the maceration duration. The commonly employed addition of commercial tannins in the order of magnitude of 100 mg/L only increases total phenols by 4 % when their initial concentration is 2,500 mg/L in a typical red wine. Commercial tannin additions can rarely compensate for the lack of phenols in underripe fruit or their deficient extraction during the winemaking process, though they might impart minor improvements (Chapter 4.1.15).

4.1.5. Impact of phenolic ripeness

In the wines referred to in table 2, extractable tannins expressed as TP_{max} increased with increasing Brix readings. However, this correlation was rather weak with $r^2 = 0.34$. This is in contrast to the still widely held view that Brix readings can be used to assess phenolic ripeness. They only indicate alcoholic ripeness, which is something quite different and poorly related with phenolic or physiological ripeness, in particular under conditions of ongoing climate change. A glance at figure 22 helps better understand in a more memorable manner how little Brix values have to do with the extractable phenolic potential of grapes. They are a bad guideline of what will become the very essence of red wine.

Figure 22: Brix readings and total extractable phenolics - a bad correlation. Data obtained from various vineyard blocks and varieties across Europe.

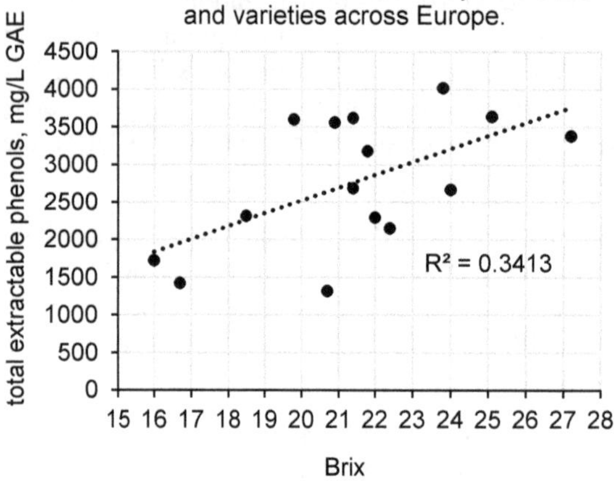

In conclusion, phenolic ripeness is more important than the berry size, the juice-to-skin ratio, and Brix readings. It determines the amount and the quality of tannins that can be extracted from the fruit. When phenolic ripeness increases, the contribution of soft skin tannins to total tannins also increases, whilst the contribution of harsh seed tannins decreases. The astringency caused by the total pool of tannins is higher when grapes are unripe. Interestingly, the main degree of polymerization of the tannins from skins and seeds was shown to be unaffected by ripening and maceration length, suggesting that other factors than polymerization affect perceived astringency (Llaudy et al. 2008). Based on these facts, the assessment of phenolic ripeness is of particular importance.

Assessment of phenolic ripeness

There is no lack of recommendations for assessing phenolic ripeness of red grapes (Cayla et al. 2002, Glories and Saucier 2000, Gunata et al. 1987, Rousseau and Delteil 2000, Harbertson et al. 2002, Celotti and Carcereri de Prati 2005, Jensen et al. 2007, Cagnasso et al. 2008, Rolle et al. 2012). They use different solvents and extraction conditions to extract the phenolic material from the fruit and quantify it by different analytical means. Since there is a lack of standardization, results are not comparable. Hence, these measurements are not presented in this context.

In practice, the phenolic ripeness of grapes is usually assessed by sensory means. For that purpose, it is useful to follow a consistent sampling protocol comprising grapes from all parts of the vineyard block and all parts of the bunch in predetermined proportions. Pulp, skins, and seeds have to be separated from the berries and tasted individually. Ripe grapes display the following characteristics:

- Berries can easily be separated from the stems without becoming damaged.
- Berries remain pressed upon light finger pressure, whilst unripe berries are elastic and return to their original shape.
- The flesh is no longer marrowy and gelatinous, but largely liquefied.
- The smell of crushed berries is reminiscent of red or black berries and does not display the smell of green bell pepper or freshly mown grass.

- Skins are evenly colored, soft, and friable, but not damaged. Upon chewing, they do not leave any dryness or herbaceous notes.
- The seeds are brown, largely lignified, and can be easily detached from the flesh. Seeds sticking to a gelatinous flesh indicate a lack of ripeness. However, attention should be paid to the impact of variety. Some varieties' seeds lignify quite easily, while others never reach full lignification, although the fruit is in full maturation.
- When seeds are bitten with the teeth, they appear crunchy, not very astringent, and release nutty flavors.
- The stems are brown and lignified, more or less like the seeds. In such a case, a minor part of the stems can be left in the fermenting must. This way of working is sometimes used to extract additional tannins from the stems, but it requires much care to not extract too much tannin.

The sensory characteristics of unripe berries will be found again in the resulting wine. Especially under cool-climate conditions, the ripeness level has a bigger impact on mouthfeel and phenolic composition than the maceration regime. Conversely, the effect of the tannin extraction methods on the sensory characteristics of the wines is more pronounced when phenolic ripeness is achieved. Wines that are harvested before complete ripeness tend to show more 'adhesive', 'puckering', 'grippy', and 'sour' characteristics on the palate, whereas wines obtained from ripe fruit display a tannin providing sensations one can describe as 'grainy', 'particulate', and 'sweet'. The sensory subqualities of tannin are discussed in chapter 5.

4.1.6. Extended macerations

Many winemakers typically ferment the must until the sugar level has dropped to between 5 to 0° Brix regardless of the phenol extraction achieved. Depending on fermentation conditions, it may take five to seven days to reach this point. Musts still containing some residual sugar are expected to continue fermentation after they are pressed. This approach yields wines with good color, fruity flavor, and a soft and round mouthfeel. They can be consumed when relatively young or after a short maturation period. Their low tannin levels do not require prolonged aging to achieve a higher quality.

The practice of extended maceration prolongs skin contact after the must has fermented to dryness. It is performed by winemakers seeking to extract more tannin and can take a few days to several weeks. For that purpose, the tank is closed, while the must is generally left undisturbed and the cap allowed to sink to the bottom. However, other winemakers may continue must homogenization by pumping over once per day or per week. This approach requires a high degree of phenolic ripeness and brown seeds to limit extraction of aggressive seed tannins.

Based on what is known about phenol extraction, extended maceration is expected to only increase tannin extraction and the formation of polymeric pigments, but not anthocyanin content. Thus, the resulting wines display higher color stability, but also lower color intensity due to losses of anthocyanins by their progressive binding onto fermentation solids or by oxidation. There are numerous studies that corroborate these expectations (Scudamore-Smith et al. 1990, Fischer 1997 a, Yokotsuka et al. 2000, Reynolds et al. 2001, Harbertson 2009, Weiand 2010, Frost et al. 2018), most of them invoking higher astringency ratings, shifts in the aroma profile towards cooked overtones, or better general quality scores in the resulting wines. Hence, this approach is often recommended for the production of full-bodied and tannic red wines with aging potential when the technical and microbial risks associated with it can be avoided. The final result will depend on cultivar (Auw et al. 1996), fruit ripeness, and the length of the maceration. Grape composition can easily overwhelm the effect of extended maceration (Zimman et al. 2002). In some areas or years, a longer hang-time can be more useful to obtain more phenolics.

In general, there is not much to gain from lengthy macerations of fruit grown under hot-climate conditions, which can produce more commercially viable wines when briefly macerated. Under cool-climate conditions, it is only advisable when complete phenolic ripeness is achieved. Otherwise the extraction of harsh seed phenols will cause an unpleasant astringency.

The critical point of extended macerations is that without any analytical measure of phenolics, the outcome is highly variable and hardly predictable. As most grapes release roughly 85 % of their phenolic material within the first 10 days of skin contact at 25° C (Table 2), the effect should not be overestimated or generalized. However, it can become significant in slowly extracting varieties such as Pinot noir (Figure 20). When it is actually significant, the tannin-anthocyanin ratio will shift to higher values. This equilibrium shift affects the sensory perception of a given amount of tannin (Chapter 5.3.2).

When phenols have already been exhaustively extracted at higher temperatures or after an intensive cap management regime, one cannot expect much more extraction within a reasonable period of time. Then, prefermentation juice runoff is a more reliable approach to concentrate tannin when producing premium red wines (Weiand 2010).

Maceration length also influences tannin composition

Regardless of whether extended macerations do or not lead to higher tannin levels, they do have an effect on tannin composition. When anthocyanins reach their maximum concentration after a few days, they start being converted into polymeric pigments providing stronger color intensity and stability in the finished wine. Further analysis of data underlying table 2 has shown that anthocyanin decay and polymeric pigment increase are strongly inversely correlated ($r^2 = 0.83$). However, these reactions are not a particular feature of extended macerations as they also occur in the wine after pressing. The most likely unique characteristics of extended macerations comprise the following features:

- The continued presence of the entire yeast lees, which would otherwise be partially depleted upon pressing and discarded with the pomace, has an impact on tannin and astringency. Post-fermentation yeast lees release macromolecules that interact with tannins, thus modifying their astringency in a way that is considered beneficial. Details are outlined in chapter 5.3.3.

- When anthocyanin extraction is finished, further pomace contact only enhances tannins. As a result, the tannin-anthocyanin ratio shifts towards higher values. This index is of crucial importance for the maturation process of red wines. It determines to what extent the wine requires or tolerates oxygen uptake during aging. At a comparable total phenol level, wines with a high tannin-anthocyanin ratio are more astringent and more oxygen-sensitive than those displaying more anthocyanins. Their maturation requires more time, because its promotion by the various means of oxygen supply (Chapter 8) is only feasible to a limited extent. Ultimately, wines obtained from extended macerations require a longer maturation period, which excludes their use for early marketing and consumption.
- Tannins extracted during extended macerations are primarily seed tannins, which can be undesirable for sensory reasons, in particular when underripe fruit is processed. The additional amounts of tannin that might be extracted by extended macerations do not necessarily improve sensory quality and market value of the wine.

4.1.7. Partial seed removal: Skin and seed tannins

Seed tannins endure a bad reputation with winemakers, and it is useful to address their properties in comparison with skin tannins. At the beginning of maceration, more skin tannins than seed tannins are extracted. Seed tannins are less extractable; they dissolve in the must more slowly and with some delay (Ribéreau-Gayon and Milhé 1970, Ribéreau-Gayon et al. 1970, Joutei and Glories 1994, Koyama et al. 2007, Lerno et al. 2017). This delay is explained by the prior requirement for hydrolysis of the lipid layer around the seeds as well as the hydration of the seed tissue itself. However, dissolution of seed tannins starts much earlier when seeds have been mechanically damaged during crushing, but that kind of technical shortcoming has become rare. A good crusher opens the berry without breaking the seeds.

Skin-derived tannins display a higher degree of polymerization than seed tannins and are strongly associated with polysaccharides (Chapter 5.3.3), which makes them convey a soft and round mouthfeel on the palate. In contrast, seed tannins show a lower mean degree of polymerization, a higher percentage of galloylation, less association with polysaccharides, and noticeably more astringency per concentration unit than skin tannins.

The concentration of skin tannins and their proportion in total tannins has been shown to positively correlate with wine grade allocation and projected bottle price (Kassara and Kennedy 2011). Nevertheless, the divergent sensory quality of skin and seed tannins is subject of ongoing debate in red winemaking. Since both seed and skin tannins are commercially available with an acceptable degree of purity, it is easy to understand the sensory difference between them by tasting a low-phenol red wine or model solution after a 200 mg/L-addition of one or the other. Whilst skin tannins display limited astringency and provide a perception of mellowness and fullness on the palate, seed tannins cause a sharp increase of drying astringency (Canals et al. 2008). This kind of astringency can have far-reaching impact on quality.

In general, the concentration and extractability of seed tannins decrease as grape maturity progresses, while skin tannin concentration increases. Thus, the so-called visual seed maturity, which occurs when grape seeds are brown, can become a critical factor and strongly impact general wine quality in bad harvests or cool climates. Tannin extraction techniques strive to favor skin tannin extraction and to reduce seed tannin extraction (Chapter 4.1.7). This is the reason why sometimes technical measures of partial seed removal are taken during maceration. Modern fermentors can be equipped with screens and some other features that offer technical options for that purpose, which were unknown some decades ago.

Seed removal unavoidably goes at the expense of total phenols, which are the essence of any red wine. Early work (Singleton and Draper 1966) provided a rough estimate of seed tannins showing that complete extraction of seeds as the sole source of tannin can provide 2,000 to 4,000 mg/L tannin measured as total phenols, and that about half of that amount would be extracted during usual red wine fermentation conditions, making up approximately half of the tannin of a typical red wine. The exact contribution of seed tannins to total tannins also depends on the variety. Thus, Pinot noir is known for its extraordinarily high percentage of seed tannins. In this context, the number of seeds per berry is more important than the amount of tannin per seed (Harbertson et al. 2002).

Figure 23: Extraction of total phenols (A 280) as affected by total seed removal during fermentative and extended maceration of two varieties: Merlot (ME) and Blaufränkisch (BF).

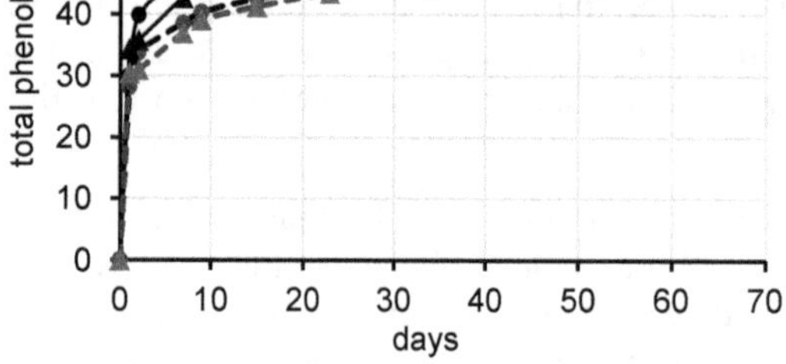

Figure 23 shows the effect of total seed removal before the start of fermentative maceration by the example of two varietal musts, Merlot and Blaufränkisch. In this trial, total seed removal was achieved by driving the crushed grapes across a sieve by means of a blade such that the seeds and juice pass through apertures of the sieve while retaining the skins. Subsequently the juice was separated from the seeds and recombined with the skins and pulp for fermentation. Under these conditions, seed tannins contributed to 13 and 16 %, respectively, of the total phenols measured after 9 days of maceration, but to 28 % on average after 50 days. With increasing maceration length, the share of seed tannins in total phenols rises disproportionately, while the extraction of skin tannins is strongly slowed down if not completed.

In general, total phenol losses after complete seed removal are in the range of 20 % when short macerations not exceeding two weeks are run, depending on cultivar, berry size, the number of seeds per berry, etc. They are even less when only partial seed removal is performed in commercial winemaking, where the must is already half fermented before seed removal can actually take place.

When tannins are measured instead of total phenols, the data take other values. After a 12-days maceration period of Merlot, tannin measured as procyanidins was only slightly lower after seed removal as compared to the reference without seed removal (Lee et al. 2008). However, the percentage decrease of tannins and the astringency resulting therefrom can be much larger when underripe fruit or fruit with a high seed-to-skin ratio is processed. Thus, partial seed removal can be a valuable option when there is a risk of extracting harsh tannins from green seeds.

The effect of partial seed removal will be even greater after longer maceration periods because seed tannin extraction is facilitated by previous seed hydration and swelling of the seeds, which requires approximately two to three weeks. Any subsequent increase of tannin and total phenols reflects the exclusive extraction of seed tannins. Thus, after one month of maceration, the amount of seed tannins can even be higher than that of skin tannins in most of the finished wines, with the exact ratio depending on the variety (Casassa et al. 2013 a, 2013 b). This behavior suggests that the maceration period should not exceed two weeks if one wishes to limit the extraction of astringent or aggressive seed tannins.

Due to the specific gustative properties of seed tannins, their deliberate enhancement by extended macerations requires a longer maturation period of the wine in order to soften their harshness via polymerization with anthocyanins and integration into polysaccharide complexes (Chapter 5.3.3). Sometimes their pronounced astringency is used to elaborate what is called structure and texture on the palate. When doing so, the ideal ratio of seed tannins to skin tannins is an open issue. The question is about personal preferences because wines focused on structure and backbone driven by prominent tannins are the opposite of those characterized by mellowness. Preference tests by consumer panels indicate that the preferred mouthfeel of premium red wines comprises rather smooth tannins and a creamy finish than a lingering astringency. In any case, excessive extraction of seed tannins excludes early commercialization and can cause the need for fining with a proteinaceous fining agent (12.1.3) to lower astringency.

Practical hint

Whenever extended maceration is run, care should be taken to thoroughly close the fermentor with a fermentation lock when fermentation is still active in order to maintain the carbon dioxide atmosphere in the headspace and to avoid oxygen uptake. Otherwise dissolved oxygen will easily stimulate bacteria growth and the appearance of volatile acidity or ethyl acetate. Furthermore, it risks to over-oxidize the must and to precipitate a part of the tannin that has been extracted with much effort and time.

4.1.8. Pre-fermentation juice runoff

As previously discussed, the extractable phenolic potential of grapes is subject to limitations, which are predetermined by viticultural factors and cannot be overcome by extended macerations. Prefermentation juice runoff, also called saignée or bleeding, is a practice to concentrate phenolics by removing a portion of juice right after crushing, thus increasing the solids-to-juice ratio and mimicking a smaller berry size. Since the anthocyanins and tannins extracted during and after fermentation are found in the skins and seeds, this practice is meant to increase their concentration in the finished wine without altering the wine's basic chemistry. As a result, these wines generally display increased flavor, astringency, color, and aroma.

From a different perspective, saignée can also be used to counteract the phenolic and aromatic dilution effect of water additions to high Brix musts meant to lower alcohol content. When it is paired with water additions, the net result is a lower Brix must that still possesses the initial solids-to-juice ratio and the potential of extractable phenols resulting therefrom (Harbertson et al. 2009). This is an interesting scope of application with regard to ongoing climate change.

Without water additions and for the sole purpose of concentrating phenolics, compositional changes induced by increasing juice removal were more linear than the sensory ratings of the resulting wines. Comparison of chemical with sensory data indicated that in wines averaging about 1,500 mg/L flavonoid phenols (as gallic acid), a difference in flavonoid content of 268 mg/L did and 165 mg/L did not make a sensorially significant difference (Singleton 1972). This is partially explained by the difficulty in rating astringency (Chapter 5.2). Moreover, sensory differences or benefits obtained by bleeding cannot always be detected by tasters or tend to diminish during aging (Gerbaux 1993, Zamora et. al. 1994, Gawel et al 2001, Fanzone et al. 2013).

In order to ensure a sensorially significant effect, it is reasonable to target a juice removal of some 10 to 20 % of the total volume. It is likely to have the greatest positive effect when working with large berries, especially if rain close to harvest has caused them to swell. After more than 30 % juice removal, cap treatment (Chapter 4.1.12) becomes difficult. When compared to a standard maceration length of seven days at 20 and 25° C for three cultivars, previous juice removal of 15 % showed a quality increase in the resulting wines comparable to that obtained by macerating for seven days at 30° C (Weiand 2010). A higher fermentation temperature can be more beneficial than juice removal.

Anthocyanins and other phenolics begin to be extracted from grapes one hour after crushing. Hence, juice that is removed prior to this time displays very low amounts of these compounds, thus allowing for the increase of their levels in the bled must on the basis of a higher solids-to-juice ratio in the latter. On the other hand, draining off juice at this early stage is cumbersome as the phase separation between liquid and solids has not yet taken place. It is easiest once fermentation has commenced and the cap has risen, but this will partially negate the concentration effect in the original must. The removal of juice must be early enough to prevent it from taking up significant amounts of tannins and anthocyanins. Nevertheless, the juice that has been removed always contains some color. It is fermented separately and made into a rosé style wine.

It is crucial to bear in mind that saignée concentrates all flavors, including harsh tannins and green vegetal aromas of unripe fruit. Hence, it should only be performed on good quality grapes.

4.1.9. Impact of maceration temperature

Fermentation temperature has a tremendous impact on phenol extraction and general red wine quality, albeit many producers are not yet fully aware of that. Very early investigations (Amerine 1955, Amerine and Ough 1957) have shown that temperatures of 21 and 27° C produced more color and flavor than lower temperatures around 15° C. This effect is further differentiated by a temperature gradient, which exists between the inner and the outer part of the fermentor, and between the cap and the liquid. Under conditions of uniform temperature of cap and liquid in Cabernet Sauvignon fermenting in a research scale fermentor, after 10 days at 35° C tannins were approximately 100 % higher and total phenols 50 % higher than at 20° C (Lerno et al. 2015).

Figure 24: Effect of temperature on total phenol extraction in Pinot noir.

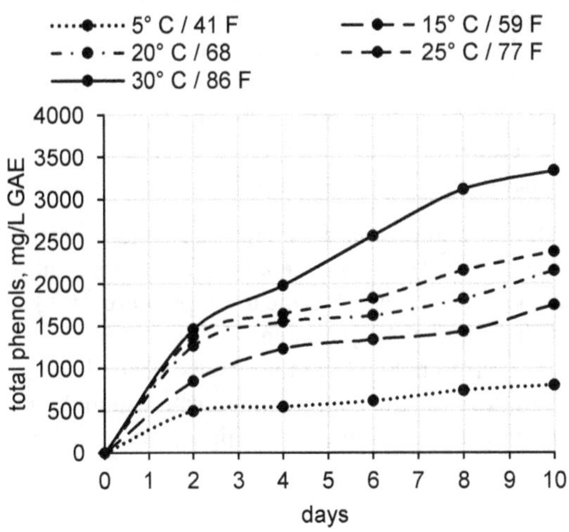

Figure 24 depicts in more detail the temperature effect in a Pinot noir must held under comparable conditions. After 10 days' maceration, total phenols extracted at a uniform and constant temperature of 30° C were almost the double those at 15° C, whilst extraction at 5° C was negligible. For skin phenolics, temperature affected the rate of extraction but not the final concentration, with increasing temperatures favoring faster extraction. For seed phenolics, increases in fermentation temperature increased both the rate of extraction and the final concentration. The must temperature appears more important than cap temperature in driving the extraction of phenolics (Lerno et al. 2015). It also seems to have more effect on phenol extraction than the pump-over or punching-down regime.

The temperature effect on anthocyanin extraction is even more pronounced. However, in contrast to total phenols consisting primarily of colorless flavan-3-ols and tannins, they reach a peak within a few days, depending on variety, after which they decrease due to their conversion into polymeric pigments (Figure 25). The higher the temperature during the skin contact period, the earlier they reach that peak, and the faster they undergo conversion into polymeric pigments (Gao et al. 1997). Additionally, increased temperature, for example 25 or 30° C instead of 20° C, was also shown to increase body and fruity (berries, cherries) flavor intensity ratings in Pinot noir (Girard et al. 1997, Fischer 1997 a, Weiand 2010) and Shiraz (Reynolds et al. 2001). By analogy, the same behavior can be expected in other varieties.

Figure 25: Effect of temperature on anthocyanin extraction in Pinot noir.

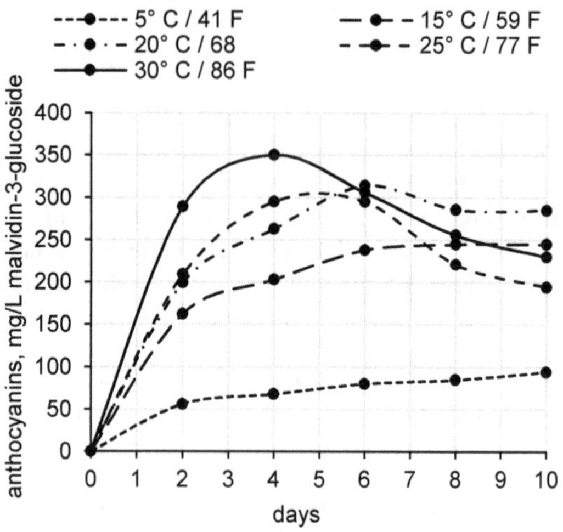

Temperature as a prime factor in phenol extraction has led to the development of thermovinification (Chapter 4.2.2), which applies a temperature above 65° C to shorten the time-consuming maceration to a few hours of pomace contact before pressing and inoculation[30].

However, temperature is not always given adequate attention when fermentation takes place during standard pomace contact. Especially under cool-climate conditions, too low a temperature can pose a serious problem and lead to heavy losses of tannins, anthocyanins, fruity flavor, and overall quality. Under these conditions, it is strongly recommended to attach to maceration temperature the importance it actually has and apply must heating to 25 to 30° C if necessary.

Under hot-climate conditions, the picture is completely different because temperature easily gets out of control during fermentation when no adequate refrigeration equipment is available. Most fermentations get stuck when a critical temperature of 33 to 35° C is reached, depending on yeast strain and alcohol content. A maximum temperature of 30° C as displayed on the tank thermometer should not be exceeded because thermometers only indicate the temperature near the tank wall, whilst the actual temperature in the inner part of the fermentor can be appreciably higher and cause the fermentation to stick.

4.1.10. Post-fermentation heating

In the context of standard macerations generally not exceeding one week or so, heating the must to 35 to 40° C (95 to 104° F) for two days after the end of fermentation is sometimes performed to reinforce tannin extraction. For that purpose, it is absolutely indispensable that the must has already fermented to dryness. Otherwise the

[30] Inoculation with an active dry yeast strain is mandatory after thermovinification because the indigenous yeast is inactivated by heat.

fermentation would get stuck and leave residual sugar behind. Furthermore, a temperature gradient frequently occurs during the process, a problem that can be best overcome by the use of a rotary fermentor. Ultimately, the tank must kept closed and the must should be cooled down before pressing to avoid excessive losses of alcohol. The heating and cooling processes require considerable energy input.

On the other hand, hot pressing is a good means to deliberately lower excessively high alcohol levels by some 0.5 to 1.0 % without relying on technically more sophisticated techniques of alcohol reduction. Whenever it is performed, it is strongly recommended to immediately cool the wine obtained in order to avoid bacterial spoilage and formation of volatile acidity, especially when the wine is in contact with atmospheric oxygen in containers not completely topped.

For Pinot noir (Gerbaux 1993, Gerbaux et al. 2003, Fischer 1997 b) and several other varieties (Weiand 2010, Lukić et al. 2017), holding the must at 35 to 38° C (95 to 100° F) for two days after a regular fermentation period yielded more tannin, aroma, and overall quality than the reference lots pressed after one week. This approach corresponds to a shortened version of extended maceration at standard temperature. Results are similar: When fruit is underripe, however, harsh seed tannins and green-vegetative aroma compounds appear under these conditions.

4.1.11 Cold soak

The rationale of cold soak is to prolong the total duration of skin contact by a preliminary extraction in aqueous medium, which precedes the extraction in alcoholic medium during active fermentation. The purpose is to favor a more selective extraction of anthocyanins, which are easily soluble in water, to the detriment of tannins and other colorless phenols that better dissolve in the presence of alcohol.

For cold soak treatment, the must is cooled with dry ice added in layers, external heat exchangers, or in double-jacketed cooling tanks to delay the beginning of spontaneous fermentation. Sulfur dioxide is also added. Subsequently it is held at a temperature ranging from 5 to 12° C (41 to 53° F) for several days before inoculation with active dry yeast.

Cap treatment by punching down or any other means (Chapter 4.1.12) must take place once or twice a day. For the rest of the time, the surface of the vessel should be covered with a plastic sheet and/or dry ice over the cap to limit oxygen uptake and the development of spoilage organisms. After pressing, the must is heated to ambient temperature and inoculated to start fermentation.

Ultimately, cold soak is primarily meant to shift the tannin-anthocyanin ratio to less tannin and more anthocyanins, accepting that the whole maceration process takes more time due to the reduced phenol extraction rate at low temperatures. It is predominantly used on Pinot noir, which is known as a cultivar displaying high levels of seed tannins and flavan-3-ols (Zou et al. 2008) in conjunction with low levels of anthocyanins. As a consequence, most research reports on cold soak focus on that cultivar. However, results obtained on either Pinot noir or other cultivars are highly conflicting.

The reason for this is the variable maceration conditions during the subsequent alcoholic extraction at higher temperatures. They blur the effect of cold soak on the resulting wine. Hence, cold soak is one of the most controversial topics in red winemaking.

Historically, cold soak was born in Burgundy, where cool temperatures at harvest often delayed the onset of fermentation by several days after crushing when no inoculation took place. In the absence of yeasts, the musts remained almost clear and showed better their color, whilst yeasts resulting from an early onset of fermentation partially mask color. This observation led to an error of judgment because low temperature slows down rather than accelerates the dissolution of coloring anthocyanins. This behavior would only change in the presence of inadmissibly high levels (> 250 mg/L) of sulfur dioxide, which promote anthocyanin extraction.

Under the cool-climate conditions of Burgundy, cold soak produced Pinot noir wines with a fruity aroma, but with less color, structure, and volume on the palate (Gerbaux 1993, Feuillat 1996). These results have been basically confirmed by later research on different varieties in other countries (Fischer et al. 1997 b, Marais 2003 a, Zini et al 2003, Köhler et al. 2005, Geffroy et al. 2013, Nel et al. 2015, Lukić et al. 2017, Ghanem et al. 2019). In these studies, no significant advantage of cold soak was seen in the finished wine, confirming the general rule that what cannot be extracted from grape solids at high temperature is even less extracted at low temperature.

Nonetheless, there are also a few reports of positive cold soak outcomes under strictly specified conditions. When compared to the traditional maceration and fermentation procedure, Pinot noir yielded more red color and a stronger fruity aroma after three to four days of cold soak at 15° C under strict anaerobic conditions. However, several pumpings through a refrigeration unit used to maintain the temperature were considered helpful for the extraction of phenolics (Cuenat et al. 1996) and suggest an impact of mechanical action on the pomace similar to that of cap management techniques (Chapter 4.1.12).

With Shiraz, a cold soak at 2° C over 10 days affected herbaceous and black currant flavor depending on the subsequent fermentation temperature. It increased color and black currant flavor in conjunction with a high (30° C) fermentation temperature, of which it is seen as a valuable complement (Reynolds et al. 2001). In a trial over five years with a less popular variety from Corsica known as Nielluccio, an eight days' cold soak at 7° C followed by seven days of fermentation at 28° C yielded slightly more total phenols, anthocyanins, and color as well as more balance and concentration on the palate in the bottled wines (Retali 2004). In trials on Tempranillo, is was shown that copigmentation (Chapter 2.2.4) plays a major role in the color increase when this is obtained by cold soak (Gordillo et al. 2010). Generally, cold-soaked warm-climate fruit generates larger color differences than cold-soaked cool-climate fruit.

The question of the reference to compare with

In summary, it appears that the hotly debated matter of cold soak has positive effects only in a few individual cases. However, to evaluate its outcome, it is crucial to specify under which conditions it is performed and to which reference it is compared. In particular, a distinction must be made between cold soak as part of the total maceration length or as its prolongation. As an example: A cold soak of 4 days followed by 5 days of fermentative maceration makes up a total of 9 days (4 +5 = 9). Unavoidably, the same cold soak length gives different results when it is followed by 9 days of fermentative maceration (4 + 9 = 13).

When cold soak entails additional skin contact time as in the second case, it generates more phenolics, color, and volume on the palate (Weiand 2009). On the contrary, when it is performed as part of the total maceration length, the resulting wine tends to display less phenolics of all kind since their extraction rate is retarded in the cold. This applies in particular to seed tannins. In contrast, the extraction of skin tannins is less retarded by low temperatures. It can be almost completed when a 2-days cold soak at 13° C is followed by an 8-days fermentative maceration, making up 10 days of total maceration length (Koyama et al. 2007).

Many results on cold soak are misleading when fermentative maceration in the reference wine is much shorter than in the cold-soak trial (de Santis and Frangipane 2010). Under these conditions, it is not surprising that the cold soak yields more tannin, color, aroma, and overall quality. Incidentally, many winemaking trials become meaningless when the experimental conditions are not normalized.

In practice, cold soak is often subject to uncontrolled microbial activities. The onset of barely noticeable fermentation activities by wild yeast stains can contribute to aroma complexity in some cases. More frequently, however, it causes elevated levels of volatile acidity and/or ethyl acetate[31], which are less desirable.

4.1.12. Methods of cap management and their impact

As an introduction, it is worth mentioning that red winemaking techniques only extract a minor part of the tannins and anthocyanins occurring in the grapes. Depending on the cultivar, this was reported to vary between 5 to 61 % and without any relationship between the total tannin per berry and the amount of tannin in the resulting wine (Harbertson et al. 2002). The complete pool of phenolic material they contain can only be extracted and quantified by means of more severe extraction techniques using adequate solvents other than the aqueous-alcoholic solution that is wine. The proportion of tannins that is not extracted but retained in the pomace depends on its adsorption by grape pulp and yeast cell wall components and their subsequent removal from the wine with the grape marc, physical sequestration, and precipitation. Cap management techniques strive to optimize tannin and color extraction in a more a less selective way taking into account the difference between bad and good tannins.

After grapes are crushed, there is a phenol concentration gradient between pomace solids and juice that causes the phenolics to dissolve in the surrounding juice. This

[31] Ethyl acetate is an ester that can be formed by spoilage yeasts and acetic acid bacteria under conditions of oxygen availability. It has a boiling point of 77.1° C (170.8 F) and an olfactory detection threshold of approximately 100 mg/L. Higher levels are reminiscent of glue and varnish. It is frequently mistaken for volatile acidity, with which it can occur simultaneously.

concentration gradient is the driving force of any extraction process. When the concentration of phenolics in the pomace solids and the surrounding must becomes equal, phenolic extraction will slow or cease, and saturation will be observed. Then mixing of the must is required to regenerate the concentration gradient.

In red wine fermentations, the extraction process is complicated by the formation of the cap that develops at the onset of active fermentation when carbon dioxide carries the skins to rise to the top of the fermentor. Cap formation has two potentially negative consequences: There is reduced contact between the liquid and the pomace solids, and heat is trapped in the floating cap. Temperature differentials between cap and liquid can easily reach 10° C (50° F) during vigorous fermentation. They are caused by a faster fermentation of the sugar in the liquid adhering to the cap than in the bulk liquid (Guymon & Crowell 1977). However, cap temperature is much less important for driving extraction of phenolics than juice temperature (Lerno et al. 2015).

Cap management overcomes spatial heterogeneity and drives phenol extraction

Due to cap formation, red wine fermentations are not spatially homogeneous throughout the bulk of the fermentation. As phenol extraction takes place in the cap, a concentration gradient in phenolics develops between the cap and the bulk liquid. When the extraction of phenolics in the cap reaches an equilibrium plateau or saturation point, no further extraction is possible until a punch-down or pump-over washes the extracted phenolics from the cap into the bulk liquid, thereby allowing further extraction back to the same concentration. These cap management techniques are the primary mode of homogenizing phenolics throughout the fermentor, though some other minimal mixing also occurs due to thermal currents arising from the temperature gradients and carbon dioxide evolution.

Concentration gradients for anthocyanins and skin tannins occur early during fermentation and diminish as fermentation progresses. In contrast, seed tannins exhibit concentration gradients throughout the fermentation. After a pump-over used for mixing, these gradients have been shown to be progressively re-established and to reach equilibrium again after approximately eight hours for skin tannins and more time for seed tannins (Lerno et al. 2017). From this it could be deduced that the must should be mixed every eight hours at a minimum. Under commercial winemaking conditions, the rapid cap formation observed sometimes suggests a higher frequency would be desirable. Otherwise the phenol concentration gradient between pomace solids and adherent liquid will be cancelled out, and valuable time is lost. The need to overcome the temperature gradient between cap and liquid is an additional reason why musts have to be mixed several times a day using one or another technique of cap management. In most wineries, it is done twice to four times a day.

During extended macerations, the spatial heterogeneity of phenolic concentration diminishes due to a loss of concentration gradients as phenolic extraction slows or ceases between mixings. Therefore, cap management techniques are limited to one intervention per day in the post-fermentation stage. More mixing events would produce large amounts of fine lees leading eventually to a binding of free phenols with isolated grape tissue cell wall compounds.

The need to overcome concentration and temperature gradients explains why cap management and the way skins are handled is one of the basic stylistic decisions a winemaker has to make. Similarly to maceration duration and temperature, it can strongly shape a wine's finished style. It also influences the possible development of detrimental micro-organisms and promotes the alcoholic fermentation by addition of oxygen. Different techniques are used for this purpose. Punching down, pumping over, and rack and return are the most important of them. Their results strongly depend on variety (Fischer et al. 2000), although for a given variety one of them can turn out to be more beneficial than another (Marais 2003 b). The picture becomes even more complex when one takes into account how they are performed with regard to the volumes involved and the timing and frequency they are applied with. Therefore, they do not allow any generalization. However, regardless of specific choice, it must always be ensured that the cap is properly wet.

Punching down

Punching down the cap below the juice surface and breaking it up is the oldest and simplest method of mixing the cap and the juice. It can be done with different equipment breaking the cap and submerging the skins under the fermenting juice. However, done by hand or even feet in its most romantic manner, it generally extracts less phenolic material than mechanical punch-down treatments (Fischer et al. 2000), though it might work well on small vats with an open top. In larger fermentors, punch-down devices consisting of pneumatically or hydraulically powered plungers are used to press down the cap into the juice. These plungers or plate-on-pole devices may or may not rotate around their own axis. They tend to be a more aggressive extraction technique than manual punch-downs or pump-overs as they might grind some seeds and facilitate extraction of harsh seed tannins.

After a punching step during active fermentation, the cap takes some four to six hours to form again. Therefore, the optimum frequency of punch-downs is four to six times in 24 hours when the quickest possible phenol extraction is sought. Otherwise the phenol concentration gradient between pomace solids and adherent liquid will be cancelled out, and valuable time is lost.

When fermentation stops, the cap starts to break apart and fall. Thus, more than one punching per day is neither required nor helpful during extended macerations.

Pumping over

Pumping over is the method by which the juice is taken from beneath the cap, usually from the bottom of the tank, and sprayed over the top of the cap to wet the skins. As the juice percolates through the skins, it extracts the phenols and other flavor compounds. Since the juice tends to always take the same way through the cap, it is important to distribute it evenly by means of deflector plates or rotating nozzles.

The sprinkling device chosen can be automated and has significant influence on phenol extraction. Juice can be sprinkled gently over the cap, slowly wetting it, or it can be shot at high pressure, breaking up the cap and redistributing the solids within it. The pump capacity should be adjusted to the must volume and be able to recycle the fermentor volume once in an hour.

Minimum pump-over frequency is twice a day with half of the must volume pumped over each time in order to overcome the phenol concentration gradient between cap and juice (Lerno et al. 2018). More frequent pump-overs can facilitate phenol extraction when the intention is to accelerate phenol extraction and to save time, and pumping over twice a small volume is more beneficial than pumping over once a big volume. However, must temperature has more effect on phenol extraction than pump-over volumes.

In comparison with punching down, mechanical action on the pomace is lower as only the juice is moved and no plate or plunger comes into contact with the grape solids. Therefore, pumping over is considered to be a more gentle way of cap management, to produce less lees of broken grape tissue, and to extract less seed tannins (Fischer et al. 2000).

Additionally and in contrast to punching down, pumping over allows for absorbing some oxygen, which plays multiple important roles in polymerizing and softening tannins, stabilizing color, and improving the capability of yeast to ferment to dryness. Deliberate oxygen supply during fermentative maceration results in wines with tannin and color characteristics of more aged wines, less astringency, and less bitterness, which may reduce the need for extended wine aging (McRae et al. 2015). However, it is reasonable to adjust the mode of oxygen supply to the phenolic strength of the must. In high-phenol musts (Chapter 3.1.4), a venturi design can be used at the delivery side of the hose to increase the oxygen uptake, however caution is advised with excessively high oxygen supply in low-phenol musts in order to avoid over-oxidation.

Furthermore, it is reasonable to limit oxygen uptake, length and frequency of pump-overs during the first two days of fermentation when the natural polyphenoloxidase enzyme is still active and able degrade anthocyanins through enzymatic oxidation. To compensate for the lower initial oxygen uptake, to promote the formation of polymeric pigments, and to improve color stability during aging, more pump-overs are recommended during the second part of fermentation (Bosso et al. 2009).

Rack and return

The rack and return method, also referred to by the French term *délestage*, removes all the juice underneath the cap to another container from where it is poured back over the cap. It is an expanded form of pumping over, which recycles only a portion of the juice with less oxygen pickup.

During délestage racking, the cap slowly falls to the bottom of the tank while the must is allowed to drain completely under the weight of the grape solids for one or two hours. Then the juice is returned to the vessel over the cap using a high-volume pump to completely and quickly flood the cap and gently break it up, thus minimizing mechanical impact and extraction of harsh phenols. The intention is to douse the pomace quickly so that the must rises faster than the grape solids during pumping. When pumping is too slow, too much oxygen remains in the still compact pomace and can promote the formation of volatile acidity. Lastly, the cap percolates up through the returned juice, which again encourages diffusion of anthocyanins and tannins without harsh extraction. This last step maximizes interaction between the cap and the must, and optimizes the effect of délestage. A refinement of this technique involves fermentors equipped with mechanisms including special screens that allow for removing a part of the seeds from the tank bottom after all the juice has been drained off.

Usually the process is repeated every day in conjunction with punching down two or three times per day. Thus, increasing amounts of seeds can be removed when this is required. The combination of partial seed removal, enhanced oxygen uptake, and low mechanical impact on the pomace solids is considered the reason why wines obtained in this way are considered to display less tannins and less astringency.

Injecting air bubbles

This technique, also known as pneumatage, reverse punch down, or in-tank pump-over, is a process of injecting pulses of compressed air during some minutes through a mixing probe in an open tank, through a racking port, or through multiple fixed injection ports near the bottom of the fermentor. As a result, the large burst of rising air bubbles brings the juice up into and over the cap, breaking it up into individual berries. Since there is no violent mechanical action or the use of moving parts that increase the risk of straining the seeds or producing excessive lees, this technique is considered gentle. Additionally, it is less labor-intensive than most other cap management regimes.

In its simplified form as described, air injection is likely to work out on small volumes, but there is a risk of the cap being only lifted without going to pieces when larger volumes are processed. Musts with a very high solids-to-juice ratio can also cause trouble. Therefore, automated devices have been developed that inject sequentially pulsing bubbles of air or any other gas. They allow adjustment of all parameters (pressure, frequency, and duration) associated with the formation of the bubbles to any stage of the fermentation curve, variety, or winemaking style. Thus, the jets of gas they release can be modulated in such a way as to produce disintegrating waves able to plunge, break, and crumble the cap completely. Some unfilled space should be left to make sure that the must does not spill over.

Oxygen uptake is minimal as the bubbles are too large to allow for significant oxygen diffusion into the must, but systematic studies on the resulting wines are still lacking.

Submerged cap

Submerged cap is a technique that uses a grid that can be locked at any desired location inside the tank to hold the cap submerged under the surface of the must. This means that the cap remains permanently, but not disrupted, in the must. The obvious advantage is that there is a continuous diffusion of phenols from the solid parts of the pomace into the liquid, which does not happen to that extent under opposite conditions of a floating cap. Furthermore, there is no mechanical operation on the cap itself. However, pump-over cycles are required to actively recycle the juice contained in the cap (Fischer 1997 b). Otherwise the diffusion would come to a standstill when the concentration equilibrium between solids and the surrounding liquid is established.

Systematic investigation on the submerged cap technique is scarce. One study (Bosso et al. 2011) showed significantly higher concentrations of anthocyanins, flavonoids, and tannins of submerged-cap wines as compared to the floating-cap standard under conditions of comparable maceration length and temperature. However, these differences only became statistically significant after the blending of the free-run wine with the press wine. In Pinot noir, the differences in the phenolic composition as compared to standard maceration using punching down caused losses of sensory quality (Gerbaux et al. 2003).

Combination of various traditional cap treatments

Ultimately, it is difficult to assess which of the aforementioned cap management techniques represents the softest or most efficient way of phenol extraction. Pumping over can be gentle or aggressive, and results depend upon a range of variables such as fermentation vessel, grape variety and ripeness, fermentation length and temperature. Juice can be sprinkled gently over the cap, slowly wetting it, or it can be shot at high pressure, breaking up the cap and redistributing the solids within it. Punching down can be run slowly or by force exerting a heavier mechanical load on the grape solids. Hence, it is reasonable not to be too dogmatic concerning particular techniques and concepts of cap management since they all have their advantages and disadvantages. Their outcome largely depends on their frequency and the way they are performed, the individual winery's equipment, and the winemaker's experience with local conditions. Differences in handling a given cap treatment can have more influence on wine style than the general differences between treatments. Last, but not least, grape maturity has more impact on final wine quality than the technique chosen to conduct maceration.

Whenever multiple techniques are available in the winery, one may adopt a varied and adaptive approach to cap management. This approach will often incorporate different techniques at different times throughout fermentation and vary according to a number of fruit variables which change each year. The decision for one or another technique is unlikely to be static and may evolve harvest on harvest as both the climate and the winemaker's preference shift. The most important is that one takes into account the effect the various techniques have on phenol extraction and, more importantly, on the sensory outcome. However, they should not be considered in isolation as their usefulness depends on a number of other variables and decisions both up and downstream.

If considerable tannin extraction from strongly pigmented and tannic fruit is not desirable, deep color can be achieved with less frequent cap treatments which, in turn, can limit the quantity and astringency of tannins. If the pomace solids have released a majority of extractable phenolics during the first part of fermentation, the effect of continuing frequent punch-downs or pump-overs will be less. If tannin extraction during the final phase is expected to be mostly from seeds, one may want to consider whether that extraction is appropriate depending on the style of wine desired.

Sensory assessment of tannins and astringency derived therefrom is a quite difficult task in must or wine at this early stage, though most winemakers rely on it. However, many of them overestimate their sensory discrimination capacity. As a consequence, there are so many overextracted or underextracted wines. Overextracted means that a wine has more tannin than it should have, and underextracted means that it does not have enough tannin. Total phenol measurements provide a valuable guidance for decision-making in this stage (Figures 20, 21 and 24). Advancement of fiberoptic sensors may soon permit real-time analyses of multiple chemical parameters during maceration.

Impact of oxygen uptake during cap management on the fermentation rate

All measures of the aforementioned cap management regimes make the must pick up variable amounts of oxygen. A total of 10 mg/L O_2 would be a good estimate, spaced over several pump-overs or punch-downs. However, it is fairly difficult to measure significant amounts of dissolved oxygen in the must because it is quickly consumed by two reactions:

- Oxidation of phenols and their subsequent polymerization with anthocyanins into polymeric pigments, resulting eventually in better color intensity and stability. These processes are basically identical to those occurring upon oxygen uptake during the post-fermentation maturation phase and are covered in detail in chapter 7. They are also expected to produce softer tannins.
- Consumption by fermenting yeast cells in a nonrespiratory pathway promoting yeast growth and triggering the synthesis of lipids and sterols, which increase the fermentation rate and act as yeast survival factors in the final stage of fermentation. This effect helps to better complete fermentation, especially in high-Brix musts (Ingledew and Kunkee 1985, Sablayrolles 1990, Julien et al. 2000, Blateyron and Sablayrolles 2001, Moenne et al. 2014). It is an important side effect of cap management.

Figure 26 shows results of an experimental fermentation set-up which allowed for the accurate addition of 10 mg/L oxygen to a red must at various points in time during fermentation. Lowest sugar levels were achieved when oxygen is added in the second quarter of fermentation. The dissolution of the gas was performed slowly over a period of one hour by a porous frit comparable to those used for micro-oxygenation. For comparison: The typical amount of oxygen a wine picks up during a pump-over is 1-3 mg/L O_2 (Chapter 8.1.3).

Figure 26: Impact of oxygen additions on yeast growth and residual sugar after five days of fermentation at 25° C of a red must with 24.5 Brix.

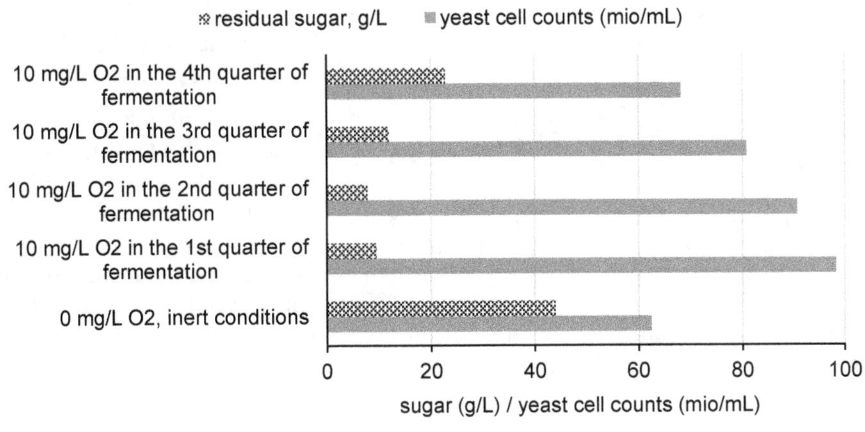

Under absolutely anaerobic conditions, which are impossible to obtain under commercial winemaking conditions, the fermentation tends to be sluggish or arrested due to limited yeast cell growth. Oxygen supply in the 1st and 2nd quarter of fermentation generally leads to highest yeast cell counts and lowest sugar levels, while oxygen uptake in the further course of fermentation has less impact. In practice, this means that high oxygen supply by punching down and pumping over is appropriate at the beginning and middle of fermentation, but progressively loses its importance for yeasts' fermentation metabolism at the end.

Mixing in rotary fermentors

Rotary fermentors can be considered a mix of floating-cap and submerged-cap vinifications, according to the frequency of their rotation cycles. They are horizontal tanks that have spirally shaped paddles along the inside, similar to a concrete mixer. They mix up the cap and juice when the tank is rotated mechanically. Their horizontal position increases the surface contact between the juice and the pomace.

Rotary fermentors are the least labor-intensive way of dealing with the cap. An additional advantage is that they make it very easy to remove the pomace from the tank by opening the door at the end of the tank and rotating it. Vertical rotary tanks have largely disappeared since they poorly mix the must whilst promoting strong extraction of harsh seed tannins.

Although horizontal rotary fermentors allow a perfect extractions of phenolics, their use is controversial. Due to the mechanical impact on the pomace during rotations, they are often accused of producing wines with increased bitterness, aggressive tannins (Suriano et al. 2012), and high levels of skin tissue fine lees. This problem can be largely overcome by carefully managing the rotation speed and frequency. It should not go beyond one rotation every five or six hours at minimum speed during active fermentation. Depending on rotation management and maceration length, this approach allows the production of soft, early-drinking red wines needing little aging as well as producing high-phenol wines with aging potential.

Figure 27: Fermentation curve of a red must (22.5 Brix) with and without oxygen in the headspace.

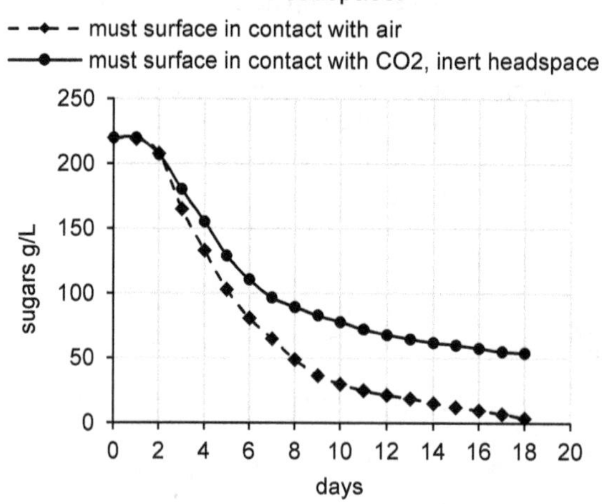

A more difficult problem to resolve is that these tanks constitute a closed system. Whilst on one hand this results in less alcohol losses and lower levels of volatile acidity, on the other hand it does not allow for any oxygen pickup. The lack of oxygen causes less formation of polymeric pigments and has also some impact on mouthfeel for a given level of tannins or total phenols.

Additionally, there are important effects on fermentation kinetics. Regardless of oxygen uptake caused by punch-downs or pump-overs, the sole contact of the must surface with air provides some oxygen with beneficial effects for the rate and completion of fermentation. An example is given in figure 27. An inert headspace makes it more difficult to complete fermentation, especially when high-Brix musts are fermented. Under real-world conditions, it is almost impossible to overdo the amount of oxygen a fermenting red must receives. The major part of it is consumed by yeast cells (Chapter 6.2).

Mixing using pressure changes

During fermentation in pressure vessels, the carbon dioxide released can be used to build up a pressure. Then an automatic valve can abruptly release the pressure causing the must to be mixed by the CO_2 bubbles escaping from the liquid. Any oxygen uptake is excluded.

This basic concept of making use of pressure changes has led to the development of more sophisticated autofermentors. They possess two superimposed fermentation chambers. The lower main chamber contains two traps into the smaller upper chamber. An elongated, perforated cylinder descends from one trap into the main chamber. As fermentation progresses, CO_2 accumulates and increases the pressure. At a certain point, the pressure forces the trap to open, and a portion of the CO_2 and the fermenting juice gushes into the upper chamber. The perforations in the cylinder prevent the pomace from escaping with the juice. In the upper chamber, the CO_2 escapes and the juice cools slightly. Eventually, the weight of the juice forces the other trap to open downward. The flush of juice back into the main chamber ruptures the cap and disperses it into the fermenting must.

Figure 28: Ganimede fermentor.

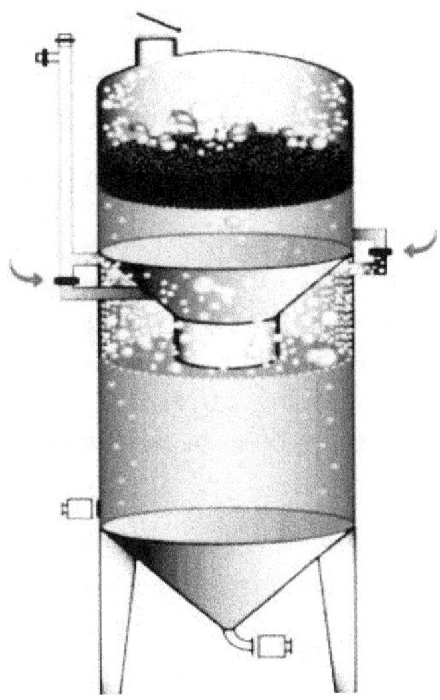

A modified version of the same principle has led to the emergence of the so-called Ganimede™ fermentor. Basically, it consists of a funnel placed inside an upright tank (Figure 28). Its operating principle is as follows:

- When the tank is filled, the cavity between the outer tank wall and the funnel remains empty because the air it contains cannot escape.
- When fermentation starts, the CO_2 produced takes the place of the air in the cavity.
- As soon as the environment is saturated with CO_2, the excess gas escapes upwards through the funnel and causes a constant mixing of the pomace in the must.
- The opening of a bypass causes a more massive action, releasing a large amount of gas accumulated in the cavity directly beneath the pomace cap, which is broken up.

- As soon as the gas vents, the cavity is flooded by the must, causing a sudden lowering of the must level. The pomace remaining above the funnel behaves more or less in the manner of a délestage racking.
- Once the bypass is closed, CO_2 accumulates again in the cavity, causing a new rising of the must level that pushes the pomace solids upwards. Thus, the whole sequence can be repeated whenever deemed appropriate.

All these pressure change systems offer a gentle and automated way of cap management with moderate phenol extraction. Like rotary fermentors, they do not allow for any oxygen uptake during fermentation. They are most appropriate for the production of fruity low-phenol red wines with strong color intended for early consumption. With regard to the Ganimede™ fermentor, tannin levels were reported to be lower and anthocyanin levels higher than after traditional vinification using punching down (Bai et al. 2013).

Transferring must from the fermentor to the press – gravity flow or pumps?

Pumps are considered to exert a detrimental mechanical effect on fermented must. Therefore, gravity flow has become increasingly popular to move the must from the fermentor to the press, and also to move wine in the later sequence of the production process. For that purpose, new wineries are frequently designed as multilevel facilities. The following advantages of gravity-fed transfers are cited:

1. a more gentle transfer of must than by the use of pumps or other mechanisms causing mechanical shear on the must solids and the additional release of harsh tannins,
2. less oxygen uptake,
3. better preservation of freshness and fruit,
4. less lees,
5. less cleaning labor required,
6. energy savings,
7. the glamor of noninterventionist winemaking and of revival of an ancient practice.

Whilst benefits in points 4 to 7 do apply without any doubt, points 1 to 3 are debatable. It is questionable whether the mechanical shear upon feeding the press with a pump is more harmful than all the punch-downs and pump-overs during maceration. It is even more questionable whether the alleged decrease of oxygen uptake makes any sense in view of the intentional oxygen supply provided during barrel aging and micro-oxygenation. It is difficult to understand how wines can be kept fresh and fruity when deliberate barrel aging is planned.

The reason for these contradictions may lie in the choice of the pumps. Slowly operating pumps and large diameters of pump head and tubing are indispensable for gentle must pumping.

4.1.13. Effect of maceration enzymes

As outlined in the previous chapters, maceration length, temperature, and cap management are the fundamental parameters controlling phenol extraction. However, in commercial winemaking conditions, the availability of fermentor capacity is an important aspect. It often reduces the time of skin contact one would desire. This is the moment when the application of pectolytic enzymes used as maceration enzymes can become beneficial.

Commercial maceration enzyme preparations are selected for their balance of pectinases, hemicellulases, and cellulases. They weaken the grape cell walls by hydrolyzing their different polysaccharides, which results in an easier diffusion of intra-cellular compounds. Thus, they accelerate they release of anthocyanins and tannins. Under a more technical aspect, their purpose is to increase press yield, to enhance color extraction, and to facilitate wine clarification.

By the end of the last century, some commercial pectolytic and macerating enzyme preparations were reported to decrease anthocyanins, polymeric pigments, and color intensity rather than to increase them. This effect was explained by a ß-glucosidase side activity, which cleaved the sugar from the anthocyanins and left the unstable anthocyanins in a colorless form (Wightman et al. 1997). In the meantime, these initial problems with deficient enzyme purification have been largely overcome, but the question still arises whether these enzymes really contribute to tannin and color extraction to a significant extent.

Reported results are conflicting. The most reasonable answer is that maceration enzymes tend to be helpful, depending on the particular commercial preparation used, its current stage of development, and the winemaking conditions in particular. A study on the different enzymatic activities (pectolytic, hemicellulolytic, cellulolytic, and glycosidasic) of 41 commercially available enzymatic preparations revealed that the differences were as large among the preparations for the same technological effect as between the preparations designated for different technological effects (Guérin et al. 1999). Hence, it is hard to generalize their effect.

Under conditions of short macerations not exceeding one week, especially at low temperatures, advantages regarding tannin and color extraction can be expected together with better self-clarification and filterability. Some enzymes labelled as color extractors mainly increased tannin (Favre et al. 2014). Hence, most studies report an increase of total phenols, tannin, and color but no increase of anthocyanin extraction (Ough and Berg 1974, Ough et al. 1975, Ducruet et al. 1997, Guerrand and Gervais 2002, Guadalupe et al. 2007, Alpeza et al. 2017), although there are also exceptions (Zimman et al. 2002, Bautista-Ortín et al. 2005). The increase of color is explained by an increase of colorless phenols and polymeric pigments resulting therefrom (Parley et al. 2001, Sacchi et al. 2005).

In sensory terms, the enzyme treatment can have a bigger effect on the mouthfeel properties than expected from its effect on phenolic composition (Nel et al. 2015). In Pinot noir, a slight increase of total phenols and color contrasted with a marked increase of undesirable bitterness and drying astringency (Gerbaux et al. 2003). This proves that chemical composition is not always a direct indication of assumed sensory characteristics. In some cases, an increase of aroma intensity regardless of phenol extraction can be observed due to secondary activities of the preparation used (Bakker et al. 1999).

Most of these results have been obtained after short macerations not exceeding one week. These are the typical conditions most wineries are subject to due to limited fermentor capacity. Under these conditions, an increase of color resulting from enzyme treatment can be perceived by eye in wines obtained from lightly colored varieties such as Pinot noir, but it becomes imperceptible in those from strongly colored varieties. In general, the benefits of enzyme treatment become negligible when extended macerations exceeding two weeks are performed. For optimizing the extraction of tannins and anthocyanins, an adequate maceration period at not too low a temperature and frequent must homogenization intervals using one of the various cap treatment procedures can be more efficient tools.

4.1.14. Effect of different yeast strains

Spontaneous fermentation by indigenous yeasts can sometimes give good results, but it is risky and requires a lot of skills and attention. Therefore, a wide selection of active dry yeast strains is available for conducting red wine fermentation.

As phenol extraction from the solid parts of the pomace is a physical process of diffusion following a concentration gradient between solids and liquid, any contribution of yeasts to the phenolic make-up of the finished wines has been seen as incomprehensible for a long time. However, more recent research has shown that yeast strains do have an impact of the phenolic and gustative profile of red wines, thus partially substantiating claims of the supply industry.

Basic requirements on yeast strains comprise a low production of volatile sulfur compounds responsible for reduction off-odors, higher alcohols, and volatile acidity. For their use in red wines, there are some additional requirements such as

- low color absorption by the yeast cell wall,
- high release of cell wall polysaccharides, in particular mannoproteins, during and after fermentation aiming at enhancing volume on the palate and lowering astringency (Chapter 5.3.3),
- high alcohol tolerance,
- facilitation of malolactic fermentation.

The contribution of yeast to wine color is a particular trait of yeast strains. On one hand, there is an average adsorption of 3 to 4 % of anthocyanins on the yeast cell wall during fermentative maceration (Morata et al. 2005), with the exact extent also depending on yeast growth and the production of yeast biomass (Balík 2006). On the other hand, yeast also contributes to wine color, as a result of the formation of strongly colored stable pigments by the release of fermentative metabolites and mannoproteins that bind to anthocyanins and protect them from precipitation (Morata et al. 2016). As a result, there can be significant differences in wine color parameters when musts are fermented with different yeast strains, especially in wines from low pigmentation grapes (Sidari et al. 2007, Bautista-Ortín et al. 2007). However, these differences only become evident when comparison is made to a reference wine fermented under identical maceration conditions. Otherwise differences in maceration would confound yeast effects.

A replicated study of Gamay fermentations with 18 yeast strains and uniform skin contact time showed differences of up to 100 mg/L in anthocyanins and of 10 absorbance units (A 280) in total phenols (Cuinier 1997). In contrast, different yeast strains were found to have little influence on anthocyanins, color, total phenols, and tannins in fermentations of Pinot noir (Lorenzini 2001, Girard et al. 2001), Sangiovese (Zini et al. 2002), and other varieties (Nicolini et al. 2003).

In contrast to anthocyanins and color, much less importance is attached to the impact of yeast strains on tannin, though such an impact does exist. In replicated fermentations of Pinot noir, five yeast strains yielded significant differences in color, tannin concentration, and tannin composition. Strain RC 212 distinguished itself by imparting increased levels of polymeric pigments, a high tannin molecular size, and a high degree of tannin polymerization (Carew et al. 2013). In fermentation trials with Cabernet Sauvignon, two Saccharomyces bayanus strains produced the most distinct wines with distinctive phenolic composition, among other differences (Blasquez-Rojas et al. 2012). In Shiraz, the yeast strains had influence on numerous compositional parameters including tannin concentration. In this case, the yeast strain explained even more variation in the final wines than fruit source, with a compelling case for yeast strain signature becoming evident (Holt et al. 2013).

How can these differences be explained in view of the fact that the diffusion of phenolics into the liquid is a merely physical process? Actually, yeasts display a native pectinase activity. As a result, they can act to some degree like maceration enzymes and facilitate the release of phenols entrapped in the grape solids. In some non-Saccharomyces strains, this activity is particularly pronounced and suggests the use of such strains for co-inoculation in conjunction with S. cerevisiae strains able to complete fermentation (Belda et al. 2016).

Significant differences in the levels of tannins and polysaccharides after two years of aging were also partially explained by variable pectinolytic activities of 10 commercially available Saccharomyces cerevisiae strains used for fermentation of a Shiraz must. They caused sensory differences in astringency and color leading to different wine styles. Strains which produced higher astringency and color also tended to yield concomitant increases in dark fruit, floral aroma, and viscosity, which are attributes known to be relevant to quality in multiple wine styles. In contrast to expectations, practical experience and literature reports (Chapter 5.3.3), polysaccharides did not exert a moderating effect on astringency in this particular study (Bindon et al. 2019). There is an obvious need for better defining which kind of polysaccharides actually soften astringency, what the time course of this effect is, and how it is influenced by filtration (Chapter 12.2)

Running yeast trials in commercial wineries

In summary, yeast strains do have an impact on anthocyanins, color, tannin, and related phenolics in the finished wine, but whether that impact becomes obvious depends on the yeast strains that are compared one with another. Therefore, they may be selected to modulate phenol retention, color, astringency, and other mouthfeel attributes. The influence they exert will also depend on the grape source and its inherent phenolic extractability. It can be strong under conditions of short macerations in low-phenol musts and lose practical importance in high-phenol ones.

There is still a lot of work to do to better understand which yeast strain is most suitable for the production of a desired wine style from a given fruit. This work will probably never end and encourage the conduct of yeast trials in one's own winery. Indeed, many wineries do so.

As a matter of fact, yeast trials are difficult to implement under commercial winemaking conditions. They require identical framework conditions such as fruit ripeness and composition, yeast nutrient status, oxygen uptake, temperature, etc. if one wants to obtain reliable results. While these conditions seem to be relatively easy to fulfill in white juice fermentations, they are much more difficult to comply with in red must fermentations. Maceration length and cap treatment regime are additional variables that must be standardized. The maceration period in particular has to be kept identical instead of pressing the lots right after fermentation to dryness. Otherwise these variables risk having more impact on the final outcome than the yeast strain itself. Furthermore it is reasonable to use a fairly high inoculation rate (0.3 g/L) to make sure that the strain under investigation becomes really dominant and responsible for the effects observed.

Considering all these frequently neglected or poorly controlled side variables, replicates are required to allow for an unambiguous statement. These replicates refer to both the experimental design and the sensory evaluation at different points of time during aging.

4.1.15. Sense and nonsense of tannin additions

Commercial tannins are preparations consisting of grape-derived tannins (Chapter 2.2.3), hydrolyzable tannins (Chapter 2.3.2), or mixtures of both. As pointed out in chapters 4.1.4 and 4.1.15, addition of such tannins can rarely compensate for a lack of tannins in underripe fruit or their deficient extraction in musts of Vitis vinifera varieties. Recommended addition rates are too low to impact measured tannin concentrations, furthermore most commercial preparations only contain 12 to 48 % tannin (Harbertson et al. 2012). The remaining part consists of impurities.

Frequently used standard additions of 0.1 to 0.2 g/L only increase tannin or total phenol content by a few percent, but can contribute to perceived astringency in some cases. Additions of oak alternatives (Chapter 11) act in a similar way, but apart from providing odorless ellagitannins to the wine, they also impart oak-derived volatile aroma compounds influencing odor perceptions.

The market launch of commercial tannins in the 1990's was accompanied by an enormous number of advertising statements and technical articles claiming that their addition would be indispensable to stabilize color, ignoring the reality that red wines with stable color had been produced long before. The rationale behind this is the actual ability of tannins to better stabilize and to enhance the color by formation of polymeric pigments (Chapter 2.2.5) that are more resistant to bleaching by SO_2 than the anthocyanins they originate from. In particular, ellagitannins extracted from oak have been shown to quickly absorb dissolved oxygen and convert it into oxygen radicals and hydrogen peroxide, which accelerate the formation of polymeric pigments (Vivas and Glories 1996). Consequently, ellagitannins are added to many musts and represent one of the largest material input costs in red winemaking in the global wine industry. However, additions of grape-derived tannins (Kovać et al. 1995) or their better extraction during maceration show a similar effect.

The actual impact of these additions depends on variety, fruit ripeness, the phenolic composition of the must, and the kind of tannin added. Hence, it is hard to generalize. In a study on Sangiovese, additions before fermentation had a stronger influence on color structure than post-fermentation additions. Wines obtained from grapes with high phenolic concentration were much less influenced by these additions than low-phenol wines (Canuti et al. 2012). Gallnut and a grape seed tannins showed the strongest impact on color (Lempéreur et al. 2002), but these tannins are also known for the harsh astringency they provide.

On the other hand and in contrast to popular expectations, there is abundant literature suggesting that the addition of these tannins to must often does not lead to large differences in the chemical, chromatic, and sensory profile of the wine (Schneider 1999, Parker et al. 2007) and sometimes even causes unwanted sensory characters in the resulting wines (Versari et al. 2013), in particular when high amounts are added (Harbertson et al. 2012). Most frequently a slightly higher astringency is observed (Bautista-Ortin et al. 2005). Similarly, the addition of oak chips to must typically shows no clear effect on color and phenols (Zimman et al. 2002), though there is much promotion claiming the opposite.

In summary, the majority of results obtained from properly conducted trials contradicts the expectations promoting the popular addition of commercial tannins to musts of red Vitis vinifera varieties. After several years of industrial experience with tannin additions under conditions of Southern France, the result was sobering: "None of the tannins met the expectations raised by their manufacturers. Gustative results are opposed to consumer expectations. The use of these additives should not be systematized" (Delteil 2000). Only in individual cases, benefits can be explained by the specific fruit or must composition and the high variability of the commercial tannin preparations (Li et al. 2018).

When tannin or oak chips additions are intended to increase astringency and structure, these additions are more beneficial post-fermentation as they allow more accurate adjustment of the quantities to the desired sensory goals when guided by previous bench trials (Chapters 9.2 and 11.2). In contrast, pre-fermentation additions are operations with an uncertain sensory outcome, though they are said to become better integrated in the wine. Generally they require 4 to 5-fold higher amounts to obtain a comparable result on the palate. Fermentation causes the larger part of the effects of pre-fermentation additions to disappear. Adsorption of the added material by yeast biomass plays a role (Barrio-Galán et al. 2015).

The choice of the maceration variables - temperature, duration, previous juice removal, mechanical treatment of the must - has much more impact on tannin level, overall quality, and differentiation of red wines than can be achieved by tannin additions to must (Delteil 2000, Schneider 2004, Sacchi et al. 2005). Only in low-tannin wines as those obtained from thermovinification the addition of commercial tannins can be expected to be a helpful means to optimize mouthfeel sensations.

Results of additions to wines post-fermentation are given in chapter 9.2.

The particular case of interspecific hybrid varieties

What is true for Vitis vinifera varieties must be evaluated in a different way for interspecific hybrid varieties. Anthocyanin concentration in these varieties is generally higher than in V. vinifera cultivars, whilst their tannin levels are much lower. Indeed, it is well known that the traditional winemaking practices aimed at tannin extraction in hybrids lead to limited tannin, very low astringency, and poor color stability. The low tannin levels are explained by deficient tannin retention in must and wine rather than by insufficient extraction from the fruit. The reason for this is binding and precipitation of tannin by proteins contained in high amounts in these grapes and, to a lesser extent, by pectins (Springer and Sacks 2014). Measures to overcome this problem are protein removal by bentonite fining plus addition of commercial tannins in amounts higher than those used in V. vinifera varieties.

4.2. Other extraction technologies

4.2.1. Carbonic maceration

In carbonic maceration, uncrushed grape bunches are held in a closed vessel previously flushed with an inert gas, usually CO_2, during one to two weeks at a minimum temperature of 20° C. Sometimes more CO_2 is added during filling of the vessel and the first one or two days afterwards. Alternatively, some winemakers place a small amount (5 to 10 %) of fermenting must on the bottom of the tank, which generates CO_2. The tank is then sealed to maintain anaerobic conditions and kept at a temperature comprised between 20 and 35° C. This is the very strict way of carbonic maceration. When the tank is only filled up with CO_2 at the moment when some of the berries at the bottom are crushed by the weight of the others above and start fermenting, the process is called 'semi-carbonic maceration'.

Under these anaerobic conditions and in the absence of yeast cells within the berries, neither fermentation nor respiration takes place, though the process is frequently referred to as 'intra-cellular fermentation'. Instead, the berries take up CO_2 from the atmosphere and use it to trigger a number of other metabolic pathways (Tesnière & Flanzy 2011). They comprise:

- A transformation of phenolic acids (Chapter 2.3.1) into odor-active compounds, in particular esters (Lorincz and Vas 1998) and the release of aroma compounds from their bound forms by the action of glycolytic enzymes (Ducruet et al. 1983).
- Approximately half of the malic acid is converted into ethanol, in contrast to malolactic fermentation that degrades it into lactic acid.
- Amino acids are generated by grape protein degradation and liberated from grape solids; they increase the nutrient status of the juice, facilitate subsequent alcoholic fermentation, and act as flavor precursors.
- A minor amount of sugar is transformed into 1.5 to 2 % alcohol by a non-fermentative pathway. This ethanol can esterify some grape components. The esters produced this way strongly contribute to the distinctive fruity aroma of these wines.

Concurrently, berries in the lower layers become gradually crushed by the pressure exerted by the weight of the upper layers. The juice released starts spontaneous fermentation, producing more CO_2 that contributes to maintain the inert conditions. In the traditional conduct of carbonic maceration, this is the moment when grape bunches and juice are removed and pressed, and the juice is inoculated to complete fermentation. However, there are deviations from that rule regarding the time of pressing. The more the alcoholic fermentation is allowed to progress before devatting, the less distinctive characteristics of carbonic maceration the finished wine will display.

This method is strongly associated with the Beaujolais region in France and employed to produce light-bodied fruity wines, which are stylistically very different from macerated wines and meant to be consumed very young. The strong impact that carbonic maceration has on fruity aroma has led some winemakers to add 30 to 50 % of whole berries or bunches to musts fermenting in the traditional way. In this case, while much of the fermentation will be performed by yeasts, the intact berries will be in an anaerobic environment, and so some intra-cellular activities will take place. Whilst this approach results in a strong aroma enhancement in white wines, it provides no lasting benefits in red wines.

Whilst there is no doubt about the strong aroma enhancement caused by carbonic maceration, the outcome on the palate is more variable. Traditionally, wines obtained from carbonic maceration represent a particular style of wine characterized by less tannins, astringency, anthocyanins, and color than wines produced by standard maceration procedures (Beelman & McArdle 1974, Köhler et al. 2005). Their low tannin levels tend to make them unsuitable for barrel aging. However, the differences regarding their phenolic make-up can become reversed in some cases (Pellegrini et al. 2000, Sun et al. 2001, Gonzáles-Arenzana et al. 2020) and even enable successful barrel aging (Salinas et al. 1996). These differences are gradual and explained by the timing of devatting. The more the pressing is delayed, the more grapes become crushed, and the more fermentative maceration takes places. As a result, there are also border line cases in which wines obtained by carbonic maceration hardly differ from those produced by standard maceration (Geoffroy et al. 2013).

4.2.2. Thermovinification

Although there are several variations of the process, thermovinification generally involves heating of the destemmed and crushed grapes to 60 to 90° C (140 to 194° F), holding the must at this temperature for a short time, and cooling it down before immediate pressing or subsequent fermentation on the skins. Scraped-surface, immersion, and tube-in-tube exchangers are commonly used for heating and cooling, with the latter being the most frequent and the simplest equipment.

Tube-in-tube exchangers are typically fed with a screw pump, and heat is provided in the form of hot water. For heating, the cold must is pumped through the inner tube, while hot water or hot must to be cooled flows in countercurrent through the outer tube. For cooling, the hot must is also pumped through the inner tube and cooled by cold must or water flowing through the outer tube.

When musts with a high percentage of skins are processed, the heat exchange between hot and cold must can become hindered in a standard tube-in-tube exchanger because the outer tube would be liable to clog. To overcome this problem, heat exchange is sometimes performed via an intermediate water loop that recirculates in the outer tube of two tube-in-tube exchangers. An alternative used in small wineries is to heat the must while recirculating on a tank via a tube-in-tube heat exchanger. This results in a lower and often heterogeneous temperature rise and is therefore not suitable for laccase[32]-affected fruit.

Heating the must rapidly disrupts the grape cell wall and immediately releases high levels of anthocyanins. As tannins require alcohol as a solvent to be released to an appreciable extent, tannin extraction during the heating step is minimal. Heat also reduces the grape microbial population and inactivates natural grape enzymes. These enzymes comprise essentially polyphenoloxidase and pectinase. Whilst the inactivation of polyphenoloxidase is a major advantage of thermovinification, pectinase activity is required to break down the grape pectin and allow for clarification. Therefore, wines obtained by thermovinification require to essential interventions:

- The addition of commercially available pectolytic enzymes as described further below.
- The inoculation with active dry yeast to induce a sound fermentation. Thus, thermovinification is a non-option for winemakers who are proponents of spontaneous fermentations.

The compositional and sensory properties of thermovinified wines strongly depend on the way thermovinification is performed. There are various implementation strategies, for example:

1. Short-time-high-temperature treatment without maceration: This is the most extreme form of thermovinification. The must is heated to 85° C[33], held there for ~2

[32] Laccase is a polyphenol oxidase (PPO) enzyme not contained in sound fruit but only produced by Botrytis cinerea on rotten grapes. It oxidizes almost any kind of phenols, thus degrading anthocyanins and color. It is inactivated by a temperature of 70° C over 30 minutes or 80° C over 15 minutes.

[33] Longer holding times at 85° C may result in "cooked" flavors.

minutes, then cooled to ~45° C by incoming must that is pre-heated, and pressed shortly after. Further cooling is required after pressing to ensure a fermentation start temperature not higher than 20° C. Compared to the traditional maceration procedures (Chapter 4.1), wines obtained by this way are characterized by a 2-fold or higher anthocyanin and color extraction (Auw et al. 1996, Lukić et al. 2017). Anthocyanin losses during primary fermentation are 20 % on average, which is of little importance due to their high initial concentrations. Since there is no alcohol present, tannin extraction is fairly limited. Hence, the wines display much less body, tannin, and astringency than those produced by traditional maceration. Therefore, they are often supplied with commercial tannins or, better, blended with macerated wines to generate different sensory characteristics. Without such tannin corrections, they are meant to be consumed young.

In several varieties, a two-hour pre-fermentation heat treatment of the grapes at 70° C and pressing before fermentation in liquid phase was shown to yield more ethyl esters, acetates, and fatty acids than fermentation in solid phase before pressing (Geffroy et al. 2015). The fresh and fruity aroma that these compounds impart is further emphasized when fermentation takes places at low temperatures after thorough juice clarification. The reduction of C6-alcohols and methoxypyrazines, key compounds responsible for herbaceous aromas reminding freshly cut grass and green capsicum, is an undeniable benefit (Fischer et al. 2000, Lukić et al. 2017) and a feature of all forms of thermovinification. This is a considerable advantage when underripe fruit is processed. It explains why thermovinification is relatively widespread in cool-climate growing areas.

2. Short-time-high-temperature treatment with warm maceration: The must is heated to 85° C for ~2 minutes, cooled to ~45° C, and held for 6-12 hours before pressing. The holding period for maceration allows for some more tannin extraction, which provides more mouthfeel and structure. Nevertheless, the wines are not very different from those obtained as specified in the previous section.
3. Medium-time-medium-temperature treatment: The must is heated to ~65° C and stored over 1-2 days allowing for passive cooling before pressing. Wines obtained in this way make the transition to those obtained by short fermentative macerations of 3 to5 days.
4. Pre-fermentation hot maceration: This approach takes advantage of both thermovinification and fermentative maceration, accumulating their effects. The must is heated to ~65° and stored for passive cooling. Inoculation with active dry yeast takes places when the temperature reaches ~20° C for subsequent fermentation on the skins. The accumulated effects of temperature plus fermentative maceration totally blur the typical sensory effects of thermovinification and allow for the production of high-phenol red wines with an aroma similar to that of macerated wines (Geffroy et al. 2013, Wang et al. 2016). In Pinot noir, this effect can be further enhanced by draining the juice after crushing and submitting only the dejuiced grapes to a high-temperature-short-time treatment at 90° C for one minute. They are then pressed warm and re-extracted with the free-run juice during fermentation (Girard et al. 1997).

In contrast to wines obtained by the previous variants of thermovinification (1 to 3), wines resulting from this combination of thermovinification and maceration are suitable for barrel-aging and longer storage. The variation of must heating and fermentation temperature opens additional options to formulate wines of targeted styles. However, this approach requires fruit of high phenolic ripeness (Chapter 4.1.5). Otherwise the intensive extraction process would give rise to the appearance of harsh seed tannins.

Whilst strategies 3 and 4 are the most energy-intensive, strategies 1 and 2 are the only ones ensuring deactivation of rot-derived laccase. Temperatures lower than 65° C in conjunction with longer holding times are not suitable for laccase deactivation; they would rather enhance detrimental laccase activity during the holding time.

Wines obtained exclusively by thermovinification contain very high levels of turbidity and fine lees. This is another reason for the addition of pectolytic enzymes. These enzymes should only be added after the heat treatment when temperature is below 60° C to avoid their inactivation by heat. An enzyme addition before the heat treatment facilitates the must flow through the heat exchanger, but cannot replace a post-heating addition in practice.

4.2.3. Flash détente

Further development of thermovinification has led to a technique called flash détente, which is the original French term of what is translated as "Flash Release" or "Thermoflash". For this process, the intact grapes are flash heated to ~85° C at atmospheric pressure, after which they are transferred to a high vacuum chamber where they are depressurized in a fraction of a second (Escudier and Moutounet 2001). The depressurization results in a quick temperature drop to approximately 30° C and the instantaneous vaporization of the water in the skins' cells. The rapid expansion of steam leads to the formation of micro fissures in the skins, to the explosion of the vacuoles in the skin cell walls, and eventually to the rupture of the berries and the fast release of tannins and anthocyanins they contain.

The steam that is formed in the vacuum chamber during this process carries the volatile aroma compounds with it. They are recovered by cold traps producing a condensate that is added back if desirable, discarded if undesirable, or undergoing a fining to remove undesirable odor compounds before being added back.

The colored juice can be fermented as such, or added back to the skins for further maceration to extract more phenolics. Wines obtained without additional maceration were found to suffer heavy losses of all phenolics during fermentation, whilst maceration subsequent to flash détente allowed to obtain polyphenol levels comparable to those of the control wines or even higher. Ultimately, total phenol levels and the tannin-anthocyanin ratio depend on the duration of the high-temperature exposure, the successive conduct of a maceration, and the length of that maceration (Escudier & Moutounet 2001, Morel-Salmi et al. 2006). As is the case with thermovinification (Chapter 4.2.2), these processing variables determine the final wine composition.

This sophisticated technique has been designed for the fast processing of large fruit volumes with the additional advantage of being able to remove green vegetal aromas from unripe fruit. Hence, it can help with color and tannin extraction or mitigate the lack of hang time under difficult growing conditions. Site characteristics of the grapes will hardly be retained, which explains why this technique is rejected by producers of premium wines.

4.2.4. Pulsed electric field

There are various attempts to develop additional extraction technologies aiming at facilitating the phenol extraction process and obtaining higher or more selective extraction rates. They make use of high hydrostatic pressure, pulsed electric fields, ultrasound, irradiation, and pulsed light. In an industry highly bound to traditions, it is likely that most of them will never reach practical implementation, but it is interesting to have an outlook of what could be possible in the future. The use of pulsed electric fields is one of them and briefly presented as an example in the following.

Pulsed electric field (PEF) is a large-scale form of electroporation, a bioscience technique. It is employed in the extraction of fruit and vegetable juices. In enology, it is a relatively recent 'emerging' non-thermal treatment of crushed grapes intended to open the cells of the berries' skin tissue to facilitate enhanced phenol extraction. Its operating principle and a tabulation of reported outcomes on red grape varieties are well reviewed by Ricci et al. 2018, and its application at near commercial scale is described by Luengo et al. 2014.

In summary, PEF involves applying high-voltage electrical pulses across two electrodes between which must flows. Voltage and electrode spacing are selected to achieve high electric field strength in the kilovolts per centimeter range. Cell membranes are both electrically insulating and thin: Accordingly high local electric field gradients – within the overall applied field - develop across them. This trans-membrane potential causes the formation of membrane pores which facilitate the extraction of phenolics during subsequent maceration. Whilst pore formation can be reversible, it becomes permanent at higher energies. The micro-scale effect of pore formation is well illustrated by Sack et al. 2009. Complete opening of the cells' pores is achieved with electric field strengths of approximately 1 to 10 kV/cm and energy inputs of 1 to 50 kJ/kg formed from multiple pulses.

Direct comparison of reported extraction data is difficult due to different PEF conditions and winemaking protocols, however the overall trend of results is quite clearly promising with reports of total phenol index (Chapter 3.1.3) increases between 13 and 63 % (Ricci et al. 2018). In part reported protocol variation is due to a lack of objective determination of full poration in many reports. However, techniques have been reported to achieve this (Sack et al. 2009, Bobinaitė et al. 2014).

The enological application of PEF has been researched for several decades, typically employing small, laboratory-scale systems, normally static rather than flow-through.

This work has yielded convincing evidence of enhanced phenol extraction and indicated the range of field strength necessary (1-10 kV/cm). Whilst this provides a convicting case for PEF scale-up, the required field strengths represent a significant engineering challenge when voltages are scaled due to the necessarily increased PEF chamber dimensions to achieve winery-level must through-puts. Such voltages are both challenging to generate and control as well as requiring careful implementation from a safety perspective. Accordingly, the continuous flow process is still under development.

Whilst phenol extraction is our primary focus, PEF is also reported to offer several associated advantages and applications. In economic terms these include a shortening of maceration times and decrease of energy inputs (compared to thermovinification) and material (such as macerating enzyme) costs. PEF has also been researched in an anti-microbial role when used upon wine itself where it can disrupt spoilage microorganisms' membranes and thus kill them, thereby offering the appealing future prospect of an alternative to SO_2 in its biocidal role and tight filtration.

5. Tannin mouthfeel properties: Causes and evaluation

The following chapter is exclusively about sensory issues. It focuses on the causes, nature, and evaluation of astringency as the primary gustative expression of tannins and red wine in the broadest sense. Whilst it is frequently perceived as a single sensation that facilitates its intensity assessment in routine sensory control, it is made up of multiple subquality sensations. These sensations and the sensory perception of overall astringency intensity are not only modulated by the amount and composition of tannin, but also by numerous wine matrix parameters such as polysaccharides, alcohol, total acidity, pH, and potassium. In particular, minor acidity corrections have a far-reaching impact on perceived astringency without any changes in tannin concentration or composition taking place. Similarly, aging on the lees is presented as a technique to enhance polysaccharide levels and lower astringency. These matrix effects explain why total phenol or even more specific tannin measurements can only roughly predict perceived astringency, and why they cannot completely replace its careful sensory evaluation. However, the reproducibility and comparability of astringency ratings are strongly affected by individual salivary flow rates and carry-over effects, whilst palate cleansers show only limited value. Hence, tastings should be spaced in time and the number of samples restricted.

5.1. The tactile nature of astringency

Tannins and the polymeric pigments they produce with anthocyanins are responsible for the oral perceptions of volume, body, structure, texture, viscosity, bitterness, and astringency to an extent that cannot be found in dry white wines. Among these perceptions, astringency is the most fundamental and the most controversial one in red wines. Although it is sometimes confused with bitterness, it is not a taste sensation, but a tactile irritation of the mucous membranes in the oral and pharyngeal cavity.

Astringency in wine can also be elicited by acids, metals, and high alcohol levels, but tannins are definitely its primary cause in red wines. Their chemical interaction with anthocyanins (Chapter 5.3.2) and polysaccharides (Chapter 5.3.3) makes them softer on the palate. Hence, anthocyanins are not only responsible for the color properties of red wines, but they also make an indirect contribution to mouthfeel qualities. This is the reason why tannins and anthocyanins have to be considered in context, and why it is more appropriate to refer to them as the tannin-anthocyanin complex.

There is general agreement on the importance of an appropriate level of astringency for quality, consumer acceptance, and differentiation of the various styles of red wine. Nevertheless, its specific sensory analysis and description is frequently unsatisfactory and replaced by the use of more general terms such as structure, body, weight, etc. The reason for this is an inadequate differentiation of the oral sensations the wine leaves behind when it passes through the mouth and throat.

Astringency is not a taste

Astringency is a perception that belongs to the group of trigeminal sensations. The designation of these sensations stems from the 5th cranial nerve, the 'nervus trigeminus'. It participates in the chemoreception in the areas of mouth, nose, and throat, where it functions primarily as a tactile nerve transmitting all tactile, heat, and pain sensations to the brain.

Besides astringency, trigeminal sensations also comprise the burning sensation of high alcohol concentrations, the prickle of dissolved carbon dioxide, the pungency of sulfur dioxide, the spiciness of pepper, and the cooling effect of menthol. This clearly shows that astringency is by no means a taste sensation which could be derived from the five primary tastes sweet, sour, bitter, salty, and umami. Conversely, it is a perception that must be quantified on its own and independently from any taste sensations. This is not altered by the fact that tannin causes astringency and bitterness simultaneously, and that these two sensations are often not sufficiently distinguished by less trained tasters. Their interaction could be designated as tannicity. It is characteristic for red wines.

From a perceptual perspective, astringency is not a single sensation or general category but rather made up of multiple subquality sensations comprising puckering, drying, rough, shrinking, adhesive, grainy, chalky, grippy, particulate, silky, etc. (Lee and Lawless 1991). Astringency is a more general term involving all these attributes to a variable extent that depends on the particular wine. It is used to facilitate communication between tasters who are less trained in the use of more particular subquality attributes, which poses major semantic challenges regarding language awareness. For wine, the spectrum of subtle differences in astringency sensations was compiled as a "red wine mouthfeel wheel" by Gawel et al. (2000).

Model solutions for sensory training of astringency and bitterness

For the purpose of sensory training, tannin solutions are not the most appropriate way of demonstrating what astringency really is because of the concomitant occurrence of bitterness. Aqueous solutions of aluminum potassium sulfate, zinc chloride, or iron-II-chloride have been shown to be more useful to unequivocally model astringency (Yang and Lawless 2005). When aluminum potassium sulfate (alum) is used, solutions of 1, 2, and 3 g/L of alum ($AlK(SO_4)_2 \cdot 12\ H_2O$) are a good starting point (Lee and Lawless 1991). Since alum tastes slightly acid, 0.3 g/L potassium bicarbonate ($KHCO_3$) per 1 g/L alum should also be added to remove sourness. Under these conditions, difference thresholds for astringency are usually around 0.2 g/L $AlK(SO_4)_2 \cdot 12\ H_2O$, depending on the tasters' sensitivity.

For bitterness, aqueous solutions of quinine sulfate or quinine chloride have proven to be conducive. When the task it to show what bitterness is, concentrations of 2 and 4 mg/L of any of these compounds should be used, but increments of for example 1.0, 1.2, 1.4, 1.6....mg/L facilitate the assessment of the tasters' highly variable sensitivity to bitterness.

Why can astringency be perceived as pleasant?

Unlike pain stimuli, the human brain does not necessarily experience trigeminal stimuli such as astringency, spiciness, or prickling as negative. Compounds that unequivocally elicit trigeminal perceptions cannot only be tolerated, but can even be particularly appreciated in food and beverages. The astringency of red wine tannin, the burning of alcohol in spirits, or the hotness of typical Indian food are common examples. However, it is debatable which intensity level of these sensations is considered positive in a given wine. The answer to this question will always be subjective. For marketing purposes, consumer preference tests provide more information than in-house ratings.

It has not yet been definitively investigated by psychobiological research why humans can perceive irritating sensations as pleasant under certain circumstances. Perhaps hormonal processes or the craving for excitement play a role. According to this tentative explanation, any as pleasant perceived effect of trigeminal sensations would be caused by a relatively weak and painless stimulus that activates the chemosensorial perception system, which in turn stimulates the whole organism. This would lead to an increase of the general activity level, which is ultimately experienced as pleasant.

Without any doubt, there are processes of learning and conditioning involved in the development of individual trigeminal preferences, just as it is known from the development of taste preferences. There is no other way to explain why wines preferred by novices are rarely markedly tannic red wines, which are more preferred by experienced wine connoisseurs. Not only can humans overcome genetically determined aversions, but they can also transform them into preferences by learning. In doing so, the acceptance of fairly astringent red wines with high tannin levels only increases after a longer period of consuming such wines. Psychophysiological processes explain why an astringency of a given intensity is perceived as positive, too strong, or too low as a function of precedent consumption experience (Burdach 1988).

5.2. The causes of astringency

Astringency is a tactile sensation in which the normal lubrication of the oral surfaces is impaired by the precipitation of salivary proteins with tannins, acids, metal ions, or solvents. Upon red wine ingestion, it is elicited when salivary proteins are partially precipitated by tannin (Kallithraka et al. 1998, Gawel 1998). This process is in accordance with what happens when red wines undergo the classical fining with protein-based fining agents such as chicken egg white and gelatin. When salivary proteins are precipitated, they lose their effect as lubricants in the oral and pharyngeal cavity. As a result, a friction occurs when various parts of the oral mucosa move along each other. This friction demonstrates the mechanical principle underlying the generation of astringency.

Basic tastes can mask one another, for example sweetness can suppress acidity, but astringency does not follow that pattern. A sweetener without any viscosity such as aspartame is not capable of decreasing astringency (Gawel 1998). In contrast, sugars used as sweetener are capable of doing so because they modify the rheological properties of the solvent and increase its viscosity. In practical winemaking, however, sweetening a red wine with sugar is not an appropriate means of sufficiently lowering an unpleasantly high astringency.

The sensation of astringency increases the longer the wine remains in the mouth (Valentová et al. 2001). It can persist several minutes after the stimulation has ceased due to swallowing or spitting out, and this all the longer the higher the tannin concentration has been. It decreases only slowly when the tannin is gradually desorbed from the oral mucosa. As a matter of principle, when tannin concentration increases, the astringency it causes increases more than its bitter taste.

Time-intensity measurements

The perception of astringency is very persistent and a highly dynamic process that changes continuously upon ingestion, especially following expectoration or swallowing. Hence, point estimates of intensity are inadequate, though they are commonly used in current quality control. For a comprehensive characterization of astringency, temporal evaluation is indispensable. This can best be achieved by computerized time-intensity measurements (Lee and Lawless 1991, Noble 1995, Valentová et al. 2002). As their designation implies, they take into account both the intensity of astringency and the period of time it persists. They are used within the scope of sensory studies, in which the sensorially perceived intensity of astringency is recorded on a chart as a function of time by use of a computer mouse. This procedure yields time-intensity curves in the manner as exemplified in figure 29.

Figure 29: Example of a time-intensity curve for astringency: a = intensity at maximum perception, b = time until maximum perception, c = total duration of the perception.

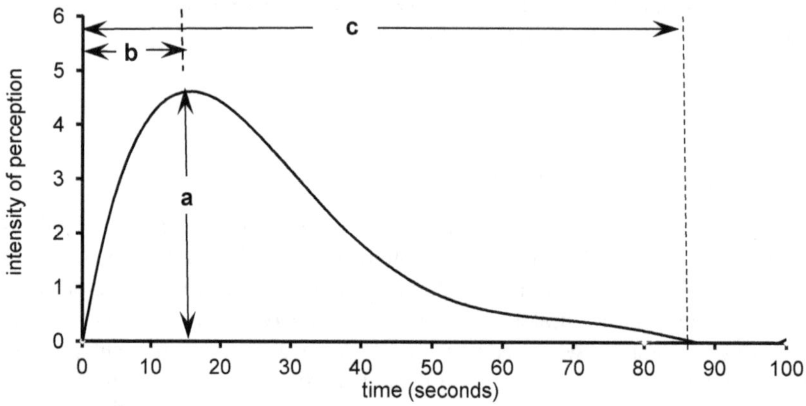

This kind of diagrams provides the following information:

- the maximum intensity (a), usually compared to a reference,
- the time until the sensation has reached its maximum (b),
- the total duration of the sensation (c),
- the intensity of response to the stimulus as the area under the curve.

As tannin concentration is increased, maximum intensity and total duration also increase. There are only small differences in time to the maximum occur. Whilst the astringent feeling can take as short as 15 seconds to fully develop, it can linger for several minutes. It builds up in intensity and becomes increasingly difficult to clear from the mouth over repeated tastings (Lee and Vickers 2009).

Judges provide characteristic time-intensity curves because the perception of astringency is highly variable between individuals and depends on their genetically predetermined salivary flow rate. Tasters with a low salivary flow rate perceive astringency more intensely, and its persistence lasts longer than for high-flow tasters (Fischer et al. 1994 a).

Differences in the individual saliva flow rate, phenomena of sensitization, and the time of desensitization required between sample ingestions explain why the astringency of one and the same red wine is perceived with highly variable intensity. Therefore, the sensory evaluation of red wine tannin and its astringency is without any doubt one of the most challenging tasks in common quality control, especially when experimental vinifications have to be evaluated. It requires a precisely specified tasting protocol when the task is to obtain reproducible data. However, this requirement is frequently neglected in current practice of technical tastings. Hence, it is not without reason that tannin measurement techniques (Chapter 3.4.3) have been developed to better predict astringency.

The carry-over effect

The irritation of the oral cavity by astringents causes saliva secretion to increase. Thus, the initial lubrication effect of saliva is restored after some time. As a result, the time interval between tasting samples is of particular importance. The basic question is how long this time must last.

The typical phenomena of adaptation, which are well known from tasting excessively sweet or sour beverages, do not apply when dealing with astringency. In contrast, intensity and temporal perception of astringency increase when a single wine is sipped repeatedly or during evaluation of several wines in one session, and increases yet further the shorter the time intervals between tastings (Guinard et al. 1986 a). This effect is called carry-over. It means that the astringent sensation imparted by a previous sample is transferred to the next sample if it has not faded away before. Thus, the remaining astringency of the previous sample adds to the astringency of the following sample and increases its intensity. The increasing perception of intensity and temporal duration of that perception upon repeated ingestion is explained by the exhaustion of salivary proteins. They require time to be reconstituted.

The carry-over effect of astringency has serious practical consequences in red wine tastings. It causes an increase of astringent perception and hence the differences in astringency to be incorrectly quantified. This problem usually arises in the context of extensive red wine tastings in a short period of time. It can easily compromise the validity of intensity ratings and triangle tests when used in the sensory evaluation of astringency[34].

The limited effect of palate cleansers

For the abovementioned reasons, there is practical interest in reducing the carry-over effect, to faster restore the lubrication effect of saliva, and to accelerate desensitization. For that purpose, water or bread rolls are frequently administered for palate cleansing when tasters are required to evaluate significant numbers of samples. However, their effect is highly debatable. Systematic studies on the usefulness of palate cleansers provided conflicting results, which are more sobering than promising.

Studies on the desensitization effects of plain unsalted crackers, water, and aqueous solutions of pectin, carboxymethylcellulose (CMC), polyvinylpyrrolidone (PVP), and proteins have shown that reducing or preventing the build-up of astringency is best achieved after ingestion of pectin solution (1 %), CMC solution (1 and 5.5 g/L), and unsalted crackers (Brannan et al. 2001, Colonna et al. 2004, Ross et al. 2007). After desensitization by one of these means, additional rinsing the mouth with water is considered indispensable to remove the residuals from these palate cleansers before tasting the next wine. However, only water is not sufficient to restore sensitivity towards astringency.

The aforementioned results have been contradicted by other studies. In one of them, milk was shown to be the best cleanser to remove astringency in comparison with still water, carbonated water, or nothing (Taladrid et al. 2018). In another one comparing the effects of water, CMC, crackers, milk, chewing wax, and nothing, there was no difference in their ability to prevent astringency build-up from occurring over repeated ingestions. On the other hand, panelists were best able to discriminate among the astringency of tannin solutions when water or nothing was used as a palate cleanser (Lee and Vickers 2009). These results were partially confirmed by comparing the effects of still mineral water, unsalted crackers, skimmed milk, drinkable sweetened yogurt, and a 2 % pectin solution. None of these palate cleansers could prevent the occurrence of astringency build-up. Regarding sample discrimination, drinkable plain sweetened yogurt, followed by a water rinse, provided the best results, whilst commonly recommended crackers showed the lowest discrimination (Vidal et al. 2016).

The inconsistency of all these results is obvious. Their lowest common denominator is that palate cleansers are either not very effective in preventing astringency build-up over repeated ingestions or show no effect at all.

Practical hints

The disappointing results of palate cleansers highlight the importance of limiting the number of samples evaluated in a single session and considering a predetermined waiting time between samples. There is no point in burdening tasters with more than 10 to 12 red wine samples in a row. When more samples are to be tasted, breaks are required to allow for desensitization and reconstitution of salivary proteins. Red wine

[34] A triangle test is a discrimination test, in which the panelists are presented with one different and two alike samples at once, and are asked to identify the odd sample. Samples are coded by randomly generated numbers.

evaluations require much more time than the tasting of white wines, which display several times less astringency.

Proteins contained in certain foods compete with salivary proteins as they also bind with tannins. Thus, more salivary proteins are preserved and enabled to decrease the perception of astringency to some extent (Madrigal-Galan and Heymann 2006). This explains the suitability of cheese dishes to accompany tannic red wines in a social environment. However, their use is not recommended in a professional tasting context as their composition and effect is highly variable and can even be counterproductive.

5.3. Sensory differentiation of astringency by wine composition

5.3.1. Impact of tannin composition and polymerization

It is obvious that astringency increases when tannin concentration increases, but this increase does not apply to all wines in the same way. For a given tannin level, the astringency depends on

- the chemical structure of the tannin and its mean degree of polymerization (Chapter 2.2.3),
- the availability of anthocyanins able to react with tannins and flavan-3-ols or having already reacted with them in the formation of polymeric pigments (Chapter 5.3.2),
- the availability of mannoproteins and other polysaccharides interacting with tannins (Chapter 5.3.3),
- the basic composition of the wine matrix with regard to alcohol content, pH, total acidity, and potassium concentration (Chapter 5.3.4).

As outlined in chapter 2.2.3, the flavonoid building blocks of tannins undergo polymerization, in particular in musts and young wines. The polymers generated in this way constitute the true tannin with a variable mean chain length, or better said, with a variable degree of polymerization. In this context, the degree of polymerization refers to the mean degree of polymerization (mDP) because small and large polymers occur simultaneously (Cheynier et al. 2006). Their nature and amount depend on the relative proportion of their precursors. In musts and very young wines, there are essentially two groups of precursors:

- monomeric and oligomeric flavan-3-ols, the latter designated as procyanidins,
- monomeric anthocyanins.

A high concentration of anthocyanins increases the amount of anthocyanin polymers with respect to other products. Excess of flavan-3-ols favors their polymerization between themselves and might result in browning. When a well-balanced ratio of flavan-3-ols and anthocyanins occurs in the early stage of winemaking, all kinds of polymers can be potentially formed, but with a focus on polymers consisting of both flavan-3-ols and anthocyanins, the so-called polymeric pigments (Fulcrand et al. 2004).

The proportions of these three polymeric phenol fractions in must influence astringency and over-all quality of the tannin in the finished wine. This elucidates why the tannin-anthocyanin ratio (Chapter 3.2.3) of young wines is as important as the absolute concentrations of these two groups of substances. Details are given in chapter 7.4.3. Furthermore, it explains why there is an interest in striving to modulate the tannin-anthocyanin ratio by different winemaking practices (Chapter 4.1), though an absolutely selective extraction is impossible.

Polymerization is a process which proceeds faster the higher the initial concentration of the molecules involved. The reason is easy to understand: Before two molecules react with each other, they first have to come together in three-dimensional space according to the law of chance. The chance of doing so increases when their number increases. Figure 30 displays the relationship between the rate of polymerization and the initial concentration by the example of catechin, expressed as the decrease of its monomeric form measured spectrophotometrically with 4-dimethylaminocinnamaldehyde (Zironi et al. 1992). In this context absolute values are less important than the curve slopes. It becomes obvious that phenols in high-phenol red wines display higher chemical dynamics and undergo faster polymerization than in low-phenol ones.

Figure 30: Polymerization of (+)-catechin as a function of its initial concentration in model solution (12 % alcohol, pH 3.7, 21° C) exposed to air and expressed as the decrease of its monomeric form.

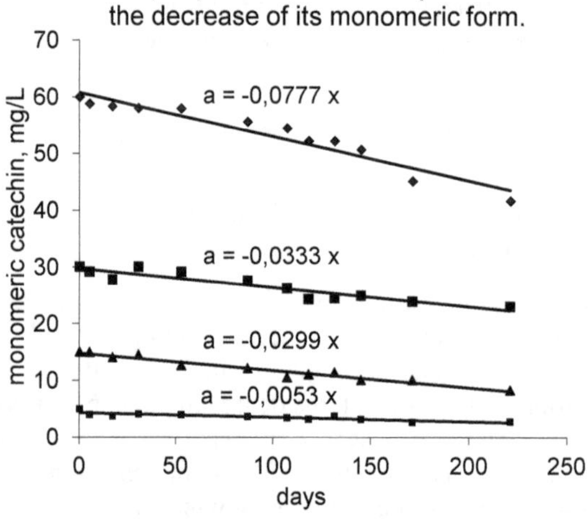

Nonetheless, polymerization is not a one-way process that would result in steadily increasing polymerization. Simultaneously, depolymerization also occurs. It is based on an acid catalyzed interflavanic bond cleavage of procyanidins, cancelling their bondings at C4→C6 and C4→C8 (Dallas et al. 1995). The products resulting from depolymerization can polymerize again in a different arrangement and form procyanidins other than those they have originated from. As a result, the tannin of red wines is not a static parameter, but rather undergoes continuing structural changes with sensory consequences.

The more polymerization proceeds, the more astringency increases, whilst bitterness decreases (Noble 1994). The dimer of two flavan-3-ol units already displays noticeable astringent properties, which intensify during further polymerization towards increasingly larger molecules (Arnold et al. 1980, Delcour et al. 1984).

Red wines are expected to lower their initial astringency during maturation and aging. Traditional enology has explained this phenomenon by an increasing polymerization of tannins, which would reduce their reactivity with salivary proteins and eventually lead to their insolubility and self-precipitation (Ribéreau-Gayon and Stonestreet

1966). However, the major polymerization reactions occur during the first year post-fermentation, whilst there are less changes of the degree of polymerization in subsequent years (Nagel and Glories 1991). On the other hand, tannin softening continues for a long period subsequently without any losses of tannin by precipitation taking place.

Trials have shown that, based on comparable concentrations, the degree of polymerization of isolated grape procyanidin fractions is the variable that most affected overall astringency and the different individual astringency subqualities, with increased drying, chalky, adhesive and puckering sensations correlating with increasing chain length (Vidal et al. 2004 b). The highly polymerized tannins predominantly precipitate together with salivary proteins, while the tannins not precipitating are low molecular weight polymers (Sarni-Manchado et al. 1999 a).

More recent studies on Cabernet Sauvignon wines of different harvests but from a narrowly limited growing area have shown that wine aging gives lower astringency scores and lower degrees of tannin polymerization. The loss of astringency associated with wine aging should be regarded as the result of an average size reduction of tannins rather than their precipitation (Chira et al. 2012). These results indicate that except for very young wines, depolymerization of tannin proceeds faster than concomitant polymerization. The net result is a decrease of the mean degree of polymerization (mDP) during wine aging.

However, under commercial winemaking conditions and for a given tannin concentration, it is misleading to derive a general statement about astringency based on mDP. On the contrary, in a set of 13 Tempranillo wines, astringency has been shown to be even more affected by the subunit composition of procyanidins, i.e. colorless or brown tannins not associated with anthocyanins (Chapter 2.2.3), than by their total concentration or their mDP. Thus, for example, the amount of epigallocatechin was negatively correlated with astringency (Quijada-Morín et al. 2012).

This picture becomes more complicated by the presence of wine constituents other than procyanidins (Cheynier et al. 2006). One of these constituents are the anthocyanins. Both their concentration and the tannin-anthocyanin ratio resulting therefrom play a role (Dallas et al. 1995, Schneider 1998, Durner et al. 2015).

5.3.2. Impact of anthocyanins

Anthocyanins are considered tasteless. Nonetheless they decrease tannin astringency some time after they have been added to red wine under experimental conditions. One of the possible explanations is their potential ability to increase the tactile sensation that is described as volume, body, softness, or mellowness. It is able to modulate and partially mask astringency, which is an opposite tactile sensation (Vidal et al. 2004 b, 2004 c, Paissoni et al. 2020). Another explanation is based on the positive charge and good solubility of anthocyanins in water. They effect a reduction of tannin hydrophobicity[35] that occurs when they are incorporated into the tannin structure, thus increasing its solubility. As a result, red wines taste softer when they contain high levels of anthocyanins.

As sensory terms are poorly defined, it is useful to specify in this context what is meant by terms such as volume, body, softness, or mellowness. It is the sensation one experiences upon tasting white wines with high levels of glycerin. For the purpose of sensory training, this sensation can be unambiguously modelled by adding 15 to 20 g/L of glycerin to a standard dry white wine. Conversely, subsequent additions of tannin to the glycerin-enhanced wine will counteract that effect.

In chemical terms, the softening impact of anthocyanins on astringency is based on their rapid reaction with colorless phenols. The reaction products are low-molecular weight copigments (Chapter 2.2.4) in young wines and high-molecular polymeric pigments (Chapter 2.2.5) in aged wines (Somers 1971, Haslam 1980).

Using the example of red wines made from Cabernet Franc, it has been shown that low levels of procyanidins associated with high concentrations of anthocyanins contribute to volume on the palate and tannin softness. As the amount of tannin increases and the ratio of anthocyanins to tannins decreases, tannin perception shifted from soft to green and then from green to hard and drying. Tannin softness as opposed to tannin hardness is related to the levels of anthocyanin-tannin adducts (polymeric pigments) and of colorless tannin polymers (procyanidins), respectively.

The formation of these two tannin subgroups is governed by the initial tannin-anthocyanin ratio. The more anthocyanins are present in the young wine, the more polymeric pigments are formed, and the softer is the tannin in the aged wine. This explains why high levels of anthocyanins are necessary not only to obtain satisfactory color intensity but also balance and mellowness. Green tannins that are characteristic of poor quality wines are associated with a lack of anthocyanins and polymeric pigments, demonstrating that anthocyanins actually participate in wine flavor (Cheynier et al. 1998).

Taking into account the importance of anthocyanins for gustative sensations, a tannin-anthocyanin index gains additional interest in quality control. It can be expressed as the total phenol-anthocyanin ratio as proposed in chapter 3.2.3. Thus, it complements the informative value of the total phenol content. Both total phenols and total anthocyanins are easy to measure (Chapter 3).

In this context, however, anthocyanins do not refer to the total anthocyanin content, but rather to its color-active ionized fraction (Chapter 2.2.4). SO_2 levels play an additional role as SO_2 bleaches anthocyanins. Using Australian red wines, it was shown that overall quality ratings increased when color intensity and the percentage of ionized, color-active anthocyanins increased. However, there was no correlation between color intensity and total anthocyanins, nor between ionized color-active anthocyanins and total anthocyanins (Somers and Evans 1974). This lack of correlation is attributable to varying levels of SO_2 and polymeric pigments.

[35] Hydrophobicity is the physical property of molecules that are repelled from water, thus decreasing their solubility in water.

A larger study (Jackson et al. 1978) on young red wines from the Beaujolais area in France revealed quite linear correlations between overall quality and taste on one hand, and total anthocyanins, colored ionized anthocyanins, uncolored anthocyanins, and color intensity on the other hand. Increasing bleaching of anthocyanins by increasing free SO_2 levels lowered the quality on the palate.

High free SO_2 levels make red wines thinner and more astringent

Anthocyanins can only occur in their ionized color-active form when they are not bleached by SO_2, which forms a colorless adduct with them (Chapter 2.2.4). In other words, any SO_2 addition lowers the level of colored anthocyanins. This can be easily observed when SO_2 is added to young red wine in a way to enhance the level of free SO_2 from 20 to 40 mg/L, for example. Any winemaker is familiarized with the loss of color associated therewith, but much less aware of the changes on the palate it causes. Upon comparing mouthfeel attributes of both the untreated and the SO_2-enhanced sample, one will readily notice that the wine with the higher SO_2 level displays less volume, mellowness, softness, etc., but more astringency.

This experience demonstrates that anthocyanins in their ionized color-active form soften astringency and contribute to volume and mellowness on the palate. As a consequence, attention should be paid to preserve them as much as possible and to not adjust free SO_2 before bottling to higher levels than required. Chapter 12.4 gives details.

The chemical binding with anthocyanins makes colorless tannins taste softer

The aforementioned results of sensory studies on real wines have been reinforced by more detailed investigations on the nature of the molecules involved. In one study (Weber et al. 2013), gel-permeation chromatography was used to break up the tannin of a red wine into two fractions, which were characterized by sensory analysis. The higher molecular fraction with a higher proportion of polymeric pigments and otherwise bound anthocyanins was less 'drying' than the low molecular fraction containing a higher proportion of brown-yellow tannins and short-chain procyanidins. It must therefore be concluded that a decrease of astringency during wine aging is due to an increasing incorporation of anthocyanins into the tannin-anthocyanin complex. Thereby, the increase of astringency observed upon polymerization of tannin in solutions low in anthocyanins (Chapter 5.3.1) is cancelled out by the attachment of anthocyanins onto the polymeric structures.

In a similar study (Vidal et al. 2004 b, 2004 c), isolated polymers of identical degree of polymerization were compared by sensory techniques. Red colored polymers with incorporated anthocyanins showed significantly less astringency than polymers obtained from a wine free from anthocyanins. The addition of highly purified anthocyanins to a model wine resulted in a lowering of its astringency and bitterness.

A sensory study (Gawel et al. 2007) on Australian Shiraz wines, anthocyanins caused a 'chalk-like" texture. In contrast, wines that elicited a 'puckering' sensation were characterized by relatively low anthocyanin levels, high concentrations of colorless tannins, and low acidity. The impact of anthocyanins on in-mouth textual properties was emphasized.

The dependence of these mouthfeel properties resulting from flavonoid composition as affected by viticultural variables has also been shown in Shiraz. Differences in the grape berry flavonoid profile were obtained by exposing bunches to varying levels of sunlight intensity through canopy manipulation. After vinification under standardized conditions, tannins and polymeric pigments in the wines were related to berry flavonoid composition. Higher concentrations of anthocyanins and skin tannins in berries, coupled with a lower extraction of seed tannins, were associated with higher wine quality. The ratio of anthocyanins x skin tannins/seed tannins was proposed as an indicator of wine flavonoid composition and wine quality (Ristić et al. 2010). Once again, the availability of high anthocyanin levels has been shown to enhance wine quality on the palate.

However, working on the same variety, these sensory effects of phenolic composition have been shown to be additionally differentiated by the climate of the growing area. In warmer climate, the effect of ripeness on mouthfeel was smaller than in cooler climates. The effect of the tannin extraction method also differed depending on climatic region and ripeness level. Whilst some mouthfeel attributes and phenolic fractions were consistently associated with region and ripeness, the effect of the tannin extraction method was less consistent (Nel et al. 2015).

Practical hint

Some thin red wines with dry tannins can benefit from additions of a few percent of young red wine, which contains at least ~500 mg/L anthocyanins. The optimum dosage must be determined by bench trials. Additionally, free SO_2 should be measured and adjusted to the level desired for bottling if necessary (Chapter 12.4). In some wines this approach can provide better sensory results than reducing astringency by laborious finings aiming at reducing tannins (Chapter 12.1.3).

5.3.3. Impact of mannoproteins and other polysaccharides

Tannins are hydrophobic molecules; the polarity of water induces them to spontaneously associate one with another. When these associations exceed a certain size, they loosely aggregate to particles of colloidal[36] nature, which are called micelles (McRae et al. 2014). These micelles are small spherical entities with a diameter ranging between 5 and 100 nm. In a further step, the micelles can react with other colloids, in particular with polysaccharides such as mannoproteins, and true proteins, forming even more complex colloidal particles (Saucier et al. 1997 a). When they reach a critical size and concentration, they can be removed by filtration to a sensorially significant extent. This phenomenon can sometimes be observed when very concentrated red wines are filtered. It does not occur in standard red wines (Chapter 12.2).

[36] A colloid is a substance consisting of particles substantially larger than molecules but too small to be visible by the unaided eye. The have at least one dimension in the size range of 1 to 1,000 nanometers (nm).

According to the understanding described above, polymeric phenols only constitute the skeleton of tannins, which are additionally differentiated by the attachment of non-phenolic colloids. In the enological context, these non-phenolic colloids are polysaccharides.

Tannins that are spatially engaged in more complex molecules are less astringent. For that reason, the chemical breakdown of the molecular structures of the phenols involved cannot explain all sensory phenomena elicited by tannins. Unfortunately, there are no analytical approaches for measuring polysaccharides that would be as easy to perform in routine control as those used for measuring total phenols, tannic phenols, etc.

Polysaccharides originate from the grapes as well as from the breakdown of yeast cell wall components. They consist of simple sugar molecules linked together to form a polymer, which is not fermentable. The sugar can be glucose, mannose, arabinose, rhamnose, galactose, or a mix of various sugars.

There are three major groups of polysaccharides, differing in their relative proportions and specific types of sugars they are made of. They comprise:

- Mannoproteins which are released into wine from yeast cell walls during fermentation and later during contact with yeast lees. They are composed of up to 250,000 mannose molecules forming a polymer to which proteins or peptides[37] are bound. They are the largest polysaccharides found in wine and display molecular weights up to 100 times higher than the average molecular weight of red wine tannin. They account for approximately 50 % of the yeast cell wall dry mass.
- Arabinogalactan proteins that are grape-derived and released during maceration. As their name suggests, the sugars they contain are arabinose and galactose.
- Rhamnogalacturonan-2-polysaccharides that are released from pectins of the grape cell walls during maceration. The sugar they are made up from comprise rhamnose, galacturonic acid, and some fructose.

Sensory properties

Polysaccharides can positively influence the integration of mouthfeel elements by interaction between structural and textural features, by increasing the wine body, suppleness, and viscosity, and by reducing the perception of astringency and bitterness. In particular the maximum intensity and total duration of astringency decrease as viscosity rises (Smith et al. 1996). Some of these macromolecules can also reduce the burning sensation of high alcohol levels, occasionally also described as palate hotness. This observation can explain why some wines appear to be less hot on the palate than others even though they have the same or higher alcohol content. Ultimately, they can provide a sense of sweetness as a result of bridging the sensory sensations between tannin, acidity, and alcohol, thus aiding in harmony and integration. All these properties explain their enological interest.

[37] Peptides are chains of between two and fifty amino acids. When they contain more than fifty amino acids, they are called proteins.

Among all polysaccharides, yeast-derived mannoproteins arouse most interest in winemaking. They can be released or extracted from yeasts and have a strong sensory impact on wine mouthfeel properties. Therefore, the terms 'polysaccharides' and 'mannoproteins' are sometimes arbitrarily interchanged or only one term is used to refer to both groups of substances, though polysaccharides have a broader meaning than mannoproteins.

Historically, the release of mannoproteins from post-fermentation yeast lees has been the primary reason of aging white wines on their lees (Fornairon-Bonnefond et al. 2002). In the meantime, this technique is also applied more and more to the barrel-aging of red wines. Alternatively, they can also be legally added as commercially available yeast derivative preparations.

In one trial the major polysaccharides present in wine were isolated regardless of their origin and split into two fractions. One fraction comprised a mixture of neutral polysaccharides, mannoproteins, and arabinogalactan-proteins, whilst the other fraction contained rhamnogalacturonan-II. After their addition to model wine, both fractions increased the 'fullness' sensation and reduced the attribute ratings associated with astringency in the model wine. The addition of a grape anthocyanin fraction also increased 'fullness', though to a lesser extent (Vidal et al. 2004 b, 2004 c, 2004 d).

Yeast cells already start releasing intracellular mannoproteins during primary fermentation. When fermentation is completed, they die within a few days or weeks. Dead yeast cells undergo autolysis, a process of cell disintegration during which they release even more mannoproteins than during fermentation. In red wines, mannoproteins bind with anthocyanins and tannins, resulting in a lowering of their astringency, a better color stability, and an increase in volume and body of the wine (Feuillat et al. 2001, Escot et al. 2001). Furthermore, they facilitate malolactic fermentation (Guilloux-Benatier et al. 1995).

For the purpose of increasing the concentrations of these molecules, several approaches are taken:

- the selection and use of yeast strains which produce high levels of mannoproteins during primary fermentation,
- the identification of yeast strains which autolyze rapidly after completion of primary fermentation,
- the periodic stirring the yeast lees,
- the addition of exogenous ß-glucanase to wines stored on their lees,
- the addition of exogenous mannoprotein preparations obtained from yeast cell walls.

Mannoproteins released or added during fermentation

Yeast strains strongly differ in their ability to release mannoproteins during alcoholic fermentation. The enhancement of these compounds is one of the primary goals in yeast selection. Comparing three yeast strains, a maximum concentration of 100 mg/L could be observed, with greater release occurring when the yeast mortality rate was high (Escot et al. 2001). Much higher levels were reported from other trials, in which yeast strains releasing high amounts of mannoproteins resulted in better sensorial and physicochemical wine characteristics than strains with a lower mannoprotein release (del Barrio-Galán et al. 2015). In general, differences between yeast strains regarding mannoprotein release are detectable on the palate when all other factors are held comparable, though these differences might decrease during aging.

Mannoprotein release was shown to significantly increase under stress conditions induced by an increase in fermentation temperature and sugar concentration in the must. Under these stress conditions, the yeast cells were subject to a higher mortality rate. However, there are also strains that release more mannoproteins when the fermentation conditions are nearer to optimal and the yeast cells more viable (Giovani et al. 2010). The initial colloid content of the must has an additional influence: The more turbid the must, the lower is the release of mannoproteins (Guilloux-Bénatier et al. 1995).

Studies comparing the effects of yeast overexpressed of mannoproteins with the addition of commercially available mannoprotein-based preparations yielded conflicting results. In one study, the additions of commercial mannoprotein preparations to the must yielded clearer effects than the inoculation with overproducing yeasts. Moreover, they enhanced sweetness and roundness of the wine (Guadalupe et al. 2007). In a follow-up study, the mannoproteins released from overproducing yeast strains significantly reduced colorless tannins (procyanidins), whilst the mannoproteins added as commercial preparations to the must reduced polymeric pigments plus colorless tannins (Guadalupe and Ayestarán 2008) or only color (Guadalupe et al. 2010). The reduction of tannin and stable color suggests a precipitation of the co-aggregates of mannoproteins-tannin and mannoproteins-pigment.

Since no further differences with regard to phenolics were observed, winemakers can use both alternatives to reduce wine astringency and increase wine smoothness and body, but have to accept one or another drawback. A reduction of tannins can be a problem for some wines, and color losses a problem for other wines. On the basis of economic criteria, mannoprotein-overexpressed yeast strains are less expensive.

However, care must be taken that these strains also meet other equally important requirements such as good fermentation capacity, high alcohol tolerance, low production of sulfurous compounds causing reduction flavor, etc. There is no yeast strain uniting only positive features, and practitioners usually do not have enough information about the characteristics of the large number of commercially available yeast strains. Therefore, increasing mannoprotein levels in the post-fermentation stage might be a more reliable approach.

Aging on the lees increases mannoproteins and makes red wine tannin softer

The release of mannoproteins continues during aging on the lees as yeast autolysis progresses. Comparable to what happens during fermentation, these macromolecules interact with tannins. As a result, astringency decreases and 'fullness' increases (Escot et al. 2001, Feuillat et al. 2001).

Mannoproteins in yeast cell walls are bound to ß-glucan[38], from which they have to be released by the action of the enzyme ß-glucanase. This enzyme is active in the post-fermentation yeast cells to a limited extent. Therefore, the release of mannoproteins is relatively slow and may require several months or years to occur. Periodic stirring of the lees increases the ß-glucanase activity, the contact of yeast cells with the surrounding liquid, and the mannoprotein concentration. The addition of commercially available ß-glucanase enzyme preparations can accelerate the process. In one trial, this enzyme addition achieved within three weeks polysaccharide concentrations which required five months in the absence of the enzyme. However, when the enzyme preparation contains ß-glucosidase impurities, losses of anthocyanins and color will occur (Palomero et al. 2007). In general, the additions of these enzyme preparations is beneficial, though their effect is variable and not always evident (del Barrio-Galán et al. 2011).

For very practical reasons, a distinction must be made between heavy lees and fine lees. Heavy lees are defined as those that settle within 24 hours after pressing. They are composed of large particles consisting of grape skin tissue fragments, residual seeds, tartrate crystals, yeasts, and tannin-protein complexes. Since these particles do not contribute to wine quality, there is little value of storing wine on heavy lees. They must be considered dirt. In contrast, the fine lees that remain in suspension consist primarily of yeast cells, which bring about the desired effect. Therefore, it is good practice to rack the wines one or two days after pressing.

When external white wine yeast lees were concentrated and added to red wine in proportions (0.8 and 5 %) that yeast lees represent in freshly fermented wine, an adsorption of color compounds and tannins by the yeast biomass occurred during the first days after addition. The spectrophotometrically measured color loss did not exceed 25 % and was not apparent by visual evaluation. The tannins adsorbed were prevalently those with the highest degree if polymerization, i.e., the most astringent ones (Chapter 5.3.1). After the initial decay of tannin, there was a retarding effect on polymerization reactions of the remaining tannin, likely by the mannoproteins released from the yeast and leading to the maintenance of low and medium molecular weight tannins in solution. Mouthfeel attributes were affected in different ways depending on the age of the yeast lees; young yeast lees yielded the highest overall balance ratings upon sensory evaluation (Rodrigues et al. 2013). These results show that the addition of external yeast lees strongly influences astringency and related mouthfeel parameters.

Practical hints

In current practice, adding white wine lees to red wine is a debatable issue. Since white wines are obtained from clarified juices, their lees consist predominantly of yeast cells without accompanying dirt designated as heavy lees. This is in contrast to red wine lees, which contain considerable amounts of non-yeast impurities. Furthermore, white wines contain much more yeast lees than red wines, in which a larger part of the yeast lees is removed with the pomace upon pressing. Seen from this perspective, adding white wine lees to red wines is an interesting means to increase their mannoprotein content. On the other hand, this approach entails a decrease of all kinds

[38] A glucan is a polysaccharide consisting of D-glucose molecules.

of red wine phenolics, body, and structure by dilution. Whenever it is put into practice, the white wine lees should be very compact to minimize the dilution effect. It is useful to check this effect on a small volume before adding the yeast lees to the wine in the industrial process. Measuring total phenols (Chapter 3.1) might provide additional support for decision-making.

There are several alternatives to the addition of lees obtained from other wines:

- The addition of industrially produced dry yeast biomass consisting in pure yeast cells of strains that have been selected for their ability of fast autolysis, high polysaccharide release, and positive sensory impact. A few such yeasts are commercially available, but are not inexpensive considering the relatively high amounts required to obtain a noticeable effect. Inevitably, they absorb some colored material and may impart yeasty aroma attributes in some cases.
- The addition of commercial dry yeast derivative preparations of variable purity, also called inactive dry yeast, has become more widespread. Some of these preparations are able to reduce the perception of green tannins and to increase softness on the palate by absorption or precipitation of the most polymerized and most astringent tannin fractions (del Barrio-Galán et al. 2012, Gonzáles-Royo et al. 2017). As in the preceding case, the sensory effect is not immediate because the release and dissolution of mannoproteins is a time-consuming process.
- The addition of commercial colloidal solutions of mannoproteins that have already been extracted from yeast biomass. Upon addition, the sensory effect is immediate and can easily be adjusted to the desired level by previous bench trials. In most wines, no effect can be observed but when there is an effect, it can be extremely powerful. Moreover, it is important to note that any sensory effects are highly variable, ranging from an increase of mellowness and fullness to an increase of sourness and astringency. Hence, no general statement on these preparations can be made. This is the result of their variable chemical composition or impurities they contain. It confirms the need of preliminary tests on each wine.
- Gum arabic consists of polysaccharides distantly related to those released from yeasts. Therefore, gum arabic preparations are often added to wine to improve volume and fullness on the palate. While they rank among the most commercialized wine additives, pure gum arabic is absolutely neutral in sensory terms when comparative trials are conducted in the appropriate way (Constantin et al. 2011). As so often in winemaking, for many strong beliefs are more convincing than scientific evidence. The only advantage of gum arabic is a better copper solubility and some protection against copper hazes.

The special case of barrel aging

The softening effect of yeast lees is most easily noticed when young red wine is split into a filtered and an unfiltered fraction before aging in new oak barrels. The fraction aged with its lees has a much lower perception of astringency and oak character, resulting in a higher perception of fruity characters. The strong reduction of astringency in new barrels can be explained to a large extent by the binding of ellagitannins and ellagic acid released from new wood onto mannoproteins. A comparable effect can be observed when certain commercially available mannoprotein preparations are added to wines with aggressive tannins. Hence, lees contact is an integrated part of barrel aging, especially in new barrels. The worst case happens when filtered wines are barrel-aged. They will remain coarse and harsh forever (Chapter 10).

Settled yeast lees are hardly helpful for the effects sought. Therefore, lees stirring (bâtonnage) is widely performed at variable intervals. There are no fixed rules for how often it has to be done. It depends on the lees settling rate, which strongly varies from one wine to another. The faster the lees settle, the more often stirring makes sense. In practice, stirring intervals range from two weeks to two months.

Lees stirring is a stylistic tool that can change the sensory balance between fruit, yeast, and wood. It enhances yeast components and reduces the perception of wood-derived flavor, whilst the effect on fruit is more variable. This variability is explained by the strong interaction between oxygen uptake caused upon stirring and the reductive properties of yeast lees. The anti-oxidative properties of yeast lees can be explained by two mechanisms, which are discussed in detail in chapter 6.2:

- They release glutathione, an S-containing cell wall peptide that is able to reduce wine compounds that have already been oxidized.
- They directly consume dissolved oxygen and prevent it from oxidizing wine compounds. This withdrawal of oxygen carries more weight than the reducing effect of glutathione.

Traditionally, stirring is considered a tool that generates an oxidative process that is detrimental to fruit intensity, but this assumption has to be revised in view of more recent knowledge on the reductive strength of suspended yeast lees. Actually, there is a complex balance between oxygen uptake by the wine and oxygen consumption by yeast lees, which can hardly be standardized. Suspended yeast lees consume dissolved oxygen at a variable rate and to a variable extent. The oxygen thus consumed ceases to be available to react with intrinsic wine constituents (Schneider et al. 2016).

In this way, the presence of suspended yeast lees limits wine oxidation. It strongly impacts the outcome of oxygen supply during barrel-aging and micro-oxygenation (Chapters 8 and 10). The more oxygen is consumed by yeast lees or the less oxygen remains to react with wine compounds, the more the tannin-anthocyanin polymerization reactions are slowed down and the more oxygen-sensitive fruit aroma is preserved. In practice, stirring the lees or adding additional lees to an over-oxidized red wine can drastically alleviate oxidation characters within a few days without relying on SO_2 additions.

Take-home message about aging sur lees

Polysaccharides stemming from both ripe fruit and yeast significantly contribute to the sensory differentiation between good and bad tannins, between ripe and green tannins.

Fermenting musts with a yeast strain tending to release high amounts of mannoproteins is beneficial for mouthfeel perceptions. This release continues in the post-fermentation stage and explains why aging on the lees is highly recommendable. It can be improved by periodic stirring and the addition of a commercially available ß-glucanase enzyme preparation.

However, after devatting, a major part of the yeast is removed upon pressing and discarded with the pomace. The remaining suspended yeast lees are nonetheless able to deplete oxygen picked up and to act as a reducing agent (Chapter 6.2). On the other hand, their total biomass is often too low to release mannoproteins to a sensorially significant extent. Thus, addition of concentrated white wine yeast less can become beneficial in some cases. This is particularly true when red wines are aged in new barrels, which supply the wine with high amounts of oxygen and oak-derived ellagitannins (Chapters 10.2 and 10.3). The color losses this approach entails are acceptable.

Aging on the lees can be replaced or improved by the post-fermentation addition of inactive dry yeasts selected for their high mannoprotein release, or by the addition of ready-to-use solutions of mannoproteins that have already been extracted from such yeasts. These ready-to-use solutions have considerable practical interest. They do not require a time-consuming extraction process as do inactive dry yeasts. Their sensory effect can be evaluated immediately upon addition. However, it is important to be aware of the large differences existing between the various commercial preparations. On a case-by-case basis, they are able to mask excessive astringency by increasing volume and fullness on the palate. Thus, in some wines they can replace fining procedures aiming at lowering astringency via a partial removal of tannins. Unfortunately, not all of these preparations show the desired sensory effect. Furthermore, for the same preparation, the effect differs from one wine to another.

Many wines do not respond to the addition of inactive dry yeast or ready-to-use mannoprotein solutions. This can be explained by the phenomenon of sensory saturation that appears when colloids occur above a certain concentration. Therefore, previous trials are indispensable on each wine. Another explanation is the basic wine matrix, which has a comparatively much greater effect on mouthfeel and taste than the polysaccharides. Increasing polysaccharides modulates the effects of the wine matrix on mouthfeel, but alcohol and acidity levels are the main drivers of mouthfeel and taste. This is shown in the following chapter.

Microorganisms like yeasts that are suspended in wine can feign a fullness on the palate, which the wine ceases to display after filtration. This phenomenon is not necessarily attributable to a decrease of polysaccharides or other colloids. It only occurs selectively upon filtration of very concentrated wines. Cut-off and adsorption specifications of the filter media play an additional role in this process. In general, fine or sterile filtration removes more colloidal material than coarse filtration.

5.3.4. Impact of alcohol, potassium, acidity, pH, and sugar

The sensory perception of tannin and its astringency is largely influenced by additional matrix variables, which comprise pH, acidity, potassium, and alcohol content.

Alcohol

Depending on the wine, model solution, and implementation of sensory studies, increasing alcohol levels decrease perceived astringency (Fontoin et al. 2008), do not affect it (Valentová et al. 2001), or increase it (Obreque-Slier et al. 2010 b). Many tasters also interpret the burning sensation caused by increased alcohol levels as astringency. The variable impact of increasing alcohol levels on astringency are shown in figure 31, which reports sensory data of two wines obtained by a trained panel consisting of wine producers.

Figure 31: Impact of increasing alcohol levels on the perception (0-5) of astringency, bitterness, burning, and sourness in a dry red wine blend (A) and a dry Merlot red wine (B).

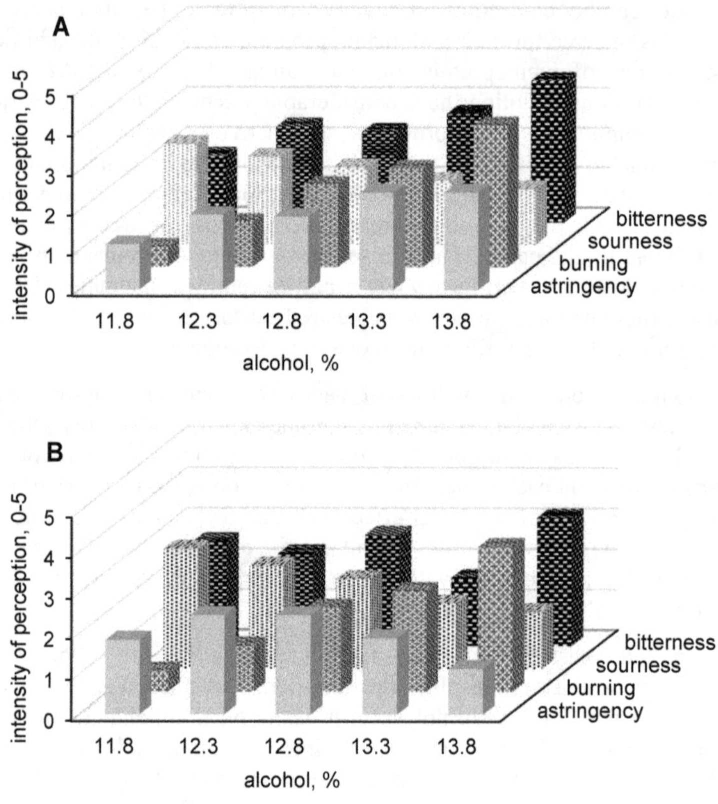

Both the astringency of tannins and the burning flavor of alcohol are tactile sensations, which affect the salivary flow rate. As a result, a large part of what is described as astringent or tannic by less trained tasters and wine novices is the direct effect of high or badly integrated alcohol levels. For assessing the changes of mouthfeel sensations caused by varying alcohol levels, more precise terms such as puckering, grippy, drying, abrasive, adhesive, smooth, fleshy, etc. are more appropriate than the more generic term of astringency (de Miglio and Pickering 2008).

Additionally, the order of wine assessment with regard to alcohol concentration significantly influences the sensory perception on the palate. Depending on whether high alcohol wines are tasted before low alcohol wines or the other way round, there are opposing effects of alcohol (12 to 16 %) on the sensory profiles. The largest differences occur when low alcohol wines are tasted first, with substantial changes in the sensory profiles of high alcohol wines. These effects are so strong that wine alcohol concentrations need to be considered when wine quality has to be professionally assessed (King et al. 2013), in particular when astringency ratings have to be performed.

With regard to bitterness ratings, there is general agreement that the perception of bitterness increases when alcohol increases (Fischer and Noble 1994 b, Vidal et al. 2004 d, de Miglio and Pickering 2008, Fontoin et al. 2008). This is largely in accordance with the bitterness ratings shown in figure 31. However, the order of samples with regard to alcohol content seems less important than for astringency ratings.

Acidity and pH

Increasing acidity reinforces the sensation of astringency, but this effect is directly due to the lower pH with goes along with increased acidity (Sarni-Manchado and Cheynier 1999 a, Fontoin et al. 2008) and not to the kind or amount of acid present. The increased astringency produced upon wine acidification is supposed to result from the pH-driven increase in the affinity of tannins for binding with salivary proteins (Guinard et al. 1986 b, Peleg et al. 1998). Another possible explanation for the inverse relationship between pH and astringency may be the reduction in salivary viscosity upon rapid alteration of salivary proteins under conditions of reduced pH (Sowalsky and Noble 1998). Be that as it may be, alcohol has a similar or stronger influence on astringency and other oral sensations elicited by dry red wines than pH (de Miglio and Pickering 2008).

Whilst the astringency of tannin strongly increases upon acidification, organic acids contained in wine and authorized for acidification elicit an astringent sensation by themselves (Lawless et al. 1996). Additionally, acidification of wine also lowers the ratings of what many tasters describe as body, fullness, and viscosity, though it might considerably reinforce the sensation of aftertaste, length, and persistence when wines are too flat.

In summary, acidity management is an efficient tool for modulating tannin-derived astringency. Chemical deacidification using potassium carbonates can be a valuable means of lowering astringency in wines deemed to tannic. Whenever it is envisaged, its extent should be based on the outcome of preliminary bench trials using additions of 0.2, 0.4, 0.6... g/L $KHCO_3$. Details and practical application are outlined in chapter 12.1.1.

Sugar

Sugar has an indirect effect on astringency. Whilst sweetness as a basic taste can suppress other basic tastes like acidity or bitterness, this does not apply to the tactile sensation of astringency. Hence, an artificial sweetener such as aspartame is not capable of decreasing astringency (Smith et al. 1996, Gawel 1998). The reason is that it fails to increase the wine viscosity. However, sucrose, glucose, and fructose modify the rheological properties of the solvent, increase its viscosity, and thus reduce astringency (Valentová et al. 2002). The attenuation of astringency by sugars seems related to increases in the salivary volume (Chapter 5.2) produced or the lubricating characteristics of the viscous sugar solutions (Lyman and Green 1990).

In practical winemaking, however, sweetening a dry red wine with sugar is not an appropriate means of sufficiently lowering an unpleasantly harsh astringency. For that purpose, sugar levels of more than 10 g/L would be required. They are off-limits for dry red wines.

Ripe and green tannins

Based on theoretical considerations, the pH value is often presented as the main driver of perceived sourness in wine. In truth, however, titratable acidity is much more important for sourness than pH (Plane et al. 1980). Whenever correlations between sourness and analytical data are run on dry wines, titratable acidity accounts for more than 75% of sourness (Schneider and Troxell 2018). This finding is substantiated by another simple observation: When a wine is diluted with water, pH remains almost constant, while TA and sour taste decrease by the dilution factor.

There is, however, a minor difference between white and red wines in that the fairly good correlation between titratable acidity and sourness is slightly less pronounced in red wines. This is attributable to the sensory quality of tannins. As outlined in the previous chapter, any tannin carries a bitter taste. Additionally, tannin in red wine made from unripe grapes carries a sour taste, which in some cases can feign up to 0.7 g/L more acidity than the wine actually has. Conversely, tannin obtained from ripe fruit can display a sweet taste, which can feign 3 to 4 g/L sugar in particular cases. Thus, the well-known effects of alcohol, acidity, and sugar on the basic taste perceptions sour and sweet in red wines are more complicated than in white wines. Figure 32 displays these interactions schematically.

Figure 32: Interaction between tannin and acidity: Influence of tannin and other compositional variables on the sour taste of red wines.

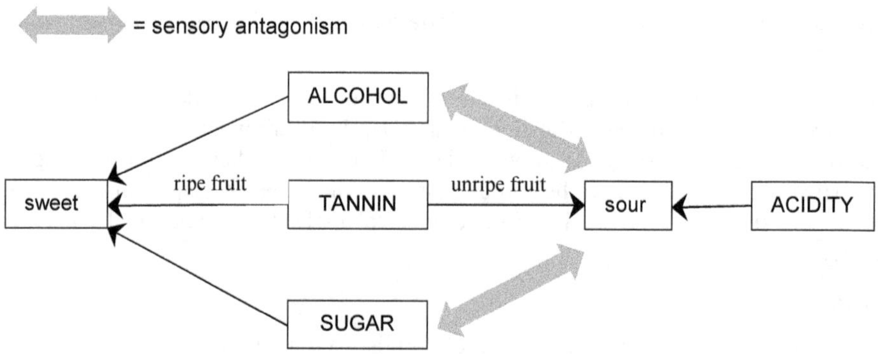

There is little knowledge about the chemical reasons that can make a tannin taste sour or sweet. Polysaccharides or their lack will probably play a role. However, it is known that such a secondary taste is not related to the degree of polymerization of tannins, and that it is not significantly influenced by the current techniques of vinification and aging.

A sour tannin appears frequently under cool-climate growing conditions, though it can also occur under hot-climate conditions after premature harvest. Interestingly enough, it will always remain sour regardless of aging. Its sour taste adds to the sourness caused by acidity. The only enological tool to provide remedy is lowering sourness via deacidification, since the sour tannin fraction cannot be selectively removed by fining. Concomitantly, that part of the astringency which is caused by too high an acidity is also removed. As already mentioned, organic acids also elicit an astringent perception (Chapter 5.2), which adds to the astringency caused by tannin.

Although complete malolactic fermentation is a standard in red winemaking, it does not always suffice in some growing regions, vintages, and cultivars to achieve a balanced mouthfeel with a well-integrated sourness. When this happens, additional chemical deacidification is advisable.

Practical hints

Owing to the sensory interaction between tannin and acidity, both parameters have to be seen in a particular context. Frequently a definitive sensory evaluation of a wine's tannin is only possible when the acidity has been set. For hot-climate growing areas strongly relying on pre-fermentation tartaric additions for pH adjustment as well as for cool-climate conditions yielding high acidity wines, minor deacidification trials with potassium carbonates should be trialed before tannin, astringency, and other mouthfeel properties are evaluated. Conversely, a trial with acidity enhancement can become advisable when a wine appears too flat and inexpressive on the palate. Many cases of too much or too less astringency can be settled with minor acidity corrections in the range of less than 1 g/L. Chapter 12.1.1 expands on these acidity corrections and how they can be evaluated by bench trials. Only when these preliminary trials do not work out, should further steps be envisaged. They comprise:

- addition of mannoproteins to better integrate tannins when the astringency is considered too strong,
- fining with proteinaceous fining agents such as gelatin to reduce tannin and astringency,
- addition of commercial tannins when the astringency is considered too low.

Potassium

Just as astringency is enhanced by acidity, it is inversely masked by a series of other wine constituents. They include anthocyanins (Chapter 5.3.2) and polysaccharides (Chapter 5.3.3), but wine potassium content also plays a major role. This can easily be understood when one tastes a wine after addition of increasing amounts of the neutral potassium salt of malic or lactic acid. In contrast to tartaric acid salts that would drop out, these salts are soluble. When they are added to wine, total acidity remains unchanged. Thus, their effect on the palate is attributable to the potassium they contain.

Potassium levels in wine are highly variable by nature. Furthermore, they are enhanced when potassium carbonates are used to reduce acidity by chemical deacidification. In practice, some of the potassium ions supplied in this way remain in solution instead of precipitating with tartaric acid, reinforce the sensation of fullness and body, and reduce the astringent perception quite independently from the acidity reduction achieved. Conversely, they are lowered by the addition of tartaric acid via their precipitation as potassium bitartrate. Cold stabilization lowers potassium levels and makes wines thinner to a degree which depends on the amount of potassium bitartrate removed by that means (Schneider and Troxell 2018).

Figure 33 shows in a schematic view the complex sensory interactions between essential wine constituents from the perspective of astringency and the basic taste sensations of sour, bitter, and sweet.

Figure 33: Sensory characteristics of tannin as influenced by other wine constituents. ⟷ = mutual masking effect

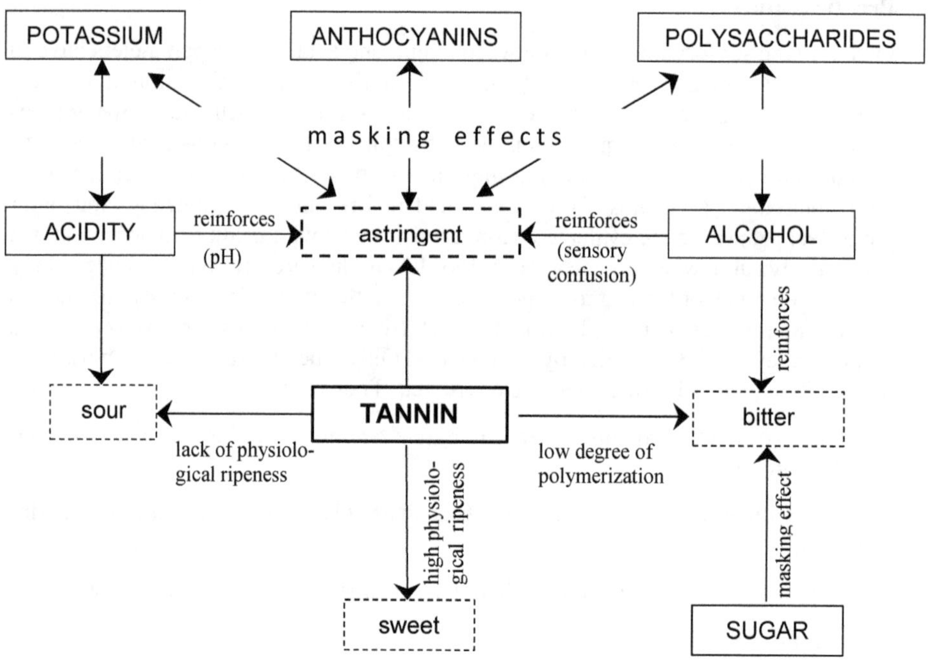

6. Oxygen consumption by oxygen scavengers

An appropriate oxygen supply is considered indispensable for the sensory integration of tannins, the softening of their astringency, and proper red wine maturation. Therefore, measures of deliberate or passive oxygenation such as barrel aging, micro-oxygenation, or periodic splashings are taken in the period between primary fermentation and bottling. The oxygen supplied is assumed to predominantly react with tannin and to be scavenged in this way. Thus, it is thought that the higher the tannin or total phenol levels, the higher the oxygen demand. However, phenolics are not the only oxygen scavengers, with suspended yeast lees of young wines being among the most important other ones. Hence, the effects of rackings and aging on the lees strongly affect the percentage of oxygen reacting with phenolics and the sensory outcome of oxygen consumption. Sulfur dioxide also interferes at a later stage of wine development. This chapter deals with the competitive behavior of the various oxygen acceptors. It gives kinetic data for the various competing reactions that occur with the wine matrix following the incorporation of oxygen. It shows that the amount of oxygen actually reacting with tannin is highly variable and barely predictable without preliminary bench trials.

6.1. General kinetics of oxygen consumption

In contrast to white wines, red wines require some oxygen uptake on their journey from their raw state immediately post-fermentation to the point where they can be bottled. This period is best described by the French term 'élevage', which means the stage in which the wines' definitive sensory characteristics are shaped prior to bottling. It often coincides with barrel aging. In this time window, oxygen is meant to react with phenolic compounds with the primary objective of softening tannin. Better color intensity and stability are additional benefits derived therefrom. Therefore, these tannin-oxygen interactions are of utmost interest in enology.

The amounts of oxygen required are highly variable and depend on the individual wine. For some wines, the oxygen unavoidably picked up during cellar operations such as fining, blending, filtration, barrel aging, etc. might suffice. This way of oxygen supply is designated as passive oxygen uptake (Chapter 8.1). When it is considered insufficient, specific means are taken to add supplementary amounts. These means are called active oxygen supply or oxygenation (Chapters 8.2 and 8.3). In both cases, the oxygen first occurs as a gas dissolved in the liquid. It is known as dissolved oxygen (DO), which can be measured by various techniques.

As it is widely known, wine can dissolve up to approximately 8 mg/L O_2 at standard air pressure and ambient atmosphere. When this happens, the wine is saturated with that gas. However, this rarely happens under commercial winemaking conditions as special efforts would have to be made to saturate a wine with oxygen. Generally wines pick up only some 1 to 4 mg/L O_2 each time they undergo a treatment such as racking, blending, fining, filtration, etc. (Chapter 8.1.3). These amounts are gradually consumed within a few days or weeks by chemical binding to wine constituents or yeast lees. After its disappearance it cannot be measured any more. This process is repeated when the wine picks up oxygen again upon a subsequent winery operation. This behavior is shown in figure 34.

Figure 34: Time course of consecutive dissolutions and subsequent consumption of oxygen.

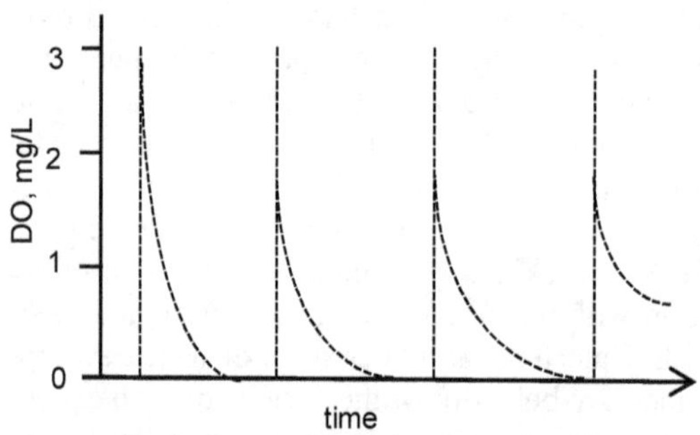

In contrast to these sporadic oxygen incorporations, oxygen can also be delivered to the wine in a continuous manner when it is stored in containers made of oxygen-permeable material such as wood or via ullage. Under these conditions, oxygen uptake and consumption occur simultaneously, resulting in low steady-state levels of measurable DO at any time.

The rate at which DO is consumed by wine is highly variable and hardly predictable. It can vary from more than 1 mg/L DO per day to less than 0.1 mg/L DO per day, but most red wines require approximately 10 days to consume one saturation concentration of oxygen (Ferreira et al. 2015). This is bad news for any winemaker. Indeed, it depends on temperature and a large number of wine compositional parameters. They include tannins, anthocyanins, heavy metals, yeast lees, free sulfur dioxide as well as the wine's history.

Figure 35: Time course of DO consumption in two filtered red wines.

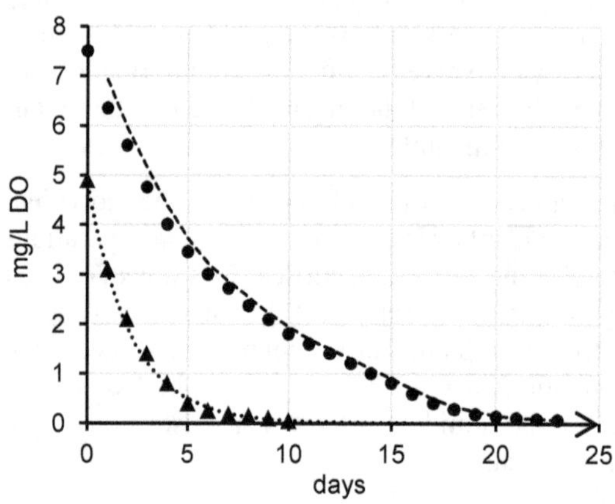

Figure 35 displays the typical course of DO consumption in two filtered red wines, which were stored under anoxic conditions after the oxygen uptake event had taken place. In these wines, the DO was consumed at an initial rate of approximately 2 mg/L per day and reduced by more than 90 % after one week. However, what is typical in these examples is not so much the time required for DO depletion, which can be more or less depending on wine composition, but rather the time course of its rate.

An outstanding feature of DO consumption in any given wine is its variable rate, which means that it is not the same over the whole DO concentration range but rather depends upon the instantaneous DO concentration and is proportional to it. Hence, when DO halves, the reaction rate also halves. As a result, the DO consumption rate can be fairly rapid in the high concentration range, whilst it slows down when the DO level approaches zero, flattening off asymptotically until no further DO can be measured. This is what is called a first-order reaction kinetics[39]. Hence, it is not admissible to attribute a characteristic DO consumption rate to a wine without considering its momentary DO level. In all wines, the change of DO over time is best fitted with an exponential function, though this might not always be apparent on superficial examination.

In order to make calculations and comparisons easier, the reaction rate can also be presented as the half-life of the reactant, in this case DO. The half-life is the period of time in which the DO concentration decreases by half of its value. It is characteristic of every wine at a given moment and is the most correct way to compare DO consumption rates between wines.

The asymptotic behavior of DO consumption explains why sufficiently sensitive DO measurement techniques allow for detecting low amounts of DO at any time when wine is stored in containers made of materials (wood, plastics) that are oxygen-permeable.

The following chapters deal with the fate of oxygen picked up by red wines and the reactions DO undergoes with wine constituents.

6.2. Oxygen consumption by post-fermentation yeast lees

Phenolic compounds such as tannins have been considered the most important oxygen acceptors in wine for a long time (Singleton 1987, Fabre 1993). Based on this assumption, micro-oxygenation (Chapter 8.3) has been developed with the objective of improving tannin quality by adding defined amounts of oxygen to young red wines stored in tanks. In doing so, the passive oxygen uptake occurring during barrel aging can be replaced and better controlled. In either case, the oxygen consumed by wine has been assumed to predominantly react with tannins and other phenolics. This can actually be the case, but more recent research has shown the importance of other oxygen acceptors than phenolics. The lower the total phenol level of red wines, the more DO consumption is caused by non-phenolic oxygen acceptors. Suspended yeast lees and sulfur dioxide are the most important of them. Alcohol and mannoproteins play a secondary role.

Figure 36 shows schematically how strongly the presence of suspended yeast lees can affect the percentage distribution of a given amount of DO reacting with intrinsic wine

---

[39] Strictly speaking, kinetics is more complex than a first-order one (Ferreira et al. 2015), but it gives a useful approximation of the realities for practical use.

constituents such as tannin, anthocyanins, etc. in a typical young red wine. As soon as a wine is filtered or clarified by any other means, considerably higher proportions of DO are available for reacting with tannins, anthocyanins, and SO_2 than before clarification. Such data can be easily obtained by comparing the specific DO consumption rate of the major wine constituents added one by one to a model solution made up of 13 % alcohol at pH 3.5 and traces of heavy metal ions.

Figure 36: Impact of suspended yeast lees on the fate of DO and the relative importance of DO scavengers in a typical young red wine.

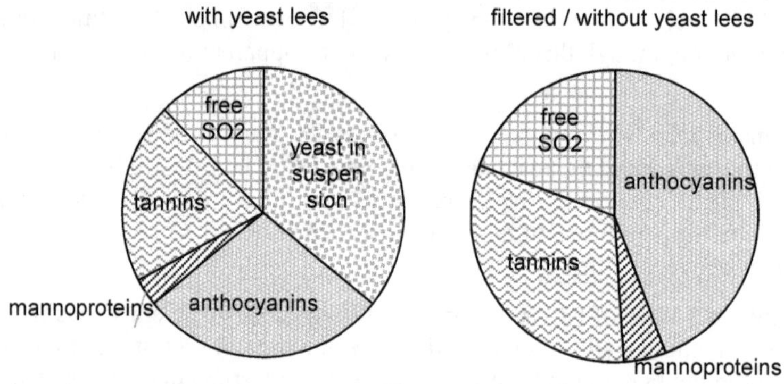

Interestingly, anthocyanins consume a major part of DO as long as they are still present in their monomeric form, and a small part is also consumed by mannoproteins probably due to reducing amino acids they contain. DO consumption by alcohol is not measurable though it occurs to a negligible extent. Heavy metals do not consume any DO at all but only act as transition metals catalyzing the transfer of oxygen to oxidizable compounds.

To summarize, in young red wines, the suspended yeast lees consume a large but variable portion of the oxygen the wine takes up or is deliberately supplied with. This oxygen is digested within the yeast cells, withdrawn from the wine, and is no longer available to react with intrinsic wine constituents and to oxidize them. This behavior of yeast lees has not as yet been fully appreciated in the wine industry, but it has a tremendous impact on the sensory outcome of micro-oxygenation, barrel aging, and other means of oxygen supply. It explains many phenomena occurring during wine aging, which cannot be explained by phenol and oxygen chemistry. Therefore it is outlined in detail hereinafter.

Respiration has been shown only to be involved in yeast lees' oxygen uptake during the first days and weeks post fermentation, as 99 % of the yeast cells die within two weeks post-fermentation. After that period and during the following months and years, DO consumed by dead yeast cells is used in a non-respiratory pathway for the oxidation of cell membrane lipids and ergosterol (Fornairon et al. 1999, Salmon et al. 2000, Rosenfeld et al. 2002, Fornairon-Bonnefond and Salmon 2003). These reactions confirm earlier findings on brewers' yeast. They are discontinued after pasteurization, but there is no direct evidence yet that enzymes are involved. They mainly occur in suspended yeast cells, whilst the effect of settled lees on DO consumption is negligible.

For the quantification of the oxygen consumption by suspended yeast lees under enological conditions, special methods have been developed (Schneider et al. 2016). Their results obtained on six commercial yeast strains (A-F) are presented in the following section. They also allow estimation of the remaining proportion of DO available for the oxidation of wine constituents.

Effect of yeast concentration

The decisive variables in this context are the oxygen consumption rate (OCR, in mg/L DO per hour) and the yeast concentration (NTU). The OCR indicates how fast suspended yeast lees remaining in suspension after primary fermentation consume DO. The yeast concentration measured as NTU (nephelometric turbidity units) strongly correlates ($r^2 = 0.98$) with the yeast cell count in model solution and white wine as long as no malolactic fermentation has taken place. In this case, 1 NTU equals $6.3 \cdot 10^4$ cells/mL per 1 NTU.

Model solution (13 % alcohol, pH 3.5) was chosen to study the effect of yeast concentration on its ability to consume DO. Thus, any interference from oxidizable wine constituents consuming DO was circumvented. The yeast lees were obtained from young white wines that had not undergone malolactic fermentation. Yeast concentrations assayed were 25, 50, 100, 200, 300, and 400 NTU. The DO consumption in each lot was almost constant during the tests. Table 3 gives the results.

Table 3: Oxygen consumption rate (mg/L·h O_2) of yeast strains in model solution as affected by yeast concentration.

Yeasts were harvested and trialed one to four weeks after the end of alcoholic fermentation. Free SO_2 = 0 mg/L; temperature = 20° C, NTU = nephelometric turbidity units.

yeast strain	yeast lees concentration						
	5 NTU	25 NTU	50 NTU	100 NTU	200 NTU	300 NTU	400 NTU
	OCR, mg/L·h						
A	0.00	0.04	0.39	0.58	0.80	1.15	1.75
B	0.00	0.04	0.44	0.64	0.80	1.04	1.21
C	0.00	0.03	0.34	0.77	1.07	1.55	2.09
D	0.00	0.04	0.51	0.57	0.60	0.68	0.74
E	0.00	0.04	0.68	0.72	0.85	0.96	1.13
F	0.00	0.03	0.27	0.38	0.53	0.61	0.67
mean	0.00	0.04	0.44	0.61	0.77	1.00	1.27

As could be expected, increasing yeast concentration leads to higher OCRs. However, this increase is not a linear one as the OCR per cell declines at higher cell densities. This pattern is believed to be caused by a steric hindrance of oxygen diffusion at the particle level as a result of the physical presence of yeast cells, making DO diffusion within the liquid the rate-limiting step.

Although all yeasts were harvested and trialed within one month after the end of alcoholic fermentation, there are remarkable differences in the OCR among yeast strains at given concentrations. Reproducible consumption rates require a minimum of 50 NTU of yeast. This yeast concentration is present in any young wine after fermentation. Above that minimum level, OCR ranged from 0.27 to 2.09 mg/L·h.

For practical considerations, it is important to interpret wine turbidity expressed as NTU. Young red wines display a turbidity in the range of 300 to 800 NTU after alcoholic and malolactic fermentation. When suspended material settles in the course of the following months, turbidity decreases to 50 to 300 NTU. After clarification by coarse filtration through diatomaceous earth, this level is further reduced to 5 to 10 NTU. A turbidity of 50 NTU indicates a clearly visible opalescence. In red wines after malolactic fermentation, approximately half of the turbidity is caused by yeast cells.

Effect of storage and yeast age

Figure 37: Impact of storage and age of the yeast on its oxygen consumption rate (OCR).

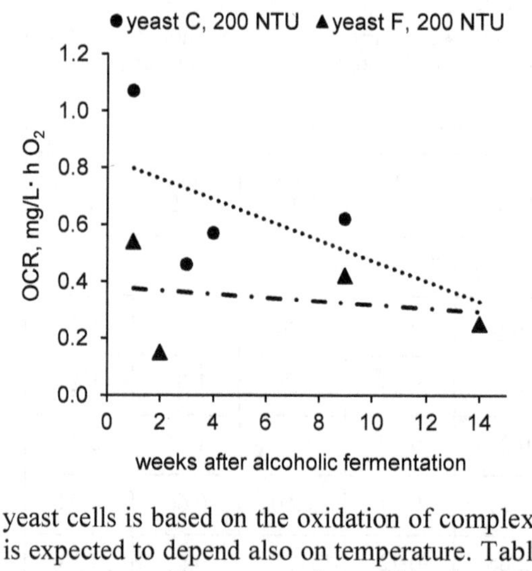

Storage and aging can be expected to have some influence on the oxygen consumption rate of post-fermentation yeast lees. Using the example of two yeast strains, figure 37 shows that there is a slight and irregular decrease during the first three months after the end of primary fermentation. After four months of aging, a residual activity of 11 to 100 % of the initial OCR has been reported (Fornairon et al. 1999), and there is still some DO consumption by yeasts even after three years of aging (Salmon 2006).

Effect of storage temperature

Since the DO consumption of yeast cells is based on the oxidation of complex yeast cell membrane compounds, it is expected to depend also on temperature. Table 4 shows that temperature effect on yeast strains with concentrations adjusted to 200 NTU in model solutions at various temperatures. It applies to each of the yeast strains to a varying degree.

Table 4: Effect of temperature on the oxygen consumption rate (OCR, in mg/L · h O_2) of different yeast strains in model solution.

Yeasts were harvested two weeks after the end of alcoholic fermentation and adjusted to 200 NTU.

yeast strain	temperature		
	5° C	15° C	20° C
	OCR, mg/L·h DO		
A	0.030	0.469	0.507
B	0.029	0.041	0.048
C	0.136	0.193	0.244
D	0.017	0.046	0.099
E	0.573	0.614	0.706
F	0.591	0.781	0.833
mean	0.278	0.357	0.406
% of OCR at 20° C	68.5	87.9	100

Under conditions of cool wine storage, the OCR of yeast cells decreases markedly. However, this effect is of limited practical importance: Oxygen consumption in real, turbid wines is not only caused by yeast cells but also by concomitant chemical binding with oxidizable wine constituents like phenols. Both oxygen consumption pathways compete one with another. Chemical oxygen binding, i.e. the oxygen consumption of filtered wines, also proceeds at a lower rate when the temperature is decreased. Hence, the yeast percentage contribution to total oxygen consumption is not necessarily affected by temperature.

Effect of free sulfur dioxide

Many or most red wines are stored with free SO_2 at variable concentrations. Therefore, another pertinent question is whether SO_2 additions would have any effect on the OCR of post-fermentation yeast cells.

Figure 38 shows OCR measurement results obtained on six yeast strains suspended at 200 NTU in model solution spiked with increasing amounts of SO_2. Clearly, free SO_2 exerts a strong suppressive impact on the yeast OCR. Depletion of DO by reactions with SO_2 cannot explain this behavior since it would increase the OCR when initial SO_2 concentrations increase. The reason why SO_2 strongly inhibits DO uptake by yeast is unknown, but inhibition of unidentified enzymes by SO_2 might be a cause.

Figure 38: Impact of free SO_2 on the oxygen consumption rate (OCR) of yeast strains in model solution at 20° C and pH 3.5.

Yeasts harvested within two weeks after the end of alcoholic fermentation and adjusted to 200 NTU. Boxplots represent measurements of six different yeast strains.

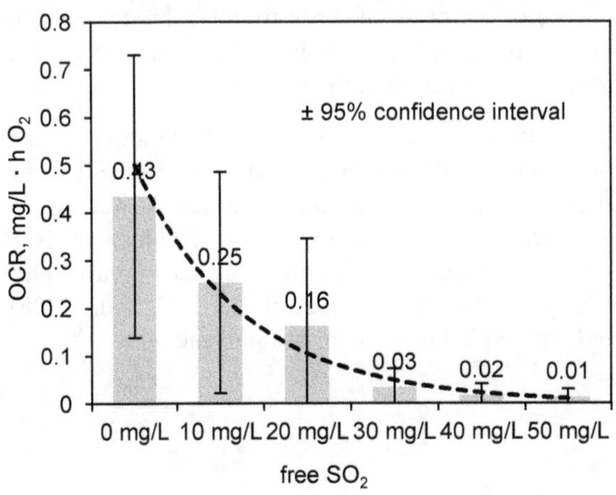

At an average wine pH of 3.5, there is a critical limited of around 20 mg/L of free SO_2, above which the OCR of yeast becomes almost insufficient to have any practical effect. In most red wines, free SO_2 does not exceed that level. As a result, suspended yeast lees can be expected to actively participate in DO consumption in these wines, thus withdrawing oxygen from reactions with phenolics and SO_2.

Effect of wine matrix

The effect of the wine matrix (red vs. white wine) was also investigated using three yeast strains. Defined amounts of these yeasts harvested one month after alcoholic fermentation were added to a sterile filtered white wine and three different red wines (A, B, and C) without any free SO_2. After another month of storage under anaerobic conditions, the wines were saturated with oxygen and the DO decrease registered. To calculate the net OCR of the yeasts, the overall DO decrease was corrected for the chemical oxygen binding by the respective wines without added yeast, which served as a control. Results are given in table 5.

Table 5: Effect of wine matrix on the oxygen consumption rate (OCR, in mg/L· h O_2) of suspended yeast lees in real wines at 20° C as affected by yeast concentration.
Data corrected for chemical oxygen binding by the wine matrix, free SO_2 = 0 mg/L; measurements taken two months after alcoholic fermentation.

yeast strain	wine	50 NTU	100 NTU	200 NTU	300 NTU	400 NTU
yeast 1	white wine	0.53	0.80	1.21	1.48	1.94
	red wine A	0.50	0.73	1.14	1.49	1.62
yeast 2	white wine	0.65	1.12	1.14	1.14	1.14
	red wine B	0.47	0.64	0.63	0.64	0.66
yeast 3	white wine	0.14	0.18	0.28	0.36	0.38
	red wine B	0.10	0.14	0.26	0.35	0.38

Again, the OCR is not enhanced by increasing yeast lees concentrations in a linear way, thus confirming results obtained in model solution. Furthermore, for comparable yeast concentrations, OCR figures are of the order of those found in model solution (Table 3). However, there is a clear and statistically significant tendency towards a decrease of yeast OCR in red wines as compared to the white wine.

This matrix effect is explained by a partial adsorption of red wine phenolics by yeasts. As a consequence, these phenolics interact with components of the yeast cell membrane responsible for the oxygen consumption. It is assumed that they lower the accessibility and reactivity of oxygen towards the sterols and unsaturated fatty acids of the cell membrane. These losses are irreversible and also observed after soluble phenolics had been separated from the yeast cells (Salmon et al. 2002, Salmon 2006). Their extent will obviously depend on the phenol concentration of the wine.

Total oxygen consumption capacity of yeast lees

Three yeast strains (A, B, and C) suspended in model solution at a concentration of 300 NTU each were used to assess their total oxygen uptake capacity. A manometric method offering conditions of potential oxygen supply of up to 80 mg/L was used for that purpose (Schneider et al. 2016). As shown in figure 39, plotting oxygen uptake against time yielded an almost linear increase of oxygen consumed at the beginning, followed by a transition to a steady state indicating depletion of the yeast oxygen uptake resources after 40 to 60 mg/L of oxygen had been consumed.

Figure 39: Total oxygen consumption capacity and consumption kinetics of yeast cells suspended at 300 NTU in model solution at 20° C.
Yeasts harvested one month after alcoholic fermentation.

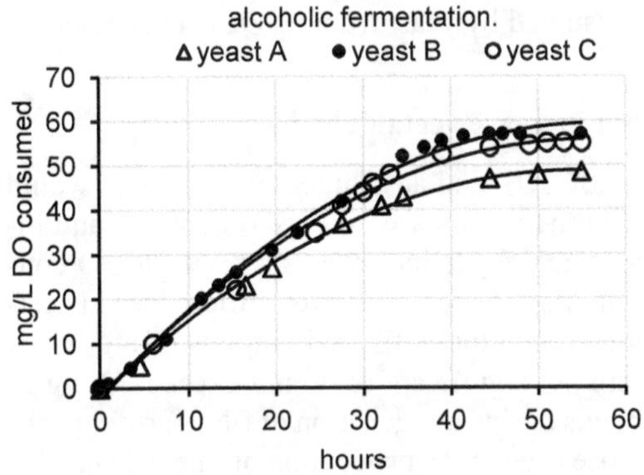

These results suggest that even minor concentrations of suspended yeast cells corresponding to approximately 50 NTU ($3.15 \cdot 10^6$ cells/mL) would suffice to consume one saturation concentration of DO (~8 mg/L DO). Accordingly under industrial winemaking conditions, the total oxygen consumption capacity of yeast lees is not a limiting factor to protect wine against oxidation.

Combined effects of lees concentration, age, and free SO_2 in real wines

Figure 40 shows a real-world example of DO consumption in a young red wine before and after filtration. The unfiltered wine still containing suspended yeast lees consumed DO more than twice as fast as after filtration.

Figure 40: Consumption of DO in a young Cabernet Sauvignon red wine before and after filtration.

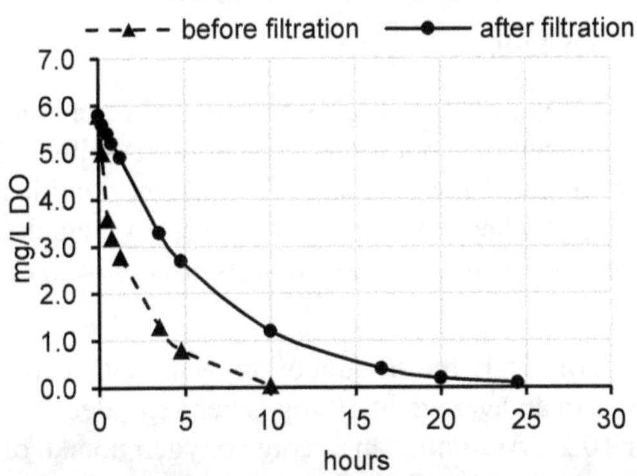

The DO decrease plots are either close to linear or exponential functions depending on the individual wine. In most of the wines, however, the plots take the form of a more or less convex curve fitting a negative exponential function. They correspond to a first-order reaction kinetics indicating a change of the OCR over time and according to the instantaneous DO concentration. Mathematical treatment of the equations of the curves registered with and without lees, respectively, allows for calculating the percentage and the absolute amount of DO consumed by the yeast lees, with the remaining part corresponding to the DO consumed by chemical oxidation of the wine matrix. In the example of wine given in figure 40, the DO consumed by the yeast lees was 58 %.

In white wines displaying variable levels of free SO_2 and yeast lees ranging from 2 to 26 weeks of age, DO consumed by yeast ranged from 0 to 47 %, but none of the aforementioned parameters such as yeast strain, yeast concentration, wine, age, free SO2 showed any significant influence on the DO consumption rate by the yeast lees they contained. The combination and interaction of the various parameters did not allow for a reliable prediction of yeast reactivity towards DO in a given wine. Moreover, the rate of chemical binding of DO is an unknown variable, rendering it impossible to deduce the percentage of DO consumed by yeast from its OCR (Schneider et al. 2016).

DO consumption by yeast lees does not increase acetaldehyde

Acetaldehyde is the key binding partner of SO_2, and any increase of its level would require higher SO_2 additions to bind it. In the worst case, total SO_2 content can ultimately exceed legal levels. Hence, any acetaldehyde increase as a result of DO consumption by yeast lees would be of crucial importance. However, differences in acetaldehyde concentration between the filtered and the unfiltered aliquots of red wines having consumed up to 7.5 mg/L DO are not significant or do not exceed ±3 mg/L. These results are in contrast to what happens during oxygenation of film-forming non-Saccharomyces yeast cultures such as those used in the production of Sherry, and also in contrast to the chemical oxidation of ethanol to acetaldehyde by the Fenton reaction (Chapter 7.2).

In conclusion, making use of the DO consumption and reducing properties of post-fermentation yeast lees does not increase acetaldehyde levels and SO_2 requirements. On the other hand, the oxygen withdrawn by yeast cells is not any more available to produce acetaldehyde by chemical oxidation and therefore not involved in the polymerization reactions of phenols (Chapter 7.2).

6.3. Practical working with yeast lees

As shown in the previous chapter, yeast lees display a highly variable oxygen consumption capacity. Under commercial winemaking conditions, the unpredictability of this particular feature makes it difficult to assess how much of the oxygen provided by micro-oxygenation, barrel aging, or any other means of oxygen supply actually reacts with tannin according to widespread winemakers' expectations or is scavenged by yeast lees.

When wines are stored in standard barrels of 225 L, the amount of oxygen they receive varies between 5 and 20 mg/L per year with an average of 10 mg/L per year, decreasing with the age of the barrels (Chapter 10.2). Assuming an average oxygen uptake of 0.05 mg/L per day in the barrel, a yeast OCR of 0.002 mg/L·h would be required to remove that oxygen at the rate it dissolves in the wine if yeast cells were the only oxygen receptor. However, a simultaneous and variable chemical binding of oxygen occurs.

Continuous oxygen supply comparable to barrel aging also takes place when red wines stored in tanks are submitted to micro-oxygenation (Chapter 8.3). Yeast lees compete for oxygen with red wine phenolics and thus hinder the maturation process (Salmon 2006), which is the basic intention of micro-oxygenation. This finding explains why young red wines withstand more oxygen uptake than aged ones after clarification. However, these relationships are further complicated by a partial adsorption of wine phenols on yeast cell membranes, which lowers the reactivity of both yeast and phenols towards oxygen (Salmon et al. 2002).

The great unknown

Table 5 reports that typical yeast concentrations of some 100 NTU in young red wines are able to consume DO at a rate of 0.1 to 1.0 mg/L·h during the first month after primary fermentation. Thereby, they tend to be only slightly less effective than in white wines where they have been shown to consume up to 47 % of total DO (Schneider et al. 2016).

These findings indicate that yeast lees can consume a substantial amount of DO provided during micro-oxygenation or other oxygen supply procedures and withdraw it from reactions with red wine phenols as intended by this treatment. However, that amount is hardly predictable. At best, it can be assumed that the more turbid the wine, the higher the OCR of the yeast lees and the less DO is available to react with wine constituents like phenolics etc. For that reason, micro-oxygenation and similar approaches of active oxygen supply (Chapters 8.2 and 8.3) require a great deal of hands-on experience and careful control to avoid over-oxidation.

Autolysis and additional properties of yeast lees

In summary, the reducing effect of post-fermentation yeast lees is based primarily on their ability to directly consume oxygen, which is thus withdrawn from wine oxidation and used for the oxidation of yeast cell compounds. These properties are much more pronounced in suspended yeast cells than in settled yeast bottoms.

The release of reducing amino acids and peptides is a secondary property of yeast lees. These compounds include in particular cysteine, homocysteine, and glutathione. They have some enological interest as anti-oxidants in white wines, but much less in red wines when compared to the direct DO uptake by yeast lees. In the specific case of glutathione that is highly advertised for white wine preservation, results on its protective effect against the risks of strong oxidation in red wines are highly conflicting (Gambutti et al. 2015, 2017).

Moreover, when yeast lees autolyze, they release colloidal polysaccharides like mannoproteins that add beneficial texture, palate weight, and creaminess to the wine. They provide a sense of sweetness as a result of bridging the sensations among phenolic compounds, acidity, and alcohol, aiding in wine harmony and integration (Chapter 5.3.3).

Autolysis is the self-destruction of yeast cells by their own enzymes. However, when yeast autolysis proceeds too far, leesy aromas evocative of toasted bread, hazelnuts, and almonds will develop. Although these aroma characteristics are not desired in most white wines, they can favorably add to the aroma of red wines. Only when they start to become too strong in some red wines intended to be consumed young and fruity, racking and clarification of that kind of wine is advisable. As long as wines are stored cool (< 20° C / 68° F), this rarely happens before the first year post-fermentation.

Other than oxygen, yeast lees also absorb further undesirable compounds that might occur in wine. These include excessive ellagitannins originating from the wood of oak chips or barrels. Ellagitannins also bind with mannoproteins released from yeast cells. This feature is of outstanding importance during barrel aging since it decreases astringency and causes a noticeable tannin softening.

Suspended yeast lees are also able to adsorb substantial amounts of copper ions. This behavior is of considerable enological interest since it allows for the removal of large amounts of residual copper left after copper fining.

The question of racking

Racking means the transfer of the wine from above its solid sediments into another container. This operation does by no means imply that the transferred wine would undergo any clarification making it less turbid than it was before, though this assumption is still widely upheld. In contrast to filtration or fining, racking does not remove turbidity inducing elements.

In former times, wines were racked off several times post-fermentation. In the meantime, this approach has changed and the number of rackings diminished. Clean yeast lees absolve the winemaker from superfluous rackings. Indeed, one way of working with yeast lees is postponing or simply skipping racking after fermentation. This approach requires some explanations.

Winemakers use to differentiate between light or secondary lees and heavy or primary lees. Heavy lees can be defined as those that settle within the first one or two days after pressing, and are composed of large particles consisting of bitartrate crystals and grape solids. There is little interest in storing wines on heavy lees because such storage can result in off-aroma and flavors. In contrast, light lees can be defined as all those that require several days or weeks to settle more or less; essentially they comprise of yeast cells. Therefore, it is indispensable to rack red wines off their heavy lees after devatting and before starting their aging with yeast lees.

However, there are exceptions calling for racking. When high amounts of compact yeast lees are left undisturbed at the bottom of the aging vessel for too long, they can start to produce reduction flavor. When this happens, time has come to rack off the lees and take further steps do deal with that kind of off-flavor, most frequently using splash-racking to provide oxygen.

Another reason for racking is the need to provide oxygen most red wines require for their maturation and harmonization. In some traditional growing areas, barrel-aged red wines are racked twice or three times during the months in winter following harvest. An additional benefit of this approach is the early removal of excessive carbon dioxide distracting on the palate. According to total phenol level and tannin structure, more oxygen supply may be useful in high-phenol red wines and can be obtained in the final stages of preparing wine for bottling by blendings, finings, filtrations, and the bottling process itself.

Barrel aging favors oxygen uptake (Chapter 10.2). During aging, wines display a higher oxidation-reduction potential in barrels than in stainless steel tanks. As a result, wines aged on their lees in barrels rarely develop reduction flavor. This is in contrast to wines stored in large-volume stainless steel tanks. Such wines are often not stored on their lees for a longer period, though this is not totally impossible and increasingly facilitated by the use of equipment for micro-oxygenation (Chapter 8.3) or periodic splash-racking (Chapter 8.2).

Promoting aging on the lees by stirring

Aging on the lees is a means to deliberately promote yeast autolysis and sensorially impact the wine with autolysis products. For that purpose, yeast strain, the volume of yeast lees, yeast stirring and its frequency, and the duration of lees contact are the most important variables from both a practical and a stylistic perspective.

Settled yeast lees are not very helpful, while periodic stirring of the lees (bâtonnage) increases the content of polysaccharides and yeast-derived amino acids. Polysaccharides, especially mannoproteins released from yeast, lower the sensory perception of astringency, sourness, fruit, and oak, while they increase the wine's body, volume, and protein stability (Chapter 5.3.3).

Stirring causes an oxygen uptake able to change the sensory balance between fruit, yeast, and wood. Although stirring is thought to be an oxidative process, it is debatable to what extent oxidation reactions take place since post-fermentation yeast lees substantially contribute to oxygen consumption, thus protecting wine against oxidation. It can be assumed that the redox regime is highly variable between wines and individual barrels, depending on the oxygen uptake through the wine surface that is controlled by the intensity and frequency of stirring, and the oxygen consumption by variable amounts of yeast lees.

This picture is further complicated by rackings and SO_2 additions. Early additions tend to keep the amount of components that bind to subsequent SO_2 additions at a higher level. Acetaldehyde is a typical example of these compounds. Moreover, early SO_2 additions also hamper the completion of malolactic fermentation usually sought for red wines, whilst they might protect against adverse microbial developments to a certain extent that depends on wine pH. The most striking effect of SO_2, however, relies on his action as a reducing agent (Chapters 6.4 and 7.4.1).

It is quite possible to run aging on the lees in barrels over more than a year without any SO_2 and without any signs of oxidation, relying only on the reducing capacity of suspended yeast lees. At the latest, SO_2 should be added when there are initial signs of heavy oxidation such as a strong smell of acetaldehyde or even visible color alterations towards brown. Browning indicates the beginning of excessive and irreversible oxidation reactions that should be prevented at any cost.

Generally, stirring frequency depends on the time span suspended yeast needs to settle. The faster it settles, the more often it should be stirred in order to prevent oxidation of the topmost liquid layer. Thus, stirring frequency varies between once per week until once per month. Instead of stirring with a propeller mixer, pumping around from the bottom to the top can also be used.

The settling rate of yeast lees strongly depends on the individual wine and its oxidation-reduction regime. Figure 41 illustrates how both SO_2 (60 mg/L) additions and oxygen uptake through the wine surface increase the yeast lees settling rate and accelerate clarification. The molecular causes of this behavior are not well investigated. However, they explain why yeast lees settling is much slower in stainless steel tanks than in wooden barrels allowing for oxygen uptake.

Figure 41: Impact of oxygen uptake and SO_2 (60 mg/L) additions on settling of post-fermentation yeast lees, measured as residual turbidity (NTU) after 24 hours in standardized containers.

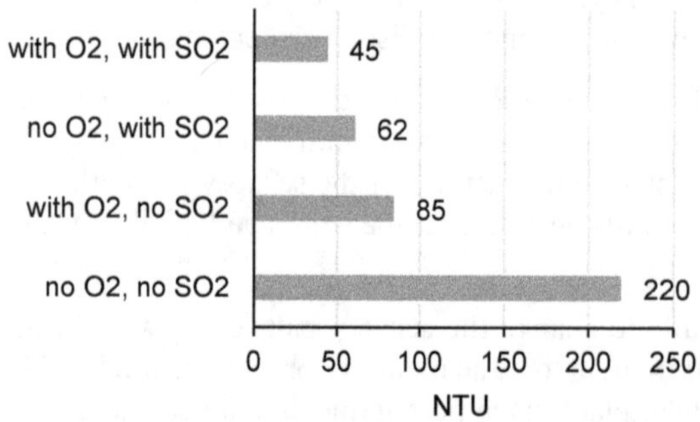

Clearly, there cannot be any fixed rules for stirring frequency. Sensory goals play a further role in scheduling stirrings. Also depending on stylistic goals, the duration of lees contact is highly variable. After the removal of heavy lees right after devatting, fine lees consisting primarily of yeast cells can remain in the wine during barrel aging over several years. In other cases, they are gradually removed by rackings.

Addition of an exogenous ß-glucanase enzyme considerably enhances yeast autolysis, the release of mannoproteins, and the sensory consequences resulting therefrom. However, it cannot perform miracles when there are not enough yeast lees available or when fermentation was run with a yeast strain that releases low levels of polysaccharides. Differences between yeast strains are considerable, and some commercial strains are marketed specifically for the purpose of enhancing polysaccharide levels during fermentation and subsequent lees contact (Chapters 4.1.14)

Stirring does not make very much sense when yeast lees are limited. As an alternative, commercial proprietary products based on lysated (broken) yeasts or yeast-derived mannoproteins are available. Their addition can complement and partially replace traditional aging on the lees. Addition of purified mannoproteins can enhance weight and volume on the palate in some wines immediately. However, the sensory impact of these preparations is highly variable. Hence, they should not be used without previous bench trials.

Furthermore, some selected dry yeast strains have become commercially available and recommended for the sole purpose of addition post-fermentation. Results have been rather modest since standard additions of these products do not provide enough biomass for proper aging on the lees. In contrast, they cause aroma losses via absorption as they have not become saturated with aroma during fermentation. Going through a fermentation cycle seems indispensable to resolve both problems.

6.4. Oxygen consumption by sulfur dioxide

The traditional use of sulfur dioxide (SO_2) in winemaking pursues three distinctive objectives:

- inhibition of fault-producing spoilage microorganisms,
- protection against excessive oxidation,
- binding with free acetaldehyde and other carbonyls and converting them into odorless addition products, thus preventing oxidative off-odors caused by the presence of free acetaldehyde characterized by a smell reminiscent of bruised apples.

This section covers the role of SO_2 as an oxygen scavenger and reducing agent during red wine aging and deliberate oxygen supply.

In wine, sulfur dioxide exists in forms other than just dissolved SO_2. First, there is a free and a bound form, with the sum of these two representing total SO_2. Bound SO_2 is basically inactive. Its amount is predetermined by wine composition or, more precisely, by the concentration of its binders. The enological interest regarding the interaction of SO_2 with oxygen and its impact on red wine maturation is focused upon free SO_2.

Three different forms of free sulfur dioxide

Free sulfur dioxide dissociates in aqueous solution such as wine. As a result, it exists in three states, which are

- molecular, gaseous SO_2,
- bisulfite (HSO_3^-), and
- sulfite (SO_3^{2-}).

These three forms are in a dynamic equilibrium[40] with one another according to the formula

$$\underset{\text{molecular}}{SO_2} + H_2O \rightleftharpoons H^+ + \underset{\text{bisulfite}}{HSO_3^-} \rightleftharpoons \underset{\text{sulfite}}{SO_3^{2-}} + H^+$$

Each form has different properties. Molecular SO_2 is dissolved in the wine as a volatile gas, has an antimicrobial effect, and is sensorially active due to its pungent smell. The bisulfite form binds with many wine components including acetaldehyde, higher aldehydes, keto acids, sugars, and phenols. These bound products are essentially non-reactive and make up the bound fraction of total SO_2. The sulfite form does not have any enological interest.

[40] A dynamic equilibrium occurs in reversible reactions, in which the rate of the forward reaction is equal to the rate of the backward reaction, meaning there is no net change of reactants and products. Hence, it is a system in a steady state. In the case of the various forms of free SO_2, when the concentration of one of the forms is lowered or removed, the other forms are slowly converted into the removed form until a new point of equilibrium is achieved.

Figure 42: Effect of wine pH on dissociation and relative proportions of forms of free sulfur dioxide.

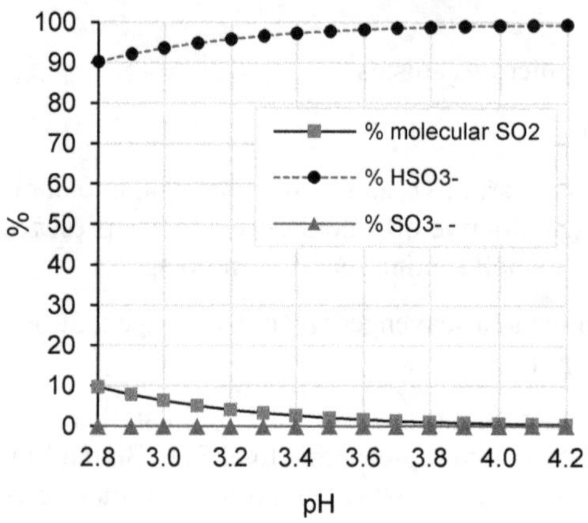

The quantities of these different forms of free sulfur dioxide depend on pH. Their relative proportions in the range of wine pH are shown in figure 42. They explain why pH has also importance in oxygen-related reactions.

This commonly known graph makes clear that

- the predominant form of free SO_2 in the pH range of wine is the bisulfite ion (HSO_3^-),
- the sulfite ion (SO_3^{2-}) concentration is very small and can essentially be ignored,
- the molecular form (SO_2), although also small at wine pH values, becomes increasingly significant with only a small decrease in pH. It is of crucial importance for wine conservation and for modulating oxygen-related reactions.

The sulfite (SO_3^{2-}) is the only form of SO_2 able to react directly with oxygen according to the formula

$$2\ SO_3^{2-} + O_2 \rightarrow 2\ SO_4^{2-}$$

However, at wine pH, the concentration of sulfite is so minute that its reaction with oxygen is infinitely slow. Therefore, sulfite is not an effective antioxidant in wine.

In contrast to sulfite, the molecular form of SO_2 is quite reactive. For example, it traps hydrogen peroxide (H_2O_2) in accordance with the formula

$$H_2O_2 + SO_2 \rightarrow H_2SO_4$$

in a spontaneous way leading to the production of sulfuric acid. This reaction is of outstanding importance and will be examined in detail in subsequent chapters.

SO_2 reacts with oxidation products rather than with DO

Older textbooks claim that the ability of SO_2 to react with oxygen is the manner by which it protects other wine constituents against oxidation. This is a simple concept easy to understand, but unfortunately it is not what actually happens. In reality, SO_2 only receives the oxygen in the form of H_2O_2 and oxidized phenols appearing at the end of a long reaction chain induced by oxygen consumption of the wine (Chapter 7.2). Thereby, it is oxidized into sulfate (sulfuric acid) according to the formula of sums

$$\begin{array}{ccccccc} SO_2 & + & H_2O & + & \frac{1}{2} O_2 & \rightarrow & H_2SO_4 \\ 64 & + & 18 & + & 16 & \rightarrow & 98 \end{array}$$

Under these conditions, 1 mg/L O_2 leads to the oxidation of 4 mg/L SO_2 according to stoichiometric[41] rules.

When SO_2 is oxidized to sulfate, free and total SO_2 are decreased simultaneously, but not necessarily to the same extent. In contrast, free SO_2 decreases somewhat less than total SO_2. The reason for this is that it is partially replaced by "fresh" SO_2 released from the pool of bound SO_2 when free SO_2 reaches low concentrations (Waterhouse et al. 2016). Hence, when DO is measured after oxygen uptake and to be related to SO_2 losses, it is indispensable to measure total SO_2 instead of free SO_2.

The sulfate produced upon SO_2 oxidation no longer contributes to measurements of either bound or total SO_2. Thus, in a certain way, this behavior can be considered as a chemical removal of SO_2. It is well known in the wine industry, for example when wine picks up oxygen during barrel aging or at bottling.

Some basic calculations: The SO_2:O_2 ratio

As will be outlined in chapter 7.2, the first step of wine oxidation produces hydrogen peroxide (H_2O_2) and oxidized phenols. The previously discussed formula stating that 1 mg/L O_2 leads to the oxidation of 4 mg/L SO_2 means that the SO_2:O_2 molar ratio is 2:1. One mole equivalent of SO_2 reacts with the hydrogen peroxide that is produced, and the second one with the oxidized phenol. This relationship translates into a SO_2:O_2 mass ratio of 4:1. Weight and mass units are values winemakers are used to work with.

Unfortunately, the SO_2:O_2 mass ratio of 4 only applies under ideal conditions. Such conditions require that the total oxygen is actually consumed by SO_2 at the end of the reaction chain, that there are no other oxygen scavengers than SO_2 present, and that the effect of oxygen consumption is entirely reversed by SO_2.

However, wine is anything but an ideal solution behaving in a chemically simple way. It contains other oxygen scavengers competing with SO_2 for oxygen. Therefore, 1 mg/L O_2 consumes less than 4 mg/L SO_2. This is because a portion of the oxygen taken up by the wine does not arrive at the SO_2 at the end of the reaction chain, but remains with other wine constituents that are oxidized irreversibly (Danilewicz 2016, Waterhouse et al. 2016). Thus, the SO_2:O_2 mass ratio becomes less than 4. As a result, the DO measured in the wine at a given moment does not allow for stoichiometrically calculating the decrease of SO_2 to expect.

Take-home summary: The higher the SO_2:O_2 ratio, the more DO is consumed by SO_2, and the better SO_2 protects the wine against oxidation.

Wine composition determines oxygen consumption by SO_2

In an attempt to evaluate where the oxygen actually ends up, a total of 10 mg/L DO was supplied to each of 10 different sterile filtered red wines containing 4 to 26 mg/L free SO_2 and stored in hermetically sealed glass vessels. The oxygen was supplied from an appropriately sized headspace containing air (Chapter 8.4). The oxygen decrease was monitored using non-invasive DO measurement through the vessel walls

[41] Stoichiometry describes the quantitative relationship between the number and mass of molecules of products and reactants in chemical reactions. It is founded on the law of conservation of mass, which means that the total mass of the reactants equals the total mass of the products.

by a fiber optic meter based on luminescence quenching. When DO was entirely consumed after two months, losses of total SO_2 were measured and related to the initial amount of DO.

Figure 43: Losses of total SO_2 (mg/L) per 1 mg/L DO in 10 red wines supplemented with 10 mg/L O_2. Initial free SO_2 levels = 4 to 26 mg/L, reaction time = 2 months.

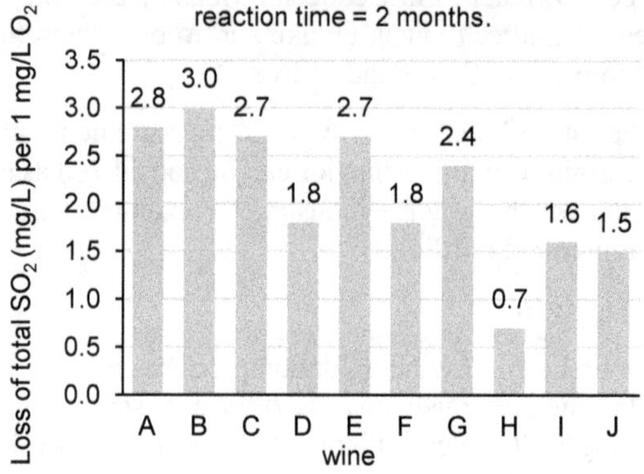

Figure 43 shows that 1 mg/L O_2 oxidizes variable amounts of SO_2, which strongly depend on the individual wine. Instead of 4 mg/L SO_2 per 1 mg/L DO (SO_2:O_2 mass ratio = 4), the average was only 2.1 (0.7 to 3.0) mg/L SO_2 per 1 mg/L DO (SO_2:O_2 ratio = 2.1) or 52.5 %, respectively. The remaining 47.5 % of oxygen was not intercepted by SO_2, but available for irreversible oxidation of other wine constituents.

This high variability of data suggests a closer look at how they are influenced by wine compositional data. Therefore, some compositional data were compared with the SO_2:O_2 ratio of the wines using correlation analyses. The correlations are given in table 6. Note that the sign preceding the correlation coefficient (r) indicates whether the correlation is positive or negative, and that the coefficient of determination (R^2) assesses how strong the relationship between two variables is[42].

Table 6: Correlation coefficients (r) and coefficients of determination (R^2) between wine compositional data and the SO_2:O_2 ratio.

Compositional parameter	r	R^2
initial free SO_2	+ 0.792	0.63
pH	- 0.512	0.26
total phenols	- 0.189	0.04
total flavonoids	- 0.350	0.12
monomeric flavonoids	- 0.281	0.08
anthocyanins	+ 0.465	0.22

Interestingly and in contrast to what one would expect, the phenolic make-up assessed as total phenols, flavonoids, and anthocyanins does not have a significant impact on the SO_2:O_2 ratio in red wines. Instead, initial free SO_2 and pH are the primary matrix variables controlling the proportion of oxygen consumed by SO_2 in clarified red wines. Of course, this picture will be completely different in the presence of yeast lees.

[42] A coefficient of determination R^2 of 1.0 indicates a perfect correlation, whilst a value of 0.0 indicates the absence of any correlation. But a value of 0.30, for example, suggests that 30 % of the dependent variable, here the SO_2:O_2 ratio, is predicted by the independent variable, the compositional parameter in this case.

Effect of initial free SO_2 levels on oxygen consumption by SO_2

As expected, the higher the initial free SO_2, the higher the $SO_2:O_2$ ratio, the more DO is consumed by SO_2, and the less DO remains to oxidize other wine constituents such as phenolics (Danilewicz and Standing 2018). This obvious relationship is illustrated in figure 44. The other way round, in wines with low free SO_2 levels or without any SO_2, a correspondingly higher proportion of oxygen reacts with the tannin-anthocyanin complex via intermediate H_2O_2. At low (< 15 mg/L) free SO_2 concentrations, these reactions become dominant. It depends on the individual wine and its stage of development if this is beneficial or detrimental. This feature has been investigated in detail during barrel aging of red wines (Pontallier 1981), where it is of outstanding importance for the evolution of tannin and its sensory expression.

Figure 44: Impact of the inital free SO_2 on the $SO_2:O_2$ mass ratio in red wines measured two months after oxygen supply.

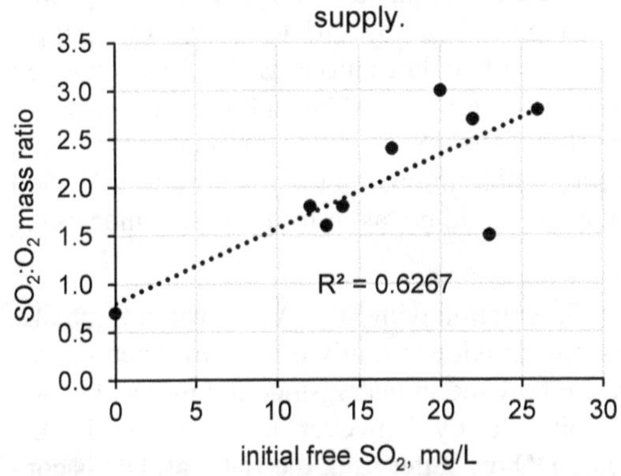

Under a more practical aspect, it is important to keep in mind that phenolics and free SO_2 in filtered red wines compete for oxygen as ultimate oxygen acceptors. To a lesser extent, even bound SO_2 can participate in this process when it is gradually released after the depletion of free SO_2 in the course of maintaining the dynamic equilibrium between the two forms. There are no other oxidizable compounds in sufficient amounts to react with all the oxygen wines take up. Iron and copper can react with oxygen, but the amounts present preclude more than a catalyst's role for them in wine.

Effect of pH on oxygen consumption by SO_2

Wine pH is another factor affecting DO consumption by SO_2. The higher the pH, the lower the $SO_2:O_2$ ratio, and the more DO reacts with wine constituents other than SO_2. This is shown in figure 45. This behavior is not very different from that observed in white wines. There are two reasons to explain it:

Figure 45: Impact of pH on the $SO_2:O_2$ ratio in red wines.

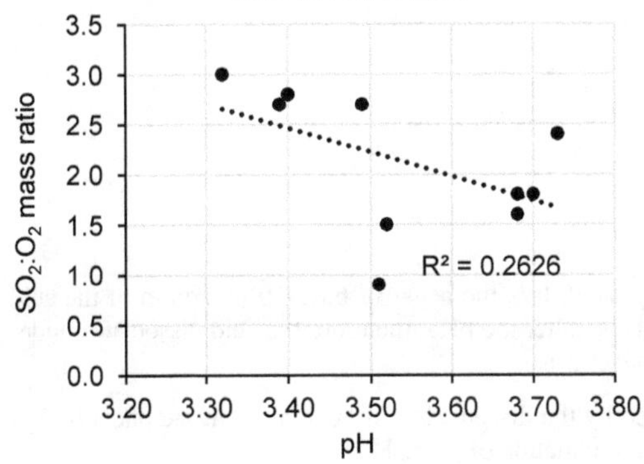

The first reason is phenol chemistry. As will be shown in the next chapter, phenolics are the first oxygen acceptors initiating an oxidation sequence. They are weak acids. Their acid strength expressed as pKa[43] is pH 9 to 10. At or above this pH, where 50 % or more of the phenols exists as phenolate[44] ions, their autoxidation is extremely rapid and much faster than the oxidation of the protonated phenols with undissociated hydroxyl groups.

Whilst wine is much more acidic than the pK's of its phenols, there is a small but calculable percentage of phenolate anions even at wine pH, and these are able to rapidly autoxidize. When the pH increases, more of the phenolics dissociate into phenolates, and more readily they bind with oxygen. At a high pH for wine, for example 4.0, there would be roughly ten times as much phenolate anions as at pH 3.0, and their oxidation rate would be ten times as fast. Therefore, unbalanced high pH wines oxidize so readily and quickly become tired and 'flabby' instead of maturing properly. The suitable lowering of pH as early as possible improves the maturation and shelf life of wines from overripe grapes, but care should be taken to not base winemaking exclusively on pH.

The second reason for the pH effect is its interaction with SO_2. As shown in figure 42, the molecular SO_2 percentage of free sulfur dioxide strongly decreases when the pH increases. This effect is widely known in the wine industry since the molecular SO_2 is the only form of SO_2 inhibiting microbial activity. However, less considered is the fact that it is also responsible for trapping DO or, better said, the H_2O_2 all DO is converted into. Consequently, with increasing pH, less molecular SO_2 is available to trap DO for a given amount of free SO_2.

Combined effects of SO_2 and pH

Since pH and SO_2 interact, the combined effects of initial free SO_2 and pH on oxygen consumption by sulfur dioxide deserve a more detailed investigation. For that purpose, a commercial red wine blend was adjusted to three different pH levels using NaOH or H_2SO_4. Each of these aliquots was adjusted to three levels of free SO_2. The nine samples obtained in that way were supplied with 7.5 mg/L DO, hermetically sealed, and stored in the dark until DO was totally consumed after two months. Then the decrease of total SO_2 was measured in comparison with the non-oxygenated references and related to the amount of DO consumed. Results are given in figure 46. The individual effects of both pH and initial free SO_2 on DO consumption by SO_2 are confirmed.

[43] The pKa value indicates the strength of an acid. It is the negative base-10 logarithm of the acid dissociation constant in aqueous solution. The smaller the pKa, the more the acid dissociates under splitting off of protons (H^+ ions) and the stronger it is.

[44] Phenolates are anions of phenols generated by the dissociation of H^+ ions from the phenolic hydroxyl groups. They are prerequisite for the autoxidation of phenols.

Figure 46: Losses of total SO_2 (mg/L) per 1 mg/L O_2 as affected by initial free SO_2 and pH in red wine.

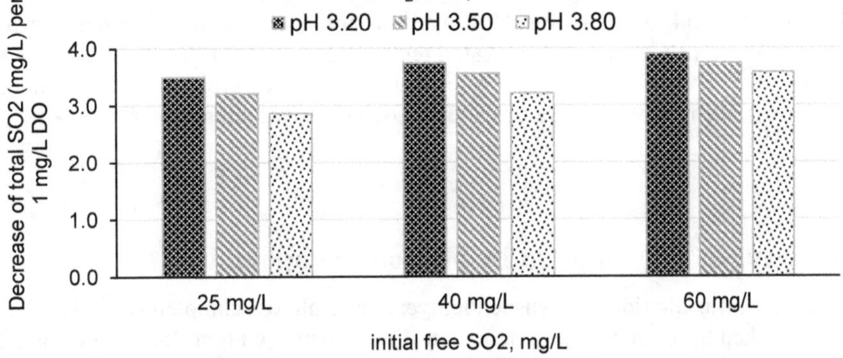

Time to summarize: High pH wines are more sensitive to oxygen they pick up during deliberate oxygenations or barrel aging. This is explained by the higher oxidizability of their phenols and the lower antioxidant effect of free SO_2 in the high pH range. However, lowering pH by acid additions for the sole purpose of better protecting wine against oxidative spoilage can be a bad idea since it considerably changes wine style. Controlling inadvertent oxygen uptake (Chapter 8.1) often turns out to be the better approach.

Practical conclusions from the role of SO_2 as an oxygen scavenger

It follows from the above that the impact a given amount of oxygen on the tannin-anthocyanin-complex and red wine evolution increases

- when free SO_2 decreases,

- when pH increases,

- when suspended yeast lees decrease by settling, filtration, etc.

Hence, the SO_2:O_2 ratio has considerable practical importance, though it is never measured under commercial winemaking conditions. Its measurement would allow assessment of the percentage of oxygen not scavenged by SO_2, to identify wines that would become easily over-oxidized during barrel aging, and to better adjust free SO_2 levels before bottling (Chapter 12.4). These considerations have particular relevance for low-phenol red wines frequently grown under cool-climate conditions. They have less importance for high-phenol red wines, which are more resistant to surplus oxygen.

Moreover, the SO_2:O_2 ratio gains in importance when barrel aging is performed at elevated temperatures. During barrel aging, there is a continuous uptake and consumption of oxygen by the wine. This oxygen turn-over and the reactions it triggers is increased when storage temperature increases, for example from 15 to 22° C. As a consequence, barrel-aging at elevated temperatures more easily leads to over-oxidation in the absence of free SO_2. Thus, it may require SO_2 additions and the maintenance of some free SO_2 to avoid oxidative spoilage. Otherwise tannins that were previously extracted with much effort and expense risk to become insoluble and dropping out. Details are discussed in section 7.3.

Storage at elevated temperatures is sometimes deliberately used to accelerate the reactions involved in red wine maturation, leading to more polymeric pigments, higher color intensity, and softer tannins. However, this approach is also associated with microbial risks, in particular when elevated temperatures coincide with the absence of free SO_2. Additionally, it requires a careful and periodic control for microbial faults. This applies even more at pH values above 3.6 that are not unusual in red wines.

6.5. Oxygen consumption by phenolic compounds

Phenolics, sulfur dioxide, and suspended yeast lees almost completely consume the oxygen picked up by red wines. Hence, in filtered wines without free sulfur dioxide the total of the oxygen supplied is inevitably and exclusively consumed by phenolics. Accordingly, such wines respond much more strongly to any oxygen uptake.

In contrast, when there are suspended yeast lees and/or SO_2 present, it is no longer possible to quantify the amount of oxygen reacting with phenolics. Indeed, the percentage oxygen consumption of each of the oxygen scavengers referred to is subject to numerous interactions and highly variable between wines. This explains so many failures in the aging of red wines when simplified or generalized standard approaches are taken.

Figure 36 shows that anthocyanins consume more oxygen than colorless tannins as long as they remain present in their monomeric form. This is because the reaction of oxygen with anthocyanins is many times faster than with all other phenolics (Vivas 1999). This explains why anthocyanins are highly sensitive to oxidation. Only after their conversion into polymeric pigments by chemical binding with colorless phenolics, they represent a much more stable form of color.

Generally, the less the phenolic fractions are polymerized, the faster they react with oxygen. Therefore, young red wines with a low mean degree of tannin polymerization and high levels of monomeric anthocyanins consume oxygen faster than aged wines.

Theoretically, red wines can consume several hundred mg/L of oxygen, but the amounts they benefit from are much lower and variable. Generally, high-phenol wines will withstand or even benefit from much more oxygen uptake during bulk maturation than low-phenol red wines. Their total phenol content is a rough measure of their capacity to take up oxygen, their ability to withstand oxygen, and their capacity to change when exposed to oxygen. It also helps explain why red wines protected from oxygen can remain coarse and astringent if made too astringent. This topic is discussed in section 7.4.

On the other hand, matching the exact oxygen requirements of a specific red wine at a given stage of development is not an easy undertaking. However, it is the actual task of what is called redox management, which means balancing reactions of oxidation and reduction. As previously discussed, the reason for the difficulties encountered are the complex interactions between the various oxygen scavengers acting as reducing agents. This explains why it is fairly difficult to precisely quantify the oxygen consumption by phenolics in absolute or relative terms. Moreover, it makes clear why red wines respond so differently to barrel aging and the various means of deliberate oxygenation (Chapters 8.2 and 8.3). Yet these techniques or their omission are the key to success in the production of great red wines. Therefore, a good understanding of what happens when red wine phenols react with oxygen is very helpful in putting enological decisions on a rational basis. The following chapter provides an introduction into the chemistry of post-fermentation redox management including practical examples.

7. Reactions between phenolics and oxygen

This chapter describes the basic chemistry of oxygen consumption by red wine phenolics and their subsequent polymerization. It explains why red wines can consume without damage fairly high amounts of oxygen that would destroy any white wine by oxidative spoilage. They even benefit from mild oxidation triggered by deliberate oxygen supply when its extent is adjusted to the type of wine and its stage of development. Regenerative polymerization of phenolics is the key to understanding the reason for this. In this process, the original phenolic hydroxyl groups lost by oxidation are regenerated without the intervention of any reducing agent. By this means, they rendered able to oxidize again and to consume more oxygen. Hydrogen peroxide and acetaldehyde appear as two key intermediate products involved in the reaction chain. They are produced upon phenol oxidation and consumed in subsequent reactions. Various competing reactions are involved, and kinetic data already introduced in chapter 6 are discussed in more detail with regard to the reactions between phenols, oxygen and the resulting downstream products. Ultimately, this section also shows that the sensory response of red wine to oxidation is more than the sum of the response of its constituents, due to the complex interplay of chemical reactions. Particular emphasis is placed on the sensory consequences of excessive oxygenation of wines that are not suitable for it.

7.1. Sensory and chemical objectives

Flavonoid phenols tend to polymerize. The polymerization reactions initiated during maceration (Chapter 4.1) continue during aging and after bottling. They are based on three types of reaction mechanisms outlined in chapters 2.2.3 and 2.2.5. They can be summarized as follows:

1. Attachment of acetaldehyde and pyruvic acid to the C4 of anthocyanins, resulting in the formation of stable colored adducts.
2. Direct polymerization of the kind "flavanol + anthocyanin" and "flavanol + flavanol" via C4→C8 and C6→C8 bondings without involvement of oxygen. The flavonoids involved can be monomeric, oligomeric, or polymeric.
3. Polymerization under the action of oxygen with formation of an ethyl bridge leading to molecular structures of the kind "flavonoid-ethyl-flavonoid". This reaction has much enological interest due to the following features and prerequisites:
   - The formation of an ethyl bridge between two phenols requires the presence of free acetaldehyde.

- Free acetaldehyde only occurs in the absence of free sulfur dioxide.
- It can be of microbial origin and stem from yeast metabolism during fermentation.
- It can also be of chemical origin and result of what is called 'coupled oxidation' of ethanol with phenolic compounds. As this reaction has much importance in enology, it is outlined in detail in the following.

Chemical generation of acetaldehyde and its reaction with phenolics

- The oxidation of ethanol to acetaldehyde requires the presence of oxygen.
- The oxygen required is supplied by means of active or passive oxygen uptake (Chapter 8).
- In the presence of oxygen, polymerization with formation of an ethyl bridge proceeds several times faster than reactions of type 1 and 2.
- Thus, oxygen accelerates maturation and aging of red wines, leading to sensory results including the generation of softer tannins that are different from those obtained when red wine is stored under anaerobic conditions. The amounts of oxygen required for that purpose are highly variable.
- The ethyl bridge formed by the action of oxygen results in the formation of polymeric tannins when it leads to the bonding of colorless flavanols (flavanol-ethyl-flavanol), and to polymeric pigments when it causes the bonding between colorless flavanols with anthocyanins (flavanol-ethyl-anthocyanin).
- Apart from the bonding of two flavonoid monomers to a dimer, polymerization also involves the bonding of dimers or oligomers to larger polymers with increasing chain length.
- The polymeric pigments formed in this way are stronger in color than the anthocyanins they have originated from. They are also more resistant against the bleaching effect of sulfur dioxide (Chapter 2.2.5).

It is essential to note that there are two kinds of polymerization, either with or without acetaldehyde generated by oxidation. The crucial question is which of these two kinds is more important, how much oxygen is required to promote overall polymerization, and, last but not least, which degree of polymerization most meets sensory expectations. These questions will be discussed in detail later in this book.

Modern enology assigns great importance to the targeted use of oxygen. However, there are also opposing voices that attach much greater importance to storage temperature and direct polymerization without oxygen and acetaldehyde. In this context, it was shown that the reactions responsible for the progressive formation of polymeric pigments are essentially anaerobic (Somers and Evans 1986). However, it is also indisputable that oxygen supply by measures of mild oxygenation increases the overall polymerization rate via the formation of intermediate acetaldehyde.

From a sensory perspective, the goal is to smoothen the initially harsh tannins, to partially convert their astringency into volume and body, and to better integrate them in the wine matrix. It will be shown in the following sections that this sensory goal is not necessarily achieved by a high degree of polymerization, rather the opposite. Surplus oxygen always leads to contrary results and over-oxidation. It occurs when the capacity of phenolics to polymerize is exhausted and when natural reducing agents are depleted. Hence, excessive oxidation changes most red wines for the worse.

Avoiding over-oxidation by maintaining the balance between oxygen supply and oxygen consumption by phenolic compounds leads to redox management being a key challenge in post-fermentation red winemaking. In this context, regenerative polymerization of phenolics plays a crucial role.

7.2. Oxidation and regenerative polymerization of phenolics

The vast majority of phenolics contained in grapes and red wine are flavonoids with two adjacent phenolic hydroxyl groups on their B ring (Figures 2 and 3). They are also called vicinal diphenols and considered the grape-derived phenols that most readily oxidize. However, as outlined previously, autoxidation of phenolics by molecular oxygen is fairly slow in wine and only possible in the high pH range. In wine, the oxidation of phenolics is catalyzed by the action of heavy metal ions such as iron (Fe^{2+}) and copper (Cu^{+}) (Waterhouse and Laurie 2006, Danilewicz 2007), which always occur in amounts that are sufficient for that purpose. The oxidation of phenols is facilitated by increasing pH, which causes their increasing dissociation into phenolate anions that oxidize more easily than the protonated phenols with their undissociated hydroxyl groups (Chapter 6.4).

When a diphenol oxidizes, it produces the corresponding quinone[45]. Thereby hydrogen peroxide (H_2O_2) is produced as a side product. Ultimately, diphenol oxidation generates two products – H_2O_2 and quinones. They initiate the formation of phenolic polymers by oxidative pathways. Therefore, their fate is of crucial importance in connection with competitive reaction kinetics that can be observed during the maturation of red wines.

Figure 47 outlines how all that proceeds in general terms. The entire reaction chain from the simple vicinal diphenols to the polymers is designated as the 'oxygen cascade'. It is the basis for understanding the ability and limitations of red wines, with their high phenolic content, to consume relatively large amounts of oxygen without oxidative spoilage occurring. It also provides the basis for a deeper understanding of redox management.

[45] Quinones are oxidized phenols in which an even number of phenolic hydroxyl groups (-C–OH) are converted into carbonyl (–C=O) groups with rearrangement of double bonds in the respective phenol ring.

Figure 47: Reactions of oxidation and regenerative polymerization of vicinal diphenols.
R = rest of the phenol molecule, of which only the B ring is shown.

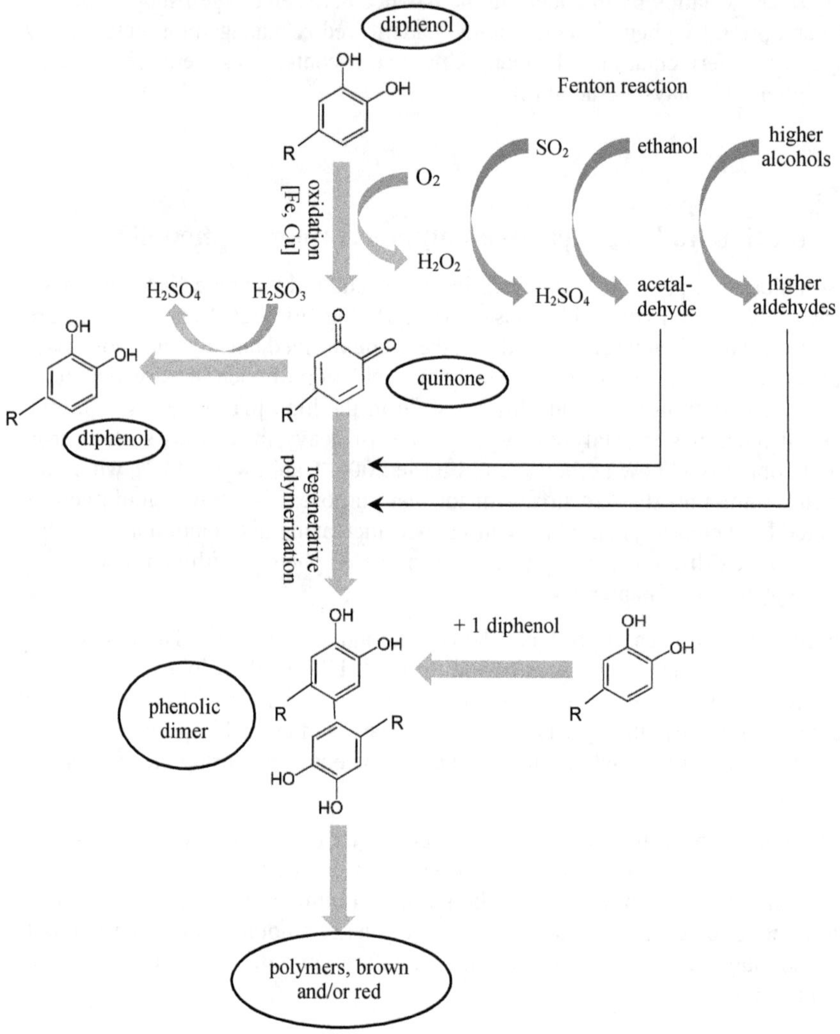

The fate of hydrogen peroxide

The H_2O_2 generated upon phenol oxidation reacts further by three pathways (Wildenradt and Singleton 1974, Singleton 1987, Kilmartin 2009, Oliveira et al. 2011), thereby oxidizing compounds that are not directly oxidizable by molecular oxygen under conditions like those in wine. These pathways are:

1. It is scavenged by molecular SO_2, which is oxidized to sulfate that is not any more measured as SO_2. This reaction only occurs when there is free SO_2 present.
2. It is reduced by ethanol. Thereby ethanol is oxidized to ethanal, also called acetaldehyde in common winery vernacular. Acetaldehyde is a major product only because of the huge concentration of ethanol in wine, which ensures that ethanol is the first oxidizable molecule encountered by H_2O_2. This explains the source of acetaldehyde during non-microbial wine oxidation. In red wines, the acetaldehyde thus produced is integrated into the polymerizing tannin structure, in which it forms the aforementioned ethyl bridge between two flavonoid moieties (Timberlake and Bridle 1976, Saucier et al. 1997 b). In white wines that contain almost no tannin, it would accumulate or bind with SO_2.
3. It oxidizes higher alcohols to higher aldehydes. When white wines oxidize, such aldehydes accumulate in the free form, which is responsible for the madeirized aroma of oxidized white wines. In red wines, they are bound to a large extent by the polymerizing tannin, thereby losing their volatility and odor activity. This is one of the reasons why red wines can consume much more oxygen than white wines without generating the typical aroma profile of oxidized or madeirized white wines. Nevertheless, the broad spectrum of compounds susceptible to oxidation by H_2O_2 is supposed to add to aroma complexity under conditions of mild oxygenation. When oxygen supply is exaggerated for a given wine, peroxide accumulation can occur and adversely impact aroma quality.

The oxidation of organic molecules by H_2O_2 that is catalyzed by iron ions in an acidic environment is called the Fenton reaction. However, in this reaction, H_2O_2 does not react directly with alcohols, but as free oxygen radicals it generates in a subsequent reaction. For the sake of simplification, the sequence of oxygen radicals is not shown in figure 47 but can be found in the specialized literature (Waterhouse and Laurie 2006, Danilewicz 2007). However, it is important to state that these radicals are the reactants that are actually responsible for the chemical and sensory outcome ascribed to intermediate H_2O_2.

The fate of quinones

As stated before, quinones represent the oxidized form of phenols. They are also unstable and react further via two predominant pathways:

1. In the presence of free SO_2 they are partially reduced back to the original phenols they were formed from. This reaction is triggered by the bisulfite (HSO_3^-) fraction of SO_2 (Chapter 6.4), which is oxidized to sulfate. Its only partial occurrence explains why a portion of the oxygen consumed remains irreversibly bound to the wine matrix, more precisely to phenols, and why 1 mg/L DO consumed oxidizes

less SO_2 than the 4 mg/L that would be expected on a stoichiometric basis (Chapter 6.4). In the absence of SO_2 it does not occur at all.

2. They combine with phenols that have not become oxidized. The result of that quinone-phenol binding is a dimer of the original phenol having been oxidized by the loss of two hydrogens, but regenerated into the reoxidizable hydroquinone (phenol) form. In other words, the original phenolic hydroxyl groups are regenerated, and the quinoidal state is canceled without the action of a reducing agent. Therefore, this reaction is designated as regenerative polymerization. Several reaction pathways of that kind have been proposed (Singleton 1987). One of them is shown in figure 48. The recombination of electrons in the aromatic ring leads to a covalent bond[46].

Figure 48: Regeneration of phenolic hydroxyl (-OH) groups via polymerization of a quinone with a phenol.

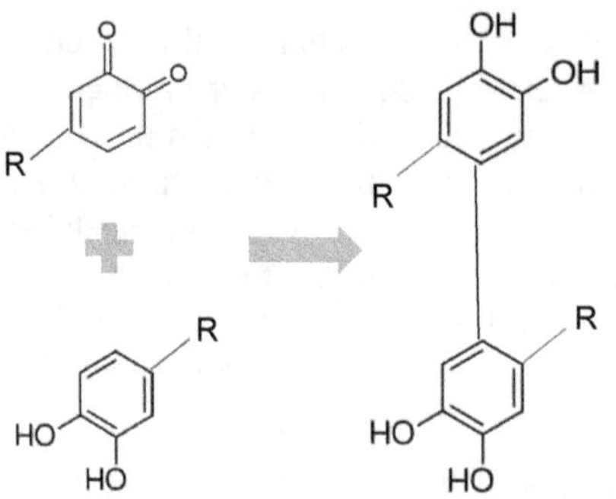

Instead of reacting with a phenol, a quinone can also react with an anthocyanin, hydrogen sulfide (H_2S), mercaptans, and sulfur-containing amino acids. The reaction with H_2S and mercaptans, albeit occurring to a minor extent in quantitative terms, explains why oxidation and the quinones resulting therefrom are able to mitigate or remove reduction flavor. However, the long-term effect of this pleasant phenomenon should not be overestimated. It has been shown that whilst oxygenation induces a decrease of free odor-active H_2S and mercaptans, it hardly affects the ability of the wine to release them again during aging. Instead, the effects obtained are the result of reversible equilibria between free, metal-complexed and oxidized forms of these compounds (Vela et al. 2018). The bottle closure and its oxygen permeability plays a decisive role in the formation of reductive off-flavors during bottle aging (Chapter 12.5).

The wonder of regenerative polymerization

The dimer discussed above does not contain an ethyl bridge. Furthermore, it is the smallest possible polymer. As a phenol it is subject to reoxidation, which leads to the corresponding quinone. When a quinone occurs as a dimer, regenerative polymerization transforms it into a trimeric phenol, and a trimeric quinone leads to a tetrameric phenol, etc. This reaction continues for as long as the polymeric phenols remain stable

[46] Reduction means the gain of electrons by a molecule through the removal of oxygen or the addition of hydrogen. In the commonly known reactions of oxidation and reduction, the reduction of one molecule occurs simultaneously with the oxidation of another molecule. Regenerative polymerization is different in that the quinone, being electrophile, can easily react with the electron-rich A ring of a flavonoid phenol. Thus, the phenol donates an electron to the quinone. As a result, both form a covalent bond by sharing an electron pair. Thereby, the hydrogens on the newly bonded carbons migrate to regenerate the quinone (=O) into the respective phenol (-OH), which can react again. Notice that the number of atoms is identical on both sides of the reaction, indicating that no reducing agent intervenes in the process of regenerative polymerization.

in solution and able to produce quinones. It explains why red wine tannin is theoretically capable of consuming unlimited amounts of oxygen. At least, the amount of oxygen incorporated per original phenol unit is many times larger than would be stoichiometrically possible without regenerative polymerization.

The vicinal quinones are partially converted in the dimer form to the equivalent of phenols with a different arrangement of the OH groups. These phenols oxidize more readily and speed up the oxidation rate as long as oxygen is available. Thus, autocatalysis (Chapter 7.3) of oxidation can occur when oxygen is not limited. Also, since these phenols are more oxidizable than the original phenol, survival of some of the original phenols is encouraged. The system favors further reaction of products before exhaustion of the original phenols and, as a result, tends to buffer or poise the wine to resist complete change by oxidation, at least to a certain extent. On the other hand, it also explains why further oxygen consumption is encouraged as long as oxygen is available.

When oxidation is very fast and when all oxidizable phenolic substrate has been converted to quinones, then there is less total oxygen consumption than if as much regenerative polymerization as possible occurs. The polymerization reaction is slower than the oxidation per se. Hence, the slower the oxidation occurs, the more regenerative polymerization augments the pool of oxidizable substrate in the wine (Singleton 1987).

Reductive strength and its limits

The enormous capability of red wine phenols to absorb oxygen explains the reductive strength of these wines. It is a measure of the amount of oxygen they are able to take up without destroying themselves. However, it is challenging to develop that qualitative relationship to a quantitative one predicting the amount of oxygen required to optimize their quality.

The better oxidizability of the phenolic polymers resulting from oxygen consumption creates a lower redox potential[47] of these molecules, also referred to as reductive strength. There are anecdotal claims suggesting that the reductive strength improved in this way increases the wine's resistance to later oxidation. However, no reliable

---

[47] The redox potential of aqueous solutions like wine is a measure of their tendency to either gain or lose electrons at an electrode, and thereby to be reduced or oxidized, when they are subjected to change by introduction of a new species such as oxygen. It is measured in millivolts (mV). The higher or more positive it is, the greater is the solution's tendency to be reduced, whilst a solution with a low reduction potential shows a tendency to be oxidized. In wine, it is an observation of its current oxidation or reduction level, which depends on pH and DO. Its measurement can provide a measure, for example, for characterizing conditions under which phenolics are stable against oxidative degradation. Its variation reflects the oxidation-reduction reactions occurring over time.

For measurements in wine, a combined electrode is used. It consists of a platinum or graphite measuring electrode and an Ag/AgCl reference electrode. Obtaining reliable results is complicated by the fact that readings show slow kinetics, do not stabilize, lack significance (Danilewicz 2012), and that the electrode calibration is disturbed due to poisoning. Therefore, redox measurements are not used in wine quality control, but they have been proven useful in monitoring changes under research conditions rather than determining their absolute values.

data are yet available that could justify such assertions. Rather, the reductive strength makes the wine hungrier for oxygen. The little known phenomenon of autocatalysis, described in the next section, is an example of this. Essentially, it is all a matter of the amount of oxygen available. Whilst it is certainly true that in most red wines the oxygen consumption rate decreases with increasing oxygen uptake and wine age (Chapter 7.3), there is no scientific evidence that oxidation makes the wine more resistant to further oxidation. Conversely, at some point the reductive strength will be exhausted in any wine.

As will be shown in the following sections, the ability of wines to resist oxidative damage is determined by their phenolic strength and the possible presence of free SO_2. Furthermore, as long as yeast lees are present, reductive strength is additionally reinforced, however they are no longer present after filtration and bottling. In bottled wines, the oxygen ingress through the bottle closure plays an additional role in wines' shelf life (Chapter 12.5). The presence of ripe tannins ensures that even graceful aging might take place. However, at the latest when a smell of acetaldehyde appears after consumption of all free SO_2, the question of agreeability is clearly answered regardless of considerations about reductive strength. In this context, wines without added sulfites (Chapter 12.4.3) might be rated in a different way.

7.3. Reaction rate, autocatalysis, and over-oxidation

In the presence of free SO_2, reduction of quinones by SO_2 competes with regenerative polymerization. When there is no or little SO_2, regenerative polymerization is in the foreground. The increase of easily oxidizable phenols it generates can enhance oxygen consumption by autocatalysis as long as oxygen is freely available.

As outlined in chapter 6.1, DO consumption in wine is not linear but follows a first order kinetics best expressed by a negative exponential function. This means that the DO consumption rate depends on the instantaneous DO concentration. When DO halves, its consumption rate also halves. Hence, the only valid way to express and to compare DO consumption rates is the DO half-life period expressed as hours or days. It is the period of time in which the DO concentration decreases by half of its value. The higher it is, the more slowly DO is consumed.

Over-simplifying one could assume that the DO consumption rate in a given wine would be the same over time or after every oxygen uptake. However this is not the case; it depends on the wine's history and it changes as a result of previous oxygen uptake and consumption. Even worse, it can increase or decrease.

In most wines, the DO consumption rate slightly decreases to the extent that the wine has already consumed increasing amounts of oxygen. As a result, the DO half-life increases. An example is given in figure 49 representing a filtered red wine stored at 20° C under conditions of constant oxygen availability ensured by several consecutive oxygen saturations.

Figure 49: DO half-life period in a filtered red wine at 20° C.

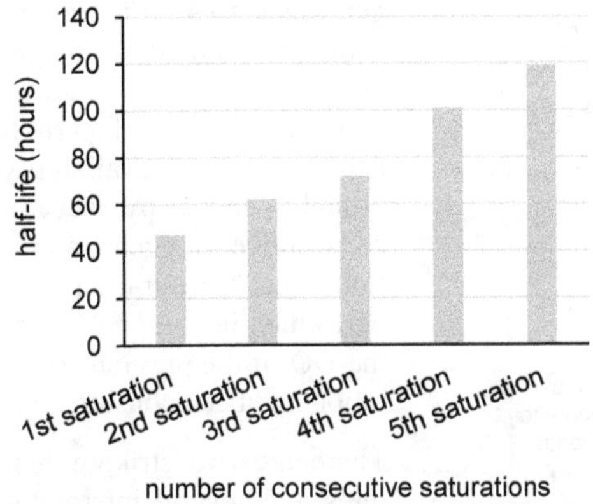

Half-life periods are not a commonly employed measure in the wine industry, which is used to work with concentration units. However, as outlined in chapter 6.1 and due to the exponential flattening of the DO consumption rate, any attempt to express this rate in mg/L per time unit must refer to a well-defined instantaneous DO concentration. This has been done to create figure 50. In this figure, the data underlying figure 49 are transformed into the DO consumption rate v observed between 2 and 1 mg/L DO. The sole reason for doing so is to facilitate understanding of what might appear complex stuff from the viewpoint of daily winery work.

It is important to understand that the concentration range between 2 and 1 mg/L DO has been chosen arbitrarily. It corresponds to a typical amount of oxygen a wine picks up upon splash racking or similar operations. It will also be used in further examples to facilitate understanding and comparisons. Moreover, those who prefer working with a per-day basis can easily convert the per-hour data. When doing so, one will state that the DO consumption rate of most wines varies between 0.1 and 1.0 g/L per day, depending on all the factors outlined previously and those following.

Figure 50: DO consumption rate v (mg/L per hour) at 20° C as affected by previous DO consumption. V given for the concentration range from 2 to 1 mg/L DO.

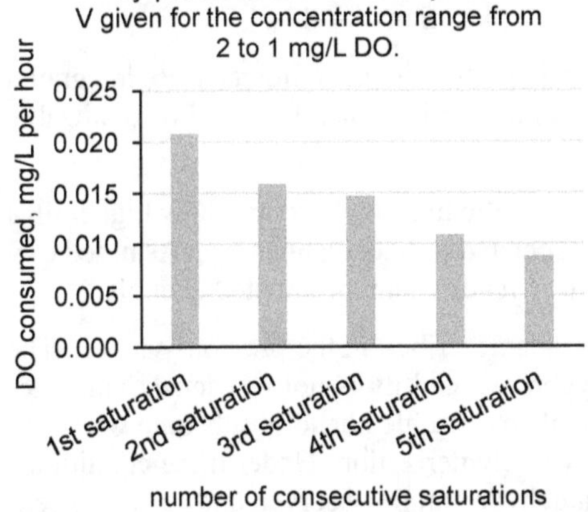

Back to practice: The DO consumption pattern shown in figure 50 partially explains why older wines consume DO more slowly than young wines, referring to yeast-free samples without microbial DO consumption. The most plausible explanation is that regenerative polymerization does not regenerate phenolic hydroxyls lost by oxidation as completely as one might expect. Thus, there is some depletion of oxidizable hydroxyls when wine picks up oxygen during the aging process. Hence, regenerative polymerization does not proceed completely.

Over-oxidation, scenario I: Autocatalysis of oxidation

However, there are also exceptions to the rule shown above, and that is the point where autocatalysis comes into play. It only occurs in a few wines, in particular in those stored with ullage. The sensory outcome can be disastrous

As a result of autocatalysis and in contrast to regular wine behavior displayed in figure 50, the DO consumption rate v increases as the amount of previously consumed oxygen increases. Figure 51 gives an example of autocatalysis in another wine stored under conditions of permanent oxygen availability brought about by several consecutive oxygen saturations. Each saturation was performed immediately after the DO of the previous saturation had been consumed.

Figure 51: Autocatalysis: DO consumption rate v (mg/L DO / h) at 20° C as affected by previous DO consumptions.
v given for the concentration range from 2 to 1 mg/L DO.

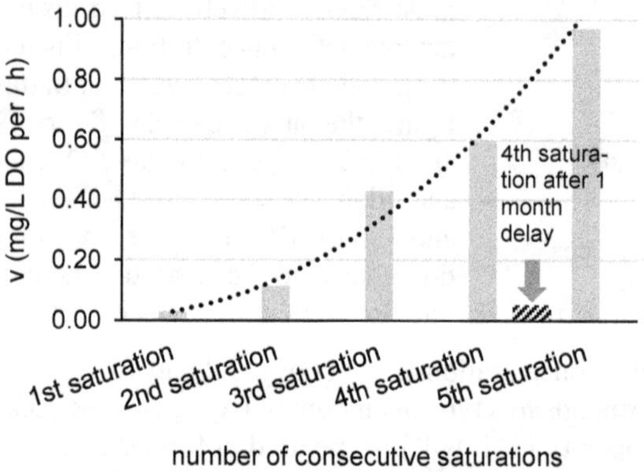

There are two striking features and several interpretations:

1. After each oxygen saturation, the DO consumption rate v increases when the fresh saturation is performed immediately after the consumption of the DO resulting from the previous saturation (grey columns). After the 5th saturation, v is 33 times greater than after the 1st saturation. The increase of the DO consumption rate is exponential, demonstrating autocatalysis as the cause. There are two explanations for this:

   - During polymerization, phenols that originally were not oxidizable are integrated into the increasing polymer, thus becoming oxidizable. The pool of oxidizable phenols increases.

   - With increasing polymerization, the phenolic hydroxyl groups show higher pKa values such that phenols are more dissociated as phenolate anions. As noted earlier, ionized phenols oxidize much more readily than protonated phenols.

   Both effects accelerate the oxygen cascade described in the previous section. The reaction runs out of control when oxygen availability is not restricted. This phenomenon can rapidly deplete the available oxidizable phenolic substrate when oxidation proceeds faster than regenerative polymerization. Under these conditions, an accumulation of peroxide and acetaldehyde can be observed and cause the sensory effects of over-oxidation.

2. When the 4th saturation was performed immediately after exhaustion of the DO incorporated by the 3rd saturation, v = 0.6 mg/L•h DO could be observed. However, when the 4th saturation was delayed until one month later, v was only 0.03 mg/L DO/h and comparable to that observed after the 1st saturation. Obviously during the interval of one month of anoxic storage, far-reaching reactions took place. They took the oxidizability of the phenolic substrate back to its original level. The causes and nature of these reactions are unknown. Acidic hydrolysis of polymers is a possible explanation.

Autocatalysis in a nutshell

Autocatalysis can occur in some wines when exaggerated and prolonged oxygen uptake takes place, i.e. when oxygen availability is unlimited. It means that the wine consumes DO faster the more it has consumed before (Petrozziello et al. 2018). As a consequence, oxygen uptake replenishing DO also increases when oxygen availability is not limited. Thus, the oxygen consumption speeds up, and oxidation begets oxidation.

There is not yet any explanation as to why some wines are more prone to autocatalysis than others under conditions of comparable oxygen availability, SO_2 status, yeast concentration, etc. In particular, there is no connection with total phenols or other macro-analytical phenol data. For that reason, careful monitoring is required when measures of deliberate oxygenation are taken.

In sensory terms, the most likely outcome is visible browning, precipitation of phenolic material, and oxidative spoilage evidenced by smell even after SO_2 additions take place to bind free acetaldehyde. Under commercial winemaking conditions, the most frequent cause of autocatalysis is storage with ullage at elevated temperatures, the latter further accelerating the reaction.

Over-oxidation, scenario II: Oxygen supply too fast

Autocatalysis is one reason why the effects of oxygen supply can run out of control, but the same outcome can happen in any wine when too much oxygen is supplied at once. This is best exemplified by the severe oxygenation conditions underlying the results depicted in figure 52. It shows the sensory outcome of oxygenation speed in a red wine blend provided with an identical amount of oxygen in two different modalities:

- Fast oxygenation by dissolving 40 mg/L O_2 all at once under pressure, corresponding to five saturation concentrations of 8 mg/L O_2 each at standard pressure.
- Slow oxygenation by dissolving five saturation concentrations in the form of 5 **x** 8 mg/L O_2 spaced in time at standard pressure. Each saturation was performed one month after depletion of DO resulting from the previous saturation.

Both lots were submitted to sensory evaluation two months after the complete depletion of DO and adjustment of free SO_2 to 30 mg/L. In this wine, the olfactory attribute “red fruits” represents grape-derived primary aromatics, whilst the term “dry herbs” depicts aroma damage caused by over-oxidation as it is better known from aged white table wines. Data are given as percentage deviation from the mean values of both lots.

Figure 52: Sensory and color impact of the oxygen supply rate on a red wine blend; comparison of slow and fast oxygenation.
Data given as percentage deviation from the mean values of both lots.
Total phenols = 1610 mg/L GAE, anthocyanins = 734 mg/L

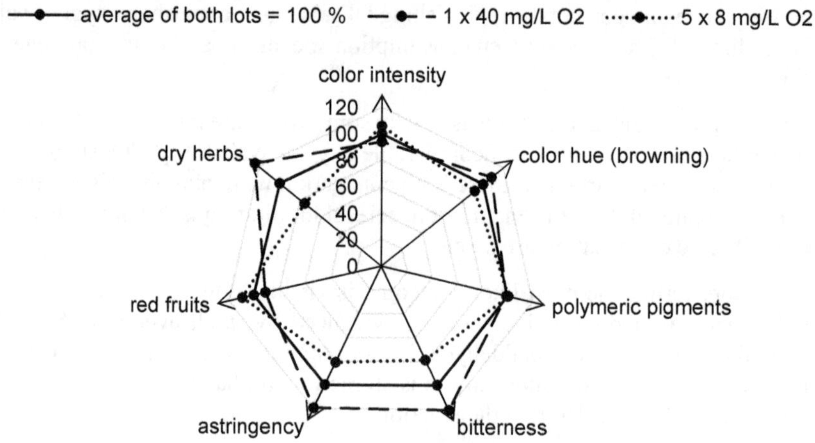

The fast oxygenation caused strong aroma deterioration expressed by a significant enhancement of the “dry herbs” term as well as an increase in astringency and bitterness. Only the slow oxygenation was able to decrease astringency and bitterness, which is in line with the outcome one expects from oxygenations. Color parameters were not significantly affected by oxygenation in this specific wine.

The increase of the “dry herbs” smell attribute after fast oxygenation is associated with a temporary accumulation of intermediate peroxide. It can also be observed in a relatively short term after addition of surplus H_2O_2 to any wine. It promotes the irreversible oxidation of higher alcohols to higher aldehydes (Figure 47). The latter are substantially involved in the aroma profile of oxidative aging and have been particularly well investigated in white wines (Escudero et al. 2000 a, 2000 b, 2002, Pons et al. 2015). The increase of astringency appears to be related to excessive tannin polymerization (Chapter 5.3.1).

The effect of temperature

Temperature has a large impact on the oxygen consumption rate of phenolics. It is another variable that can cause oxygenation effects to run out of control when it is too high (Chapter 8.3).

As stated previously, the DO consumption rate of a given wine sample is not constant, but depends on the instantaneous DO concentration, with higher concentrations increasing the rate exponentially. Figure 53 demonstrates that a similar pattern applies to temperature.

At 5° C, the DO consumption rate almost approaches the zero value, whilst it increases by a factor of 5 when the temperature increases from 10 to 20° C. This behavior is quite logical because the DO consumption by phenols involves two reactions, one with each phenol, which means that the temperature dependence must be theoretically squared.

Figure 53: Effect of temperature on the DO consumption rate of a filtered red wine.
v given for the concentration range of 2 to 1 mg/L DO.

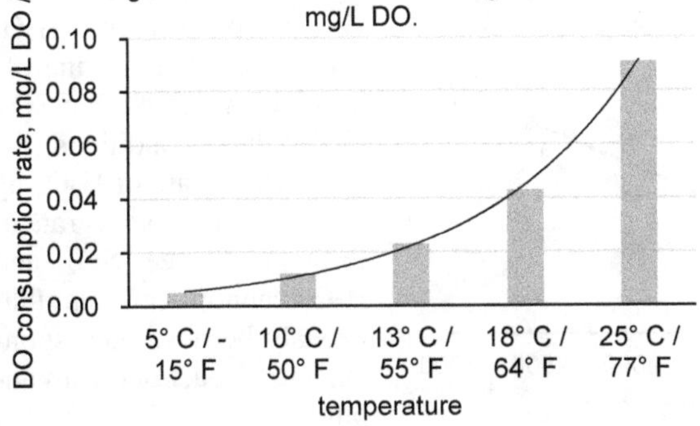

The effects of a high initial DO level and a high temperature are synergistic with regard to autocatalytic effects. This pattern has particular importance when measures of active oxygen supply such as micro-oxygenation are taken or when wine is stored under conditions of unlimited oxygen availability in only partially filled containers.

Generation, accumulation, and breakdown of hydrogen peroxide

After this foray into the dangers and risks of poorly performed oxygen supply measures, an additional look behind the scenes of chemistry helps to better understand what happens in the wine when it consumes oxygen.

When free SO_2 is present, it spontaneously traps the very largest portion of peroxide generated upon phenol oxidation. This portion has been found to be 87 % on average in wines containing free SO_2 (Schneider 2019), with the exact extent depending on initial free SO_2 and pH (Figure 46). However, in the absence of SO_2 as is typically the case during oxygenation of most young red wines, H_2O_2 is able to temporarily accumulate to measurable concentrations. Based on an experimental approach derived from enzymatic analysis and using the reaction system of $NADH_2/NAD^+$–horseradish peroxidase[48], instantaneous concentrations of H_2O_2 produced upon aerial oxygenation

[48] Horseradish peroxidase (HRP) is an enzyme commonly used in enzymatic analysis and other biochemistry applications to catalyze oxidations by H_2O_2. $NADH_2$ is nicotinamide adenine dinucleotide (reduced), whilst NAD^+ is its oxidized form whose absorbance is measured at 340 nm and converted into mg/L H_2O_2 by use of a calibration curve. Other substrates than $NADH_2$, for example Amplex red, are feasible and have been used more recently for the same purpose (Héritier et al 2016). However, in contrast to statements made in older literature (Vivas and Glories 1996), such measurements cannot be performed without the use of a peroxidase enzyme.

The direct proof and quantification of intermediate H_2O_2 in wine have been a long-cherished dream in enological research, but its execution and the interpretation of the data obtained require a great deal of care. Indeed, most enzymatic assays are performed in a medium buffered in the almost neutral pH range around pH 7 in order to optimize enzyme activity and substrate turnover. In contrast, enzymatic H_2O_2 measurements have to be carried out at the natural wine pH and in a short time period, which requires the use of unusually high amounts of enzyme (HRP) activity. Otherwise phenol oxidation would strongly accelerate and generate unrealistically high amounts of H_2O_2 (Chapters 6.4 and 7.2).

Furthermore, in wines containing free SO_2, there will be no molecular SO_2 in the neutral pH range able to scavenge H_2O_2 as it does at wine pH. As a result, H_2O_2 would be detected in any wine, albeit it does not accumulate in the presence of free SO_2 at wine pH due to its immediate scavenging.

(~ 6 mg/L DO) of red wines at ambient temperature were measured. Figure 54 shows that the formation of intermediate H_2O_2 correlates well with total phenol contents under comparable conditions of oxygenation. The amounts reported are net concentrations resulting from the equilibrium between formation and degradation of H_2O_2. However, no H_2O_2 can be detected at wine pH in the presence of free SO_2. Under these conditions, the degradation rate of H_2O_2 by SO_2 is greater than its rate of formation. Free SO_2 and H_2O_2 cannot occur simultaneously because they spontaneously react one with another.

Figure 54: Instantaneous concentrations of intermediate hydrogen peroxide (H_2O_2) during aerial oxygenation of red wines devoid of free SO_2.

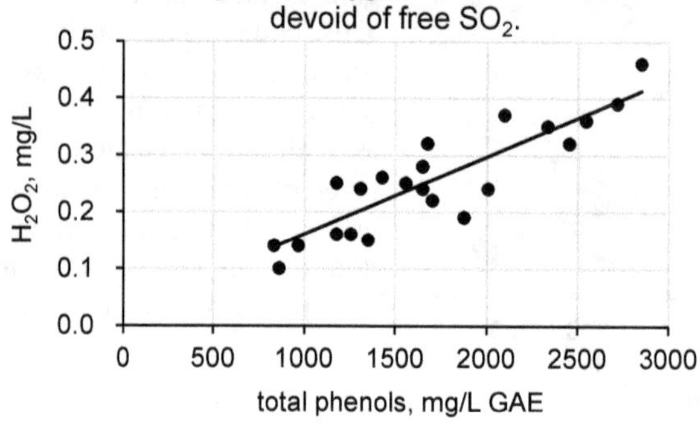

Other studies (Héritier et al. 2016) have shown that the formation of peroxide even better correlates with the concentrations of anthocyanins and polymeric pigments. These are the most easily oxidizable phenols with vicinal hydroxyl groups best meeting the requirements for the formation of peroxide and quinones according to the reactions outlined in figure 47. Under identical conditions, white wines produce less peroxide in agreement with their lower levels of these phenols, whilst iron enhances peroxide formation.

Summarizing, H_2O_2 is an unavoidable by-product of wine oxidation. When there is no free SO_2 available, H_2O_2 acts as a strong oxidant via the hydroxyl radical ($OH^{\cdot}$) it disintegrates into. As a consequence, there is a nonspecific oxidation of all organic wine constituents in the order of their concentration. This is the Fenton reaction. Due to its high concentration, ethanol is preferentially oxidized. The product resulting therefrom is acetaldehyde (Waterhouse and Laurie 2006).

Producing acetaldehyde is the primary purpose of micro-oxygenating (Chapter 8.3.1) of young red wines before any post-fermentation SO_2 addition. At a very low rate, this reaction also takes place in the presence of free SO_2. Peroxide scavenging is the primary purpose of SO_2 in mitigating the oxidation process of wine.

Acetaldehyde combines with SO_2 to an odorless adduct. Hence, as long as there is sufficient free SO_2 available, no free acetaldehyde or the typical smell reminiscent of bruised apples it elicits will be found. However, higher aldehydes and other oxidation products do not react that easily with SO_2. They are a major problem in white wine aging (Schneider 2019) and to some extent also in overly aged red wines. Accordingly, controlling the Fenton reaction is one of the core tasks of wine redox management (Elias and Waterhouse 2010).

Ultimately, the wine constituents involved in the reduction of peroxide are SO_2, ethanol, higher alcohols, and phenols themselves. They behave in a competitive manner. When little or no free SO_2 is available, the peroxide inevitably oxidizes more intrinsic wine constituents with sensory consequences. According to their concentration, phenols are the second most reactant with peroxide in red wines. They are instrumental in both the production and consumption of H_2O_2.

This can be observed when H_2O_2 is added to red wines containing free SO_2 with the purpose of reducing it, for example, from 40 to 20 mg/L. Instead of oxidizing the stoichiometric amount of SO_2 as observed in aqueous solution, it only removes a fraction of it. The remainder of H_2O_2, about half, is predominantly scavenged by phenols. It increases when total phenols and pH increase. In contrast, it is less than 10 % in white wines with their low total phenol contents (Schneider 2019). This behavior is consistent with the behavior of oxygen shown in Figure 46, which also leads to lower losses of SO_2 than one would stoichiometrically expect. It cannot be otherwise as all DO in wine is converted to H_2O_2 before it reacts further.

Figure 55: Decrease of hydrogen peroxide added to a red wine devoid of SO_2.

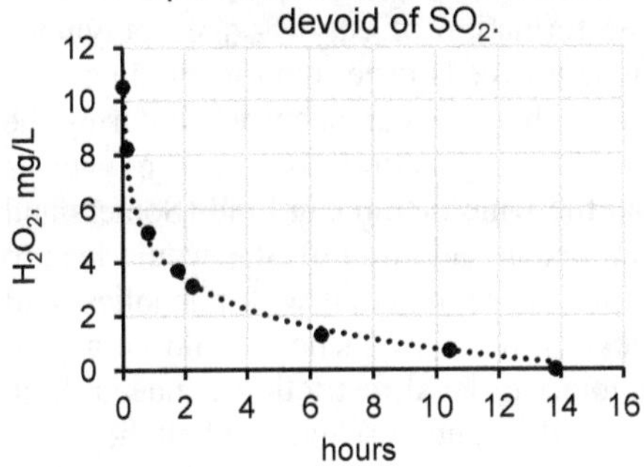

Figure 55 illustrates that standard red wines are able to consume relatively high amounts of H_2O_2 within a few hours. This consumption rate explains why intermediate H_2O_2 produced upon red wine oxygenation does not accumulate to levels higher than those shown in figure 54. White wine phenols are not able to consume so much H_2O_2 due to their low concentration. As a consequence, more of the peroxide they generate is diverted to aroma-active compounds, which become oxidized. This is one of the reasons why red wine aroma is more resistant to oxidative spoilage than white wine aroma.

However, even when red wines consume oxygen before or after bottling, there will be a moment when H_2O_2 is partially trapped by other compounds than phenols, ethanol, or SO_2. When the oxidation products resulting from these compounds are volatile, then the moment has come when significant aroma changes can be observed. These changes may be rated positively or negatively, desired or undesired, depending on the concentrations and odor activities of the molecules involved. At best, they are referred to as maturation. In contrast to the off-odor caused by free acetaldehyde, they cannot be removed by SO_2 when they are unpleasant.

Aroma changes caused by DO and H_2O_2 consumption explain why red wine redox management involves much more than stabilizing color, improving mouthfeel, and smoothening tannins. It has also far-reaching impact on aroma quality. Exaggerated oxygen supply to red wines or autocatalytic oxidation generate a transitory accumulation of peroxide with detrimental impact on aroma. In contrast, increasing rates of oxygen ingress through bottle closures increase the intensity of red fruit aroma notes in high-phenol wines, apart from decreasing the content of tannins reactive toward salivary proteins (Gambuti et al. 2016). The reason is the much slower oxygen uptake rate in bottled wines.

Generation, accumulation, and depletion of acetaldehyde

In red wines, acetaldehyde produced by oxidation of ethanol gradually binds with tannins. The same applies to acetaldehyde that has been formed microbially during primary fermentation. In doing so, it promotes phenol polymerization by creating ethyl bridges of the general kind "phenol-ethyl-phenol". This process generates phenolic polymers of the type "flavanol-ethyl-anthocyanin", "flavanol-ethyl-flavanol", "flavanol-ethyl-flavanol-ethyl-anthocyanin", and many other ones of variable chain length. Their relative proportions depend on the initial tannin-anthocyanin ratio. In this process, acetaldehyde can be replaced by pyruvic acid and glyoxylic acid, and direct bondings between two flavonoid moieties without an ethyl bridge also occur (Timberlake and Bridle 1976). Through the reaction of different single molecules via different binding forms, a vast number of distinct polymers is created. Details are outlined in section 2.2.3.

The purpose of micro-oxygenation and other means of oxygen supply is to promote the generation of acetaldehyde and, thus the formation of ethyl-bridged polymers. However, when the generation of acetaldehyde proceeds faster than its binding with phenols, it can accumulate to elevated levels. Such a case can occur when during the oxygenation of red wine, for example by micro-oxygenation, too much oxygen is added and more acetaldehyde is formed than the wine tannin can bind (Somers and Wescombe 1987). When this happens, the immediate addition of SO_2 would lead to its spontaneous binding with acetaldehyde, prevent its reaction with phenolics, and stabilize its concentration on the elevated level. The consequence would be a high amount of bound and total SO_2, which is subject to legal restrictions. Indeed, high acetaldehyde levels are largely responsible for high bound SO_2 levels, albeit they have their primary cause in sluggish fermentations. Conversely, when red wine is stored under exclusion of air, there is a gradual degradation of the acetaldehyde by binding to the tannin (Pontallier 1981, Pontallier and Ribéreau-Gayon 1983). As a consequence, SO_2 requirements are lowered when free SO_2 is to be achieved.

Figure 56 Decrease of acetaldehyde spiked at different levels in three red wines stored under anoxic conditions without SO_2.

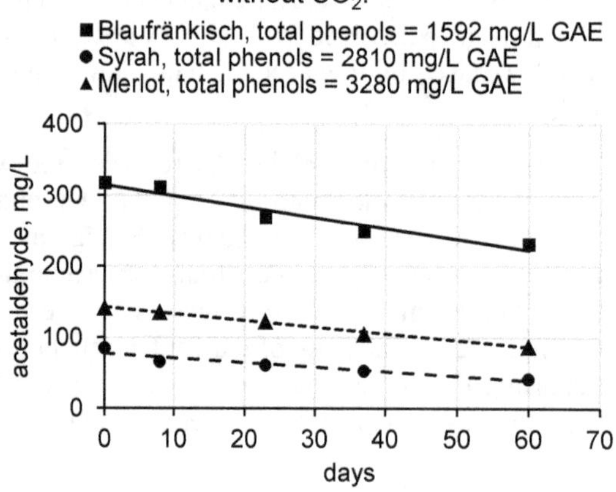

Figure 56 shows the course of such a reaction in three red wines, which were spiked with variable amounts of acetaldehyde. The average decrease of acetaldehyde is approximately 1 mg/L per day, which is as much or more than the average increase induced by a correctly performed oxygenation. Under such conditions, generation and binding of acetaldehyde proceed simultaneously, with a constant net concentration as the final result. Only when over-oxidation occurs, an increase of acetaldehyde can be observed. Therefore, the typical smell it elicits is used as a hint to modulate oxygen supply by micro-oxidation (Chapter 8.3) or other means.

A strong accumulation of acetaldehyde under oxidative conditions (no SO_2, strong O_2 uptake) leads to the following consequences:

- a rapid depletion of anthocyanins and their conversion in polymeric and other pigments,
- color changes from clear red to dark blue-mauve,
- a decrease of total phenols and tannins caused by a strong increase of their polymerization level and subsequent precipitation as insoluble compounds,
- a dry and aggressive astringency of the remaining tannins as the most obvious sensory outcome in a thinner wine lacking body.

These phenomena can be observed in some wineries when certain red wines are stored in barrels without monitoring. In this way, part of the tannin that has been laboriously extracted during maceration is lost forever. Precipitated tannins are not able to redissolve under wine conditions.

On the other hand, the enological effect that is achieved by production of acetaldehyde via the Fenton reaction can also be obtained by its direct addition in pure form. In addition, aroma damage caused by peroxide accumulation is avoided (Aleixandre-Tudo et al. 2016). However, this is not a legal procedure in winemaking.

SO_2 additions interrupt polymerization

Regenerative polymerization follows oxidation with a time lag. When it is induced by too high an oxygen uptake in a short period, it continues long after DO consumption has been completed. As a result, anthocyanins and low-molecular tannins also continue to decrease despite the absence of DO, while polymeric pigments increase. These reactions proceed further until intermediate peroxide and acetaldehyde generated in the oxidation step are exhausted. They can also be removed when SO_2 is added. Hence, SO_2 additions stop polymerization. Table 7 gives an example.

Table 7: Impact of SO_2 addition on polymerization after consumption of 8 mg/L DO in a filtered red wine blend.

	reference without O_2, without SO_2	time delay of SO_2 addition after completion of DO consumption	
		0 days	30 days
polymeric pigments (A_{520}^{SO2}) [1]	1.34	1.37	1.46
anthocyanins, mg/L [2]	944	752	724
low-molecular tannins, mg/L [3]	405	290	245
total phenols, mg/L GAE [4]	1890	1870	1840

[1] measured as residual color at 520 nm using the SO_2 bleaching method (Chapter 3.2.2)
[2] measured by the SO_2 bleaching method (Chapter 3.2.2)
[3] measured spectrophotometrically with 4-dimethylaminocinnamaldehyde (Chapter 3.4.2)
[4] measured with the Folin-Ciocalteu assay (Chapter 3.1.2)

Delayed polymerization effects can be treacherous and hazardous in real-world winemaking because they are not expected after oxygen availability has been stopped. When they occur, the most noticeable effect is a precipitation of insoluble tannins or browning of lightly colored wines.

Summary: Oxidation and regenerative polymerization of phenolics

- In regenerative polymerization (quinone + phenol), the OH groups of the phenols lost through oxidation are restored.
- The regenerated phenols are again available for oxidation. Thus, the phenolic substance of red wines can bind much more oxygen than simple stoichiometry would allow for.
- Therefore, the ability of red wines to chemically consume oxygen has no fixed end point and is unlimited as long as the polymers remain in solution.
- However, its practical limit is the phenol content (possibly flocculation, browning) and the oxidative destruction of aroma by formation of higher aldehydes and related volatile oxidation products. High-phenol red wines can easily consume more than 100 mg/L O_2 and thereby gain in quality.
- During the oxidation of phenols, intermediate peroxide is formed, which can change the aroma through the Fenton reaction, unless it is reduced by SO_2 (→ sulfate), ethanol (→ acetaldehyde), or phenols themselves.
- The resulting acetaldehyde disappears by binding to tannin via the ethyl bridge.
- The reactions are largely suppressed after SO_2 addition.

Practical hints

It is undoubtedly true that red wines can be stored and aged under much more oxidative conditions than white wines. This is due to their tannin, which is able to consume large amounts of oxygen and thus protect the wine from the adverse consequences of oxidation. The tannin even requires a certain consumption of oxygen to facilitate its polymerization and to optimize its sensory expression on the palate.

Nevertheless, oxygen supply to red wines has its natural limits when the magnitudes and the relationships between oxygen supply and oxygen consumption are neglected. Red wine is not all red wine. The phenol content is the basic factor that determines its oxygen requirements, resistance to oxidation, its capacity to age well, and its shelf life.

Over-oxidation is manifested by aroma losses and a smell reminiscent of bruised apples (free acetaldehyde) or dry herbs (higher aldehydes), loss of color, browning, or clouding due to flocculated phenolic polymers. When this happens, the oxidation cascade should be interrupted immediately by adding 30 to 50 mg/L SO_2; employed alone, further storage under anoxic conditions will not be sufficient as a remedy. However, the damage caused in the meantime will be irreversible. Only an off-odor caused by free acetaldehyde can be removed by binding it to SO_2.

7.4. Phenols, oxygen, and SO_2 interacting: Sensory results

7.4.1. The anti-oxidant action of sulfur dioxide in clarified red wines

From a sensory perspective, the most obvious action of sulfur dioxide is its capability of immediately binding free acetaldehyde, thus removing its typical smell reminiscent of bruised apples that can easily depreciate wine aroma. Acetaldehyde binds almost spontaneously with SO_2 as long as free SO_2 is available. Hence, free acetaldehyde and free SO_2 exclude one another. These sensory considerations arise at the latest before bottling when free SO_2 is adjusted in most wines with the main purpose of avoiding aroma spoilage caused by free acetaldehyde. In contrast, when wines undergo the maturation and stabilization process between fermentations and bottling, the anti-oxidant action of sulfur dioxide is what is more important. Its importance depends on the individual wine and on its momentary stage of development.

When a young red wine still containing suspended yeast lees picks up oxygen, a variable proportion of that oxygen is consumed by dead yeast cells and irrevocably withdrawn from reactions with the intrinsic wine constituents (Chapter 6.2). Post-fermentation yeast lees act as a reducing agent to which little attention is paid in current winemaking practice. After wine clarification the oxygen essentially finds two reaction partners, which are phenolic compounds (tannins and anthocyanins) and possibly sulfur dioxide.

According to widespread enological doctrines, sulfur dioxide only plays a minor role as a reducing agent in red wines as phenolics assume that role. However, this requires that the wine contains enough phenolic material able to do so. This might hold true for many high-phenol red wines, which can improve upon consumption of several saturation concentrations of oxygen slowly supplied by micro-oxygenation or barrel aging even in the absence of SO_2. The optimal amount is hardly predictable but can be determined by sensory monitoring, which allows for terminating the process at the right moment.

However, there are also other wines that do not display enough phenolic strength and in which the role of SO_2 as an oxygen acceptor becomes inevitably more important. Such wines are obviously low-phenol ones and more sensitive to oxygen consumption. Studying their behavior upon oxygen uptake is extremely instructive as it helps better understand the reactions occurring in high-phenol red wines under comparable conditions. Creating outstanding low-phenol red wines is a great challenge for any winemaker, and whoever masters that challenge will easily cope with high-phenol ones. In any case, the differences between both kinds of wine are fluid. The sensory results of excessive oxygen supply will be illustrated by means of some extreme examples.

Figure 57 reports results of a filtered young red wine with low total phenol content, which was treated in three different ways:

- adjustment of free SO_2 to 35 mg/L and subsequent consumption of 16 mg/L O_2 supplied over a period of two months,
- identical supply of 16 mg/L O_2 without previous SO_2 addition,
- untreated reference,

Sensory and chromatic evaluation took place two months after complete oxygen consumption and re-adjustment of all aliquots to 35 mg/L free SO_2. Data in figure 57 are reported as percentage deviation from the untreated reference. By analogy with the wine represented in figure 52, the aroma attribute "dry herbs" represents aroma damage caused by over-oxidation, whilst the term "red fruits" depicts grape-derived varietal aroma.

Figure 57: Sensory and color impact of oxygen supply to a young filtered low-phenol red wine (Pinot noir) in the presence vs. absence of free SO_2.
Data given as percentage deviation from the untreated reference.
Total phenols = 1080 mg/L GAE, anthocyanins = 164 mg/L.

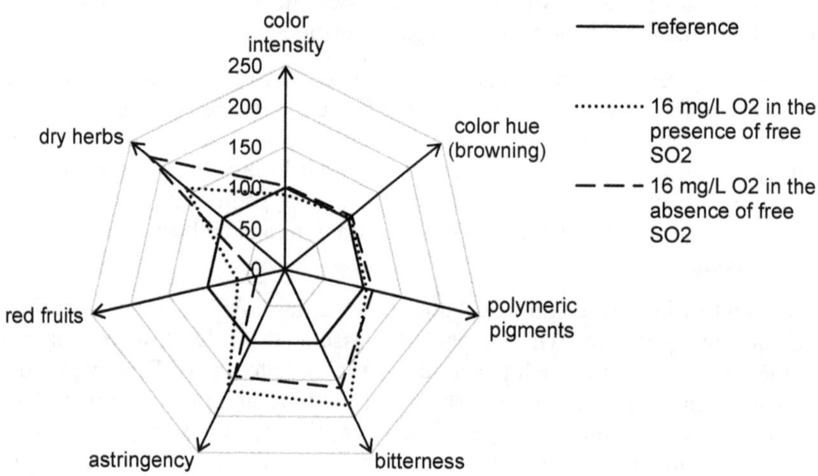

In this red wine with only 1080 mg/L GAE total phenols, the oxygen supply of 16 mg/L O_2 induced a strong decay of fruity varietal aroma and a corresponding increase of the "dry herbs" smell attribute characteristic for over-oxidation. On the palate, both astringency and bitterness also increased upon oxygenation. These detrimental effects could not be reversed by subsequent adjustment of free SO_2. Color changes were not noticeable related to a comparable level of free SO_2.

As a result, it can be stated that the oxygen supply of 16 mg/L O_2 was definitively too much regardless of the previous SO_2 addition. Nevertheless, the detrimental aroma was less pronounced when the oxygen was supplied in the presence of free SO_2, indicating some protective anti-oxidative effect of SO_2. It plays an important role by complementing the modest effect of phenolics as oxygen scavengers in this kind of rosé-like red wine in that it partially traps intermediate peroxide.

The same procedure was applied to another filtered young red wine displaying a higher level of total phenols altogether with a many times higher anthocyanin content. Figure 58 displays the results obtained.

Figure 58: Sensory and color impact of oxygen supply to a young red wine (Merlot) with intermediate total phenol level in the presence vs. absence of free SO_2.
Data given as percentage deviation from the untreated reference.
Total phenols = 1890 mg/L GAE, anthocyanins = 964 mg/L

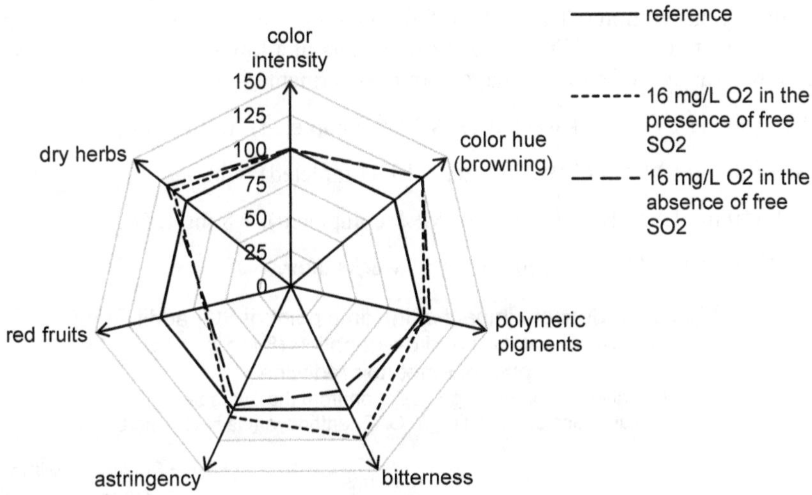

In comparison with the low-phenol red wine represented in figure 57, oxygenation under identical conditions and with the same amount of oxygen produced less losses of varietal fruity aroma. Furthermore, the presence of free SO_2 had no impact upon these losses. Whilst color intensity showed no changes after subsequent adjustment of free SO_2 to an identical level, color hue displayed higher scores in both oxygenated lots and could be explained by losses of red color due to oxidative degradation of anthocyanins. There was no increase of polymeric pigments, whilst astringency was not affected by any of the oxygenation procedures. With a moderate to medium total phenol level (1890 mg/L GAE) and a total phenol-anthocyanin ratio of 1.96, this wine had not enough tannin to elicit a noticeable astringency that could have been affected.

In both red wines shown in figures 57 and 58, oxygenation did not result in any improvement on the palate by a decrease or better integration of astringency, though the oxygen was supplied slowly over a period of two months. However, with 17 mg/L O_2 the absolute amount of oxygen provided was too high for these red wines with relatively low total phenols levels, be it with or without previous SO_2 addition. It would have been less harmful if the wines had not been filtered previously. Nevertheless, it excludes successful barrel aging of such wines as the oxygen uptake during barrel aging and subsequent steps tends to be even higher. On the contrary, storage in tanks instead of barrels and an early SO_2 addition after fermentations will help to better protect them against adverse effects of the oxygen they take up in any case during common cellar operations (Chapter 8.1.3). In this context, the time of the first SO_2 addition post-fermentation is to be examined a bit closer.

Timing of post-fermentation SO_2 additions

For several reasons, Pinot noir is a particularly suitable cultivar for studying the interactions between total phenols, oxygen, and SO_2: It tends to display low color and low anthocyanin levels in conjunction with a high total phenol-anthocyanin ratio, although total phenol contents can be quite moderate since tannins are time-consuming to extract during maceration (Chapter 4.1). This phenolic composition makes it a relatively oxygen-sensitive variety. Figure 59 shows results of a low-phenol Pinot noir submitted to four variants of redox management post-fermentation:

- SO_2 (60 mg/L) early (1 week after MLF[49] completed), without O_2,
- SO_2 (60 mg/L) late (11 weeks after MLF completed), without O_2,
- SO_2 (60 mg/L) early (1 week after MLF completed), + 8 mg/L O_2,
- O_2 (8 mg/L) early + SO_2 (60 mg/L) 11 weeks later.

Figure 59: Sensory impact of the time point of SO_2 addition (60 mg/L) post-fermentation and oxygenation (8 mg/L O_2) on a low-phenol Pinot noir red wine.
Data given as percentage deviation from the average of all lots.
Total phenols = 1850 mg/L GAE, anthocyanins = 157 mg/L.

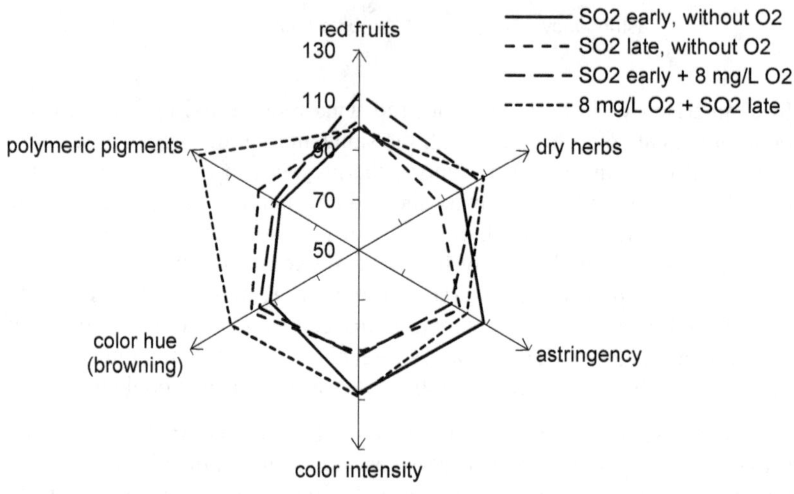

Oxygenation was performed as micro-oxygenation over two months. Subsequently all lots were adjusted to 35 mg/L free SO_2, filtered, and submitted to sensory and chromatic evaluation. Data are given as percentage deviation from the mean value of all lots. The following features can be observed:

- Storage under very oxidative conditions (O_2 early before SO_2 late) yielded the highest level of polymeric pigments, color intensity, and browning accompanied by a

[49] MLF = malolactic fermentation

modest expression of varietal aroma described as "red fruits" as a result of over-oxidation. Astringency was not significantly affected.

- The very reductive storage (SO_2 early, without O_2) also produced a low intensity of fruity varietal aroma but the highest astringency scoring. This behavior is characteristic for most cultivars stored under such conditions. Furthermore, polymeric pigments and browning were lowest, whilst color intensity was surprisingly highest. The contrast between high color and low polymeric pigments known to enhance red color is a contradiction in most varieties but fairly frequent in Pinot noir characterized by low levels of oxygen-sensitive anthocyanins. Early SO_2 addition protects them against oxidative degradation.
- Regular storage conditions (SO_2 early + O_2 afterwards) used in many traditional Pinot noir growing regions resulted in highest intensity of fruity varietal aroma in conjunction with lowest astringency and lowest color intensity.
- Delayed SO_2 addition post-fermentation without oxygen uptake (SO_2 late, without O_2) yielded a wine with intermediate scorings of most parameters evaluated.

It is obvious that the chronological sequence of post-fermentation SO_2 addition and oxygenation has a significant impact on the sensory profile. It can become a valuable instrument for the differentiation of types of wine of the same variety. Its influence increases as the contents of total phenols and suspended yeast lees decrease, and vice-versa.

The phenomena and relationships illustrated above apply to all grape varieties, though with some fluctuations. The total phenol content is the most decisive criterion for estimating how much oxygen a red wine can withstand or benefit from. Low-phenol red wines (< 2,000 mg/L GAE total phenols) rarely gain in quality with deliberate oxygen supply that exceeds the amounts they take up in any case during common post-fermentation cellar treatments (Chapter 8.1.3). In this context it is of little relevance whether their low phenol content results from varietal characteristics, origin, climate, growing conditions, or too short the maceration period. However, such wines exist, and they are not necessarily inexpensive.

In contrast, most high-phenol red wines benefit from retarded SO_2 additions as long as microbial risks are kept under control. Under these conditions, they can even benefit from extensive oxygen uptake as it occurs during barrel aging over one or several years (Chapter 10) without requiring any SO_2 additions. Many great red wines are produced in this way. However, they call for regular sensory monitoring so that one can intervene in time before irreversible over-oxidation occurs.

Sensory signs of over-oxidation during bulk storage

The typical signs of early and sometimes irreversible over-oxidation include:

- a strong smell of acetaldehyde,
- a decrease of fruit aroma,
- losses of anthocyanins without an equivalent increase of polymeric pigments,
- losses of color or the appearance in conjunction with brick-red or brown color hues,

- dry tannins and a meager body on the palate as a result of excessive polymerization or cloudiness caused by tannin precipitation.

When these signs are observed, the time has come to add SO_2, making sure that some free SO_2 will be left after a couple of days. Additionally the wine should not be exposed to any more oxygen.

From an analytical perspective, these changes are frequently reflected in a loss of all phenolic fractions.

Oxidations under reductive conditions

As shown above, the quality of an oxidation depends on favoring a stabilization process, which relies on acetaldehyde production and consumption, and by limiting phenomena of quality degradation as much as possible. The latter can be observed in two cases:

- The oxygen supply rate is too high with regard to the entirety of oxygen scavengers the wine momentarily contains. Then excess peroxide and oxygen radicals resulting therefrom affect the most fragile wine compounds.
- The total amount of oxygen supplied is too high with special regard to its total phenols content.

Adjusting oxygen additions to the wine phenolic profile as well as the utilization of other reducing substances allows us to limit the unwanted effects of oxygen and over-oxidation.

Red wines with medium to high levels of total phenols (> 2,500 mg/L GAE) are able to consume appreciable amounts of oxygen (20-50 mg/L O_2) in a beneficial way without previous SO_2 addition as long as they still contain clearly visible amounts of suspended yeast lees (> 100 NTU). The partial consumption of oxygen by non-phenolic substances such as yeast lees and, to a minor degree, also by reducing amino acids they release helps protect tannins, color, and aroma against the effects of over-oxidation. In this context, post-fermentative yeast lees take on the role of a redox buffer. Therefore, measures such as micro or macro-oxygenations as well as aging in barrels are preferably taken when the wine still contains yeast lees. This approach corresponds to what is designated as "oxygenation under reductive conditions" (Vivas 1999).

Practical hints

Red wines with total phenols not exceeding 2,000 mg/L GAE and designed to represent the fruity style should display moderate levels of free SO_2 ranging from 20 to 30 mg/L before they are subject to oxygen uptake by splash-rackings, barrel aging, filtration, or cold stabilization. This applies in particular when they have already been filtered. Higher free SO_2 levels during bulk storage are counter-productive as they hamper the maturation process (Pontallier and Ribéreau-Gayon 1983). The primary role of free SO_2 in this kind of red wine consists of scavenging intermediate peroxide, which would be responsible for the degradation of fruity varietal aromatics. Light red wines with low total phenol levels usually do not benefit from more oxygen than they inevitably pick up upon rackings, blendings, and filtrations.

On the other hand, moderately oxidative storage conditions enhance color intensity and stability. Preservation of fruity aromas and color optimization leads to a conflict of interests, which must be pondered. Extreme measures of redox management are always detrimental to overall quality.

An alternative option for harmonization of oxygen-sensitive red wines is a short-term storage at elevated temperature under exclusion of air (Somers and Pocock 1990). When doing so, temperature should not exceed 30° C (86° F). Under these conditions, warmth fosters direct polymerization of phenols without acetaldehyde bridging (Chapter 2.2.3). However, this process of non-oxidative polymerization integrates less anthocyanins and tends to generate long-chain polymers that can become more astringent than before. Furthermore, this approach involves some microbial risks, thus increasing the requirements on regular sensory controls.

7.4.2. Total phenols and oxygen sensitivity after wine clarification

In the previous section little differentiation was made as to whether oxygen is consumed in the presence of yeast lees or by clarified wines, although this can make a crucial difference in the sensory outcome.

After filtration, phenolics and sulfur dioxide are the primary oxygen acceptors. However, in high-phenol red wines, SO_2 loses its importance as a reducing agent since phenolics assume that role in an increasing or complete manner. Extremely phenolic wines with more than 4,000 mg/L GAE total phenols can be sensorially stable and long-lasting without any signs of oxidation even though they have never been supplied with any SO_2. In truth, such wines are rare, and care should be taken to check total phenols before one opts for bottling without free SO_2 (Chapter 12.4.3). The total phenol content determines whether the pendulum should swing more in one or the other direction between oxidative or reductive storage.

Screening the sensory response of filtered or even bottled wines to oxygen consumption is an instructive undertaking to better understand the importance of total phenol level. For that purpose, a selection of recently bottled red wines of various cultivars, origins, and total phenol levels were provided with 0, 10, and 20 mg/L O_2 from an air-containing headspace that was sized accordingly (Chapter 8.4). This approach is based on the fact that 1 mL of air contains 21 % oxygen, with 1 mL oxygen weighing 1.43 mg. When the oxygen provided was completely consumed after three months' storage of the tightly resealed bottles, sensory evaluation by profile analysis[50] of almost all conceivable sensory attributes took place.

[50] Profile analysis provides a detailed and relatively precise picture of a wine's sensory profile regardless of any hedonic quality considerations. It uses trained experts to quantify the intensity of perception of characteristic sensory attributes with the help of intensity scales. Samples references made of natural or chemical substances are used to standardize taster perception with regard to quality and intensity and to define the upper and lower end of the scales. In contrast to quality ratings, hedonic attributes are not considered. Sensory data obtained undergo statistical treatment to check for significance of differences between samples and for panelists' reliability and reproducibility.

As a consequence of the heterogeneous phenolic composition of the wines included in this experiment, the gustative and olfactory changes induced by the oxygen supply could obviously not show any clear trend. Therefore, the average of the relative standard deviations across treatments of all sensory parameters were calculated for each wine. It provides a measure of how strongly a wine responds to oxygen in taste and smell. Hence, it is designated as 'oxygen sensitivity' in this context. It does not give any information as to whether the oxygen treatment changed a wine for better or worse.

Figure 60: Relationship between total phenol content and sensory oxygen sensitivity in red wines after consumption of 10 and 20 mg/L O_2. Oxygen sensitivity = average of the relative changes across treatments of all sensory parameters.

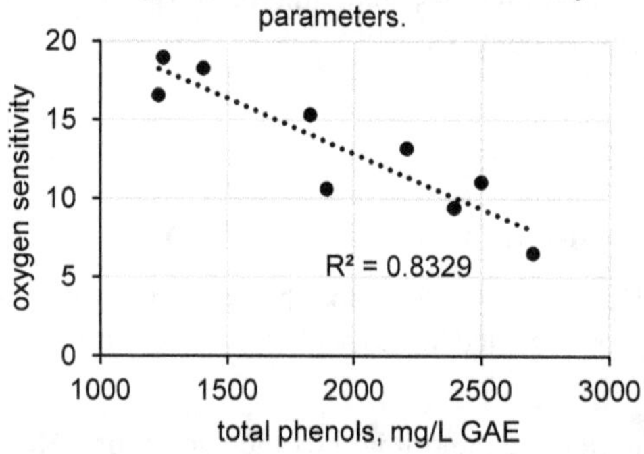

Figure 60 shows the relationship between total phenol content and oxygen sensitivity evaluated in this way. There was a significant negative correlation with r = -0.94 (R^2 = 0.83). The result was unambiguous: The higher the total phenol content, the less red wine responds to a given amount of oxygen and the more oxygen it requires for maturation and harmonization. Furthermore, it stresses the meaningfulness of total phenols as an analytical parameter that is easy to measure.

As has been shown in this chapter, understanding the interactions between total phenol content, oxygen, and sulfur dioxide is essential for a successful redox management of red wines. The craft skills of redox management consist of finding the perfect balance between

- reactions of oxidation by oxygen supply and delayed SO_2 additions, and
- reactions of reduction accomplished by phenolics, yeast lees, and SO_2.

Ultimately, redox management is inextricably linked with tannin management. The way it is performed is the logical consequence of grape quality, harvest decisions, and vinification options taken on the crush pad. It can be backed up by analytical data which provide support to sensory evaluation, but cannot replace it.

7.4.3. Why some red wines do not improve upon deliberate oxygen supply

In contrast to white wines, red wines are stored and processed under conditions that allow a significantly higher oxygen uptake. The oxidative polymerization of tannins and anthocyanins it induces is considered to enhance and to stabilize color, to smoothen tannins, to generate more fullness on the palate, and to create a more complex aroma profile. This is a general doctrine. Furthermore, it is widely acknowledged that the amount of oxygen a red wine requires depends on its phenolic strength, which can be roughly assessed by total phenol measurements.

As outlined throughout this book, total phenols in young red wines are basically made up by tannins and anthocyanins, which react among themselves via polymerization to produce polymeric pigments (tannin + anthocyanin), higher molecular weight tannins (tannin + tannin), or oligomeric anthocyanin combinations (anthocyanin + anthocyanin). Obviously, it depends on the initial tannin-anthocyanin ratio which one of these three types of reactions is predominant. They lead to variable sensory consequences when the oxygen uptake exceeds a critical limit. A large part of these differences can be explained by anthocyanin chemistry, which differs in some respects from the chemical properties of uncolored flavonoids.

Some more enological details about anthocyanins

As indicated in figure 36, anthocyanins are strong DO consumers and the most readily oxidizable flavonoid phenols. However, most of them are not vicinal phenols. Therefore, they do not generate intermediate peroxide when they oxidize. Instead, their oxidation is caused by scavenging peroxide produced upon oxidation of other phenols. This reaction causes rupture of their heterocyclic C ring yielding colorless compounds (Jurd 1966).

Peroxide is known to destroy fruity varietal aroma. As a result, anthocyanins protect aroma compounds and colorless tannins from oxidation to some extent. The deficiency of anthocyanins in lightly colored wines explains why the aroma of such wines is more sensitive to oxidative degradation. In contrast, their abundance in strongly colored wines contributes to the aroma stability of these wines.

Furthermore, since anthocyanins are not vicinal diphenols, they also cannot oxidize to quinones able to daisy chain via regenerative polymerization (Chapter 7.2). Thus, they terminate the polymer chain. As a consequence, the higher the anthocyanin proportion in the tannin-anthocyanin ratio, the smaller the resulting polymers. The tannin-anthocyanin complex of strongly colored wines tends to show lower degrees of polymerization.

The practical significance of the tannin-anthocyanin ratio

This said and returning to the three kinds of polymerization identified above, the aging behavior of red wines can be tentatively classified into three categories according to table 8. It must be pointed out in this context that this is really a very superficial classification, which by its nature does not take into account additional storage variables such as SO_2 regime and temperature.

Table 8: Types of phenol polymerization in red wines as affected by their tannin-anthocyanin ratio.

Kind of polymerization	Tannin-anthocyanin ratio	Type of wine	Sensory results of increased oxygen consumption
tannin + anthocyanin	balanced	Red wines displaying a balanced tannin-anthocyanin ratio occurring in most cultivars.	Rather stable by smell, taste, and color during aging, relatively resistant to oxidation and well suited for aging.
tannin + tannin	high	Red wines with light color and medium to strong tannins, e.g. Pinot noir.	Tendency to oxidative aging ('dry herbs') by smell, increasing astringency, browning.
anthocyanin + anthocyanin	low	Strongly colored red wines lacking tannins as obtained from many hybrid and teinturier cultivars and some V. vinifera varieties.	Rather stable by smell but decrease of color, anthocyanins, and volume on the palate.

Whilst total phenol content is the most crucial parameter determining how much oxygen is required or provided during processing and aging of red wines, its informative value is further complemented by the tannin-anthocyanin ratio. There are findings that suggest that the initial tannin-anthocyanin ratio of young wines is even more important for the final tannin composition than oxygen uptake (Garrido-Bañuelos et al. 2019). Therefore, this issue deserves a more detailed consideration with regard to the different ratios that can occur.

High tannin-anthocyanin ratio

A high tannin-anthocyanin ratio occurs when there are not enough anthocyanins available, so that tannins react preferentially with themselves to produce brown polymers displaying a harsh and drying astringency. This applies in particular under oxidative storage conditions or when there is a surplus of seed-derived tannins, which polymerize preferentially between themselves. This can become an issue when lightly colored varieties such as Pinot noir are aged in barrels.

Low tannin-anthocyanin ratio

In contrast, when there is a lack of tannins altogether with a simultaneous surplus of anthocyanins, the latter tend to react with each other or to produce colorless oxidation products. Such wines do not improve by aging. Conversely, their astringency and bitterness even increase when their chemical age[51] increases under conditions of oxidative aging, whilst there is a strong positive correlation between original anthocyanins

[51] Chemical age is an analytical index putting into relation anthocyanins and polymeric pigments using spectral measurements (Somers and Evans 1977).

and overall quality in that kind of wines (Jackson et al. 1978, Schneider 2005 a). Hence, enological measures that promote polymerization such as strongly delayed SO_2 addition, micro-oxygenation, barrel aging, and post-bottling maturation generally do not improve overall quality of these wines but rather impair it.

The relatively low tannin levels of these wines are not able to incorporate substantial amounts of oxygen without losses of quality. Nevertheless, such wines can be treacherous when they are young. At that stage, they might display considerable volume on the palate provided by their high anthocyanin levels (Chapter 5.3.2). After these anthocyanins have undergone accelerated degradation under the semi-aerobic conditions of barrel aging, there is not enough tannin available to take over their role in providing volume and fullness. Lastly, one must accept that wines with more anthocyanins than tannins do generally not improve by aging. They are better suited to be consumed as young and fruity.

Balanced tannin-anthocyanin ratio

Ultimately, a balanced, intermediate tannin-anthocyanin ratio is what most winemakers strive for. Anthocyanins make tannins taste softer. Furthermore, they improve their solubility as they combine with them and increase the amount of polymeric pigments that are retained in solution (Singleton and Trousdale 1992). As a result of this reaction, they decrease during storage with a half-life of approximately one year. Therefore, it is advantageous to start red wine aging under semi-oxidative conditions, such as in barrels, with an initial anthocyanin content that is sufficiently high and adapted to the tannin content. Under these conditions, the reaction 'tannin + anthocyanin' is predominant; the wine develops well during maturation and aging.

Since tannins are difficult to quantify in day-to-day winery operations, this ratio can be replaced approximately by the total phenol-anthocyanin ratio (Chapter 3.2.3). It complements the informative value of total phenol measurements and contributes to better adjust aging conditions and oxygen supply than total phenol data alone.

8. Means of oxygen supply

Most red wines are deliberately supplied with oxygen with the primary objective of modifying the chemical constitution of their tannin-anthocyanin complex in such a way that it elicits the desired mouthfeel properties. In doing so, we must distinguish between passive oxygen uptake and active oxygen incorporation. Passive oxygen uptake can occur through the wine surface, through certain container walls, and during wine treatments inherent to any winery operation. It is the result of ambient oxygenation and often underestimated in its extent. In can be complemented by techniques of active oxygen incorporation, of which micro-oxygenation is the most opinion-leading one. It is presented in detail without neglecting its empirical basis and other ways of active oxygen supply, some of them being closer to macro-oxygenation. Ranges of oxygen mass incorporation are given for each approach, and further sensory results are presented. Since these results are difficult to predict, a method for determining the optimum oxygen supply rate for each wine is proposed. It is based on bench trials and their sensory evaluation.

8.1. Passive oxygen uptake

8.1.1. Passive oxygen uptake through the wine surface

Oxygen supply is commonly associated with specific measures such as micro-oxygenation and barrel aging, but the reality is more complex. We must distinguish between techniques of active, deliberate oxygen supply and processes of passive oxygen uptake. The latter occurs in any winery without people always being aware or caring about it. Albeit difficult to quantify, it substantially contributes to the amount of oxygen wines take up post-fermentation.

One of the ways wines can take up appreciable amounts of oxygen is through the wine surface, which can be undisturbed or agitated. An undisturbed, static surface occurs every time that wines are stored in containers that are not completely topped. Upon stirring of such a wine, the surface becomes agitated. Agitated surfaces can also be observed when wine flows through hoses, filters, and pumps containing air pockets when they have not been carefully vented.

All gases dissolve in liquids to a certain degree. When a liquid becomes saturated with a gas, no more of that gas can be dissolved in the liquid. At normal pressures, wine can absorb up to 8 mg/L O_2 at 20° C (68° F) and even 11 mg/L O_2 at 10° C (50° F). Table 9 shows how oxygen solubility increases when temperature decreases. These data represent the saturation concentrations of dissolved oxygen (DO) at different temperatures. They are average values since they slightly depend on alcohol content.

Table 9: Solubility of oxygen in wine as affected by temperature.

Temperature	Solubility, mg/L O_2
0° C / 32° F	14.5
10° C / 50° F	11
20° C / 68° F	8
30° C / 86° F	7

Although a cold wine can hold more oxygen in solution than a warm one, this general rule does not provide any information about the speed at which wine takes up oxygen from the atmosphere. Only rough estimates of this essential value are available. The reason is that oxygen dissolved in the wine is gradually consumed by chemical binding. After its disappearance, it cannot be measured any more.

The importance of liquid movement

For precise measurement of the oxygen uptake through the surface, a wine-like model solution (13 % ethanol, pH 3.5 with tartaric acid, 1 mg/L Cu^{2+}) was sparged with nitrogen to remove all DO, supplemented with 200 mg/L ascorbic acid, and stored at different temperatures in open wide-necked bottles of 1,000 mL with a defined surface area. Under these conditions, oxygen picked up is spontaneously consumed by ascorbic acid in a stoichiometric relationship. The decrease of ascorbic acid over time allows direct calculation of the amount of oxygen (in mg O_2/m^2•h) that passed through the liquid surface. Figure 61 reports some of the results.

Figure 61: Diffusion of atmospheric oxygen through the surface of a wine model solution at normal pressure. Effect of surface agitation and temperature.

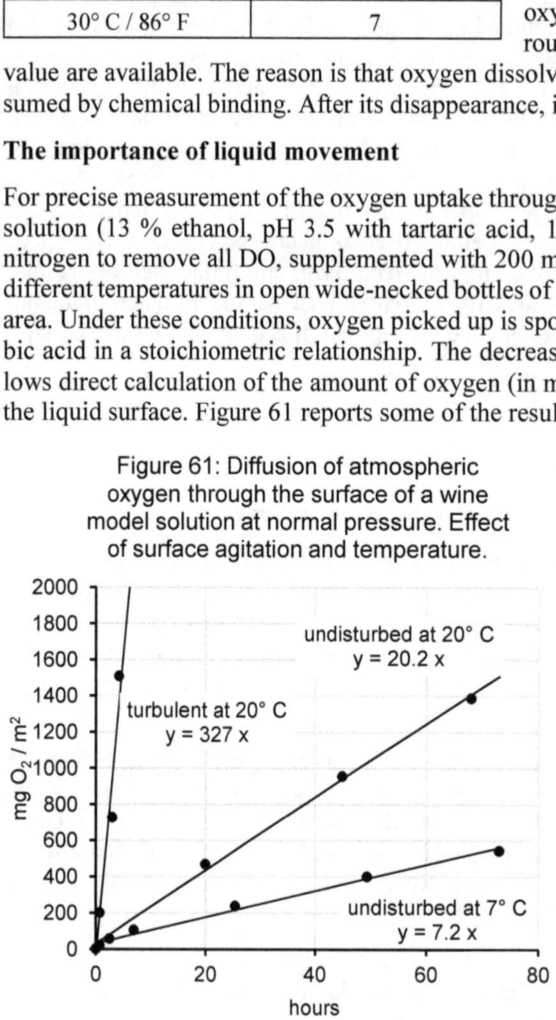

The undisturbed liquid stored at 20° C picks up 20.2 mg O_2/m^2•h or 484.8 mg O_2/m^2•day. This value increases 16-fold when the same solution at 20° C undergoes constant stirring to generate an agitated surface. The latter value is illustrative and of course depends upon the intensity of stirring and surface movement.

The impact of temperature is also crucial. The better oxygen solubility at low temperatures might lead to the careless assumption that oxygen diffusion from the atmosphere into the liquid also increases under these conditions. However, direct comparison of the diffusion rates of the undisturbed aliquots at 7 and 20° C proves the opposite (Figure 61): Decreasing temperature from 20° C to 7° C lowers the diffusion rate 2.8 times from 20.2 to 7.2 mg $O_2/m^2 \cdot h$. This result might appear surprising at the first glance but has a simple physical explanation.

The uptake of a gas by a liquid proceeds in two steps:

1. Passive diffusion of the gas from the atmosphere through the liquid surface into the upper boundary layer of the liquid. This process is almost spontaneous and only active over a small distance. At a given pressure, its rate depends on temperature, the surface tension of the liquid, and the level of saturation of the gas in the liquid. It is described by Henry's law[52]. Dissolved carbon dioxide lowers the oxygen diffusion rate through the liquid surface (Devatine et al. 2011).
2. Transportation of the gas into the liquid's interior. This process follows Fick's law[53]. Its driving force is the continuous renewal of the boundary layer. This can be caused by thermal convection or by mechanically induced agitation. In both cases, low-oxygen liquid is transported from below towards the surface, where is absorbs more oxygen. This step becomes more and more rate-determining with increasing distance from the surface.

As a basic principle, the diffusion of a gas within the liquid decreases when the temperature decreases. Although the solubility of oxygen in wine is higher in the cold, low temperature slows down its distribution into the undisturbed wine and, consequently, all downstream reactions of oxidation. However, when coldness and liquid movement interact one with another, oxygen uptake reaches extreme values. This is in particular the case when bitartrate crystal stabilization is performed using the contact process without previously purging the container with an inert gas. It is a typical situation in which many red wines are over-oxidized.

Conversely, oxygen uptake in the cold is reduced when the liquid surface is undisturbed as oxygen diffusion within the liquid slows down. In this way, the effect of a higher oxygen solubility at low temperature is cancelled out. Hence, wines not completely topped and stored in the cold oxidize more slowly, though oxygen solubility is higher. In real-world winemaking, the opposite is often observed: Red wines stored with ullage rapidly oxidize when they are stored at ambient or elevated temperatures.

Table 10: Oxygen transfer coefficients (in mg $O_2/m^2 \cdot h$) through the liquid surface of wine model solution.

~ : depending on the agitation intensity

	20° C / 68°	7 °C / 45° F
stagnant surface	20.2	7.2
agitated surface	~ 327	~ 950

The time has come to rethink traditional thoughts telling us that wine stored in the cold picks up more oxygen and oxidizes faster. It is a doctrine that's only valid when wine is kept moving in a manner allowing even dispersion of dissolved oxygen in the whole liquid volume.

[52] Henry's law states that the amount of dissolved gas in a liquid is proportional to its partial pressure.

[53] Fick's law relates the rate of diffusion to the gradient of concentration. It postulates that the flux goes from regions of high concentrations to regions of low concentrations, with a magnitude that is proportional to the concentration gradient.

The data set used to obtain figure 61 allowed calculation of the oxygen transfer coefficients through the liquid surface in mg O_2/m^2•h. For a better overview, some of them are summarized in table 10.

Surface area is more important than headspace volume

In practical cellar work, it is vital to be aware of high and barely controllable amounts of oxygen a wine can pick up from the headspace when the container is not completely topped and its headspace not purged with inert gas. This oxygen uptake is a diffusion process exclusively controlled by temperature and the wine surface area. The headspace height or the missing wine volume are not decisive.

Expressed as mg/L O_2, wine stored or processed under an air-containing headspace in small containers has to withstand a many times higher oxygen load than wine in large containers. In the same line of thinking, short, wide tanks allow for more oxygen uptake as they display more exposed wine surface than tall slender tanks.

The considerations above take us to the notion of what is called "specific surface area", expressed in cm^2/L or $inch^2/gal$. Its meaning becomes understandable by a simple example: A surface area of 1 m^2 above 1,000 liters imparts ten times more oxygen (in mg/L) to the wine than above 10,000 liters.

Headspace blanketing with inert gas is less perfect than assumed

In the hope of minimizing wine-air contact and to avoid microbial wine spoilage, headspace inertization with an inert gas attempts to drive out headspace oxygen and maintain an inert gas layer above the wine surface. In order to prevent the growth of aerobic microorganisms on and near the wine surface, the oxygen concentration must be reduced from 20.9 % found in air to 0.5 % or less at the wine surface. This is not an easy task because effective blanketing requires more than just shooting some inert gas into the headspace until it feels well. Highly turbulent gas flow rates create a churning effect causing the gas to mix with the air. Thus most of the gas is lost to the outside of the vessel, while the purity of the gas inside becomes diluted. Therefore, effective headspace inertization requires an appreciable amount of inert gas corresponding to 3-5 times the headspace volume. Since these technical details are not complied with in most winery settings, one can assume that a red wine also picks up some oxygen despite of headspace blanketing when it is stored with ullage.

Historical solutions

In past times, a layer of olive oil or paraffin was sometime dispensed on top of the wine stored in half full barrels or amphorae. This was thought to protect wine from oxygen pickup and actually can do so to a certain extent, but it creates other problems. The use of more recent antimicrobial agents such as allyl isothiocyanate disks floating on the wine is prohibited in most countries, does not impede oxygen pickup, and would impart unpleasant off-odors.

Variable capacity tanks

Stainless-steel variable capacity tanks with a floating lid are generally used to store smaller wine volumes of up to 2,000 liters (529 gallons). Their lid is simply inserted into the tank and then allowed to float on the wine surface. The lid is equipped with an inflatable bladder around its circumference. This bladder is inflated using a PE tube connected to a hand air pump, and is thus supposed to create an airtight seal on the inner circumference of the tank to protect the wine from air. A pressure gauge indicates the bladder pressure, which can be adjusted at any time to maintain the seal's tightness. The pump has a release valve to deflate the bladder when the lid is to be removed.

This apparently perfect solution has been conceived for storage of variable wine volumes without oxygen uptake from the headspace through the liquid surface. It would even work for that purpose if the commonly employed vinyl bladders and the air pump gasket did not have a common problem in forming a good seal in the long term. Therefore, these tanks usually do allow for some oxygen uptake during extended storage. The blind spot is that one cannot predict how much.

8.1.2. Passive oxygen uptake through container walls

Wineries use storage containers made of different materials displaying variable oxygen permeation rates. When they are used consciously, they facilitate meeting the different oxygen requirements of red wines. However, storage containers are often used quite carelessly without taking into account their actual oxygen permeability.

Wooden containers of the size of a standard barrel provide the wine with considerable amounts of 10 to 20 mg/L oxygen per year (Vivas 1999, del Alamo-Sanza and Nevares 2014). This oxygen uptake is driven by the partial pressure difference of the atmospheric oxygen outside the barrel and the low partial pressure of oxygen inside, a low pressure originated by the continuous consumption of oxygen by the wine.

Materials such as stainless steel and glass are the other extreme and absolutely impermeable to oxygen. This qualifies them for long-term storage of red wines ready to be bottled on condition that the containers are really topped to the brim. The various plastic materials display large differences with regard to oxygen diffusion. The diverse polyethylenes (PE) are highly permeable to atmospheric oxygen with permeability rates depending on the PE specification, while the oxygen diffusion rate through the so-called barrier polymers is lower by a factor of approximately 1,000. Porous plastic tanks made of HDPE (high density polyethylene) used as maturation vessels are claimed to replicate the oxygen permeation rate of wooden barrels. Details are given in chapter 8.3.7.

The specific oxygen transmission rates (OTR) through the different container materials are fairly well documented for specific conditions, but it is useless to know them since the actual oxygen transmission is affected by additional variables such as

- The wall thickness in mm as well as the wood porosity and moistness in the case of barrels,
- The coating of the inside of the wall with dirt, yeasts, bitartrate crystals, etc. As a consequence, the oxygen uptake during storage in old barrels is about half (5-10

mg/L O_2 per year) of that occurring in a new one (10-20 mg/l O_2 per year) (Chapter 10.2). The intensity of the cleaning between wines plays an additional role.

- The wine volume that absorbs the oxygen provided by a given container wall surface area. This leads to the surface-volume ratio. It increases exponentially with decreasing volume and it easy to calculate for cubic containers, but more difficult to quantify for barrels. Figure 62 illustrates this relationship for barrels. Under otherwise identical conditions, the oxygen uptake (in mg/L O_2) in a barrel of 300 L is only 90 % and in a barrel of 1,200 L only 60 % of that in a standard barrel containing 225 L. The release of flavor-active wood compounds into the wine is reduced to the same extent.

Figure 62: Inner wall surface area size of barrels as affected by their volume.

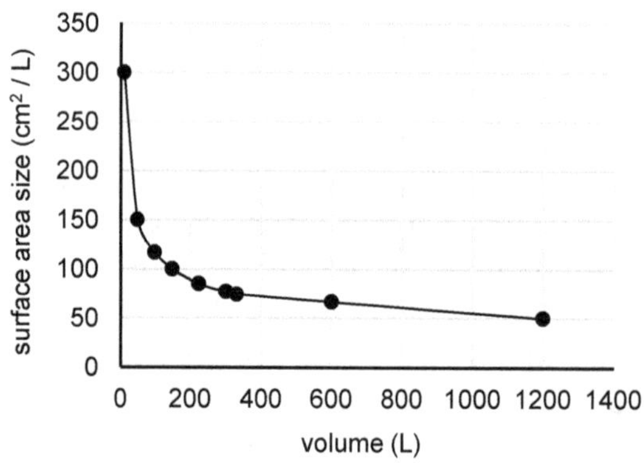

Barrels larger than standard barrels are used in some wine growing regions with the purpose of reducing oxygen uptake and the flavor impact of wood. Conversely, storing wine in a small container with a high OTR easily causes over-oxidation, especially when there is additional oxygen uptake from an unnoticed ullage. Therefore, such containers are not suitable for storage of oxygen-sensitive wines.

8.1.3. Passive oxygen uptake upon wine treatments

Apart from container headspace and diffusion through container material, wines also pick up oxygen when they are treated in the winery by racking, fining, filtration, stirring etc. because there are variable amounts of air in the machinery and pipes. However, whilst the wine industry collects significant product and process data, it cares little about the accidental oxygen uptake that comes with these routine winery operations. In contrast to the practices of the brewing industry, numerous producers do not have the slightes idea of the magnitudes involved, although suitable instruments are available to measure DO. Recording DO for various cellar activities and identifying where oxygen uptake is likely to occur would be a worthwile undertaking in many winery settings.

Table 11 shows the large range of oxygen uptake measured during different treatments and storage conditions (Vidal et al. 2001, 2003, 2004 a, Schneider 2005 b, Valade et al. 2006, 2007, Castellari et al. 2004, Calderón et al. 2014).

Table 11: Oxygen uptake caused by various wine treatments in small and medium-sized wineries.

Treatment	O_2 uptake, mg/L
transfer / racking into top-filled tanks	2 - 3
transfer / racking into bottom-filled tanks	0.1 - 1
transfer with loose suction hose	5 - 7
centrifugation	0.3 - 2
kieselguhr filtration	0.1 - 2
pad filtration	0.1 - 1
cross-flow filtration	0.2 – 1.5
membrane filtration	0.1 – 1.5
electrodialysis	0.3 – 1.3
stirring for fining or blending	1 - 2
cold stabilization using the contact process	1 - 3
transportation in untopped truck containers	3 - 5
storage in large wooden casks, per year	5 - 10
storage in barrels (225 L), per year	20 - 40
bottling	0.5 - 3.0

The large variance of oxygen uptake for each treatment rules out the specification of more precise data or average values. It can be explained by three phenomena:

1. The conditions under which the various equipments are used are highly variable. The dissolution of oxygen occurs preferentially at the beginning of a transfer or treatment when the pipes are still filled with air, and at the end of the process when the pipes are drained. At these stages, highest oxygen uptake takes place when a wine is filtered due to a larger liquid surface, turbulent wine flow, and filter pressure. This applies in particular when oversized equipments are used.
2. The wine volume plays a crucial role: The global oxygen uptake during a treatment depends on the wine volume in which the oxygen is dispersed. Thus, small wine volumes are exposed to a higher oxygen uptake, expressed as mg/L O_2 measured immediately after treatment, than large wine volumes. Small-scale winery operations experience higher oxygen uptake rates than large-scale facilities.
3. The oxygen pickup also depends on carbon dioxide dissolved in the wine at the moment when the treatment takes place (Dévatine et al. 2011). When a container is filled from the top, more oxygen is picked up when the wine has already lost the largest portion of its carbon dioxide resulting from fermentation. In contrast, less oxygen is picked up by young wines because they release CO_2 upon entering the receiving tank, which purges the headspace and removes a part of the oxygen it contains. When young wines with significant CO_2 are transferred by filling the

container from the bottom, there is almost no measurable oxygen uptake because oxygen is largely stripped by escaping CO_2 or rapidly consumed by post-fermentation yeast lees.

Oxygen pickup upon one single treatment might be negligible. However, the key question concerns cumulative oxygen pickup during many consecutive treatments. In this context, practical experience and the cellar staff's craftsmanship are essential. They can be complemented by DO measurements.

Monitoring process-related oxygen uptake

The measurement of the process-related oxygen uptake for each treatment such as pumping over, filtration, or suction of air, etc. can provide valuable information when it is performed on a stage-by-stage basis. Stage-by-stage control means that values are measured at consecutive places within the process chain, for example in the starting container, after a loose suction port, at the filter outlet, and in the receiving tank during its filling. This kind of measurement has to be performed on the spot in the winery and in a timely manner. Modern portable DO-meters with a precision of better than 0.1 mg/L DO provide valuable support for real-time measurements (Nevares et al. 2009). The total amount of oxygen is not to be underestimated, though it is highly variable and depending on the winery and the way winery staff is used to work.

Sampling, transportation, and storage of samples frequently distort DO readings. Storing or shipping samples to an external lab is even less suitable as DO decreases over time by chemical binding.

The time factor plays a particularly important role in the recording of process-related oxygen uptake. The oxygenation of the wine takes place in two phases:

1. The dissolution of atmospheric oxygen in the wine. The oxygen dissolved as a gas (DO) can be measured analytically, but it has not yet triggered any sensory consequences.

2. The binding of DO to wine constituents. The bound oxygen has disappeared and cannot be measured any more. The sensory consequences of oxygen uptake arise.

Two different reaction models are derived from this behavior, which lead to two different situations in practice:

1. The dissolution of oxygen in the wine is faster than its binding; an increase in DO is observed. This case occurs during process-related oxygen uptake such as pump-overs.

2. The dissolution of oxygen is slower than its binding; DO can be hardly measured. Such a case is will occur during micro-oxygenation, storage of wine with ullage or in containers made of oxygen-permeable materials such as wooden barrels and maturation tanks.

The analytically measured DO concentration always corresponds only to the momentary difference between oxygen uptake and binding. However, it does not provide any information about the oxygen uptake that the wine has experienced in the course of its history. It is therefore poorly suited for process control if an active oxygen supply is brought about during storage with ullage, in maturation tanks, or wooden barrels.

Among the measures of active oxygenation, micro-oxygenation (Chapter 8.3) holds a special position. Using pure oxygen, it allows one to adjust the oxygen supply to the binding of DO. Thus over-oxidation due to accumulation of DO, peroxide, and acetaldehyde (Chapter 7.3) can be avoided, provided that the process is regularly monitored. The DO concentration measured during micro-oxygenation should in no case exceed 0.5 mg/L.

8.2. Active oxygen supply by macro-oxygenation

Oxygen pickup upon current wine treatments such as transfers, finings, filtrations, etc. goes together with losses of carbon dioxide contained in young wines, in particular in those stored in stainless steel tanks. CO_2 is objectionable in almost all red wines and has to be removed when in excess, at latest before bottling. Wine treatments strongly contribute to its removal.

Luckily, gentle wine processing as required for oxygen exclusion and preservation of volatile aromatics and possibly some CO_2 in fruit-driven white wines is not a major concern in red wines. They have already lost their highly volatile aroma compounds during fermentative maceration or will lose them during barrel aging. Hence, red wines are well able to withstand or even require rugged treatments, which would be considered 'wine torturing' apt to spoil any sensitive white wine. Additionally, they provide some oxygen red wines require and white wines react negatively to. These facts demonstrate why there is a crucial difference between red and white wine handling post fermentation.

Until the CO_2 in red wine is depleted to an acceptable level, the passive oxygen uptake occurring upon treatments under the typical conditions of small and medium-sized wineries with their limited wine volumes can satisfy the oxygen demand of simple low-phenol red wines. Under these conditions, additional pump-overs, i.e. splash rackings, are a reliable method of removing any remaining CO_2 and introducing additional oxygen.

In large-scale operations, the CO_2 released during pump-overs leads to a strong reduction of the oxygen content in the headspace of large, tall tanks. In addition, the oxygen actually absorbed is distributed in larger wine volumes, which reduces the oxygen supply per volume unit.

If the passive oxygen intake is not sufficient, methods of active oxygen supply by macro-oxygenation can be used. They served well before the advent of MOX, and they continue to do so when used selectively and with an awareness of the oxygen required and the oxygen effectively supplied. They are listed in table 12, including their advantages and disadvantages.

Table 12: Techniques of active oxygen supply by macro-oxygenation, their advantages and drawbacks.	
Procedure	Effects, advantages and drawbacks
Racking / transfer into top-filled containers (splashing); higher O_2 uptake when pipe outlet is equipped with a deflector plate.	High O_2 uptake when draining via open intermediate vessel. High O_2 uptake for all other transfer operations in connection with filling from the top. But: Low O_2 uptake by filling from the top during the 1st racking by CO_2 release from young wines. Sensory results easy to control by tasting after one week.
Sucking air through the loosened suction port of the pump.	Rather high O_2 uptake, but variable and difficult to regulate. Sensory results easy to control by tasting after one week.
Sucking air through a porous sinter tube (aeration tube) on the suction side of the pump.	Rather high O_2 uptake, but variable and difficult to regulate. Sensory results difficult to predict.
Sucking air via a Venturi tube.	Rather high O_2 uptake, depending on pump pressure and wine CO_2 level. Sensory results difficult to predict.

Microbial risks of inadvertent oxygen uptake and active oxygen supply

It is part of standard enological knowledge that increasing storage temperatures promote the growth and activity of all kinds of microorganisms. In connection with dissolved oxygen this concerns especially bacteria of the genus Acetobacter. Increased contents of volatile acidity and ethyl acetate can be the consequence.

In contrast to Acetobacter, the development of Brettanomyces (hereinafter called Brett) yeasts does not rely on the availability of oxygen. In contrast, they feed on a wide range of substrates, with residual glucose and fructose as their favorites, but also on unfermentable sugars such as those released from oak as well as on grape-derived phenolic acids. However, oxygen uptake during barrel aging indirectly favors Brett growth as it decreases SO_2. Thus Brett grows more easily in zones that are poorly protected by SO_2 such as near the wine surface, around the bung hole, and in the pores of the wood. Indeed, Brett is quite sensitive to molecular SO_2. Experience has shown that 0.4 mg/L molecular SO_2 is required to prevent its growth and 0.6 mg/L molecular SO_2 to block its activity. At lower levels there is a risk of contamination, but its occurrence cannot be generalized.

The aforementioned levels of molecular SO_2 are also sufficient to prevent Acetobacter activity, but they are difficult to achieve in high pH wines, in which they would require much more than 60 mg/L free SO_2. At best, the presence of free SO_2 only slightly reduces the risk of their occurrence. Regular topping is the best way to keep Acetobacter under control.

Practical hints

As shown in chapter 7, the interaction of high oxygen uptake, high temperature and high pH has chemical consequences that easily cause detrimental over-oxidation. Additionally, it increases the risk of microbial deviations. Sufficient microbiological safety of unfiltered red wines during macro-oxygenation measures such as deliberately storing with ullage, splashing, or sucking air by venturi designs is only achievable at temperatures below ~10° C (50° F). Low cellar temperatures might slow down wine maturation, but they provide more safety. Good cellar hygiene and frequent topping are additional steps to protect wines from microbial faults.

8.3. Active oxygen supply by micro-oxygenation

8.3.1. Principles

Micro-oxygenation, also known as micro-bullage, is a technique used to add defined amounts of oxygen to wine. For a better understanding of what the prefix 'micro' means in this context, it must be compared to other regimes of oxygen supply. In total there are three of them, which differ in the amount of oxygen supplied and the time period in which it is supplied:

- Macro-oxygenation is defined as the addition of relatively high amounts of oxygen ranging from 1 to 8 mg/L O_2 all at once or in a short period of time not exceeding a couple of hours. It is specifically practiced during primary fermentation to strengthen yeast cell viability and to encourage the formation of polymeric pigments, and during splash-racking of young wines. It generates a temporary accumulation of DO.
- Micro-oxygenation is a process in which amounts of 5 to 50 mg/L O_2 per month are supplied to young red wines in a controlled and more or less continuous manner with the purpose of triggering regenerative polymerization and improving body, structure, color, and fruitiness. It can take one to several months. The O_2 supply rate should be lower than the O_2 consumption rate to avoid noticeable accumulation of DO.
- Nano-oxygenation is the very limited exposure to oxygen that typically occurs post-bottling through closure oxygen transmission. Depending on the closure type, it provides 0.1 to 10 mg/L O_2 per year. No accumulation of DO takes place. It can lead to a harmonization of tannins, in particular in high-phenol wines (Gambuti et al. 2016).

During storage in wooden barrels, the wine experiences a continuous absorption of oxygen from the almost unavoidable ullage around the bung hole and through the wood (Chapter 10.1). The same is true for maturation tanks made of HDPE. In contrast to fining, filtration, and transfers by filling of containers from the top, where a relatively large amount of oxygen is supplied and accumulates in a short time, the oxygen uptake in these vessels is slower than its chemical binding. Therefore, under these storage conditions there is rarely an accumulation of DO above 0.1 mg/L.

After replacing traditional barrels with airtight stainless steel tanks, it was noticed that the red wines matured less or at least differently and that the tannin tasted less harmonious. This observation was first made in an area in southwest France, which produces very tannin-rich red wines from the Tannat variety, and triggered the birth of micro-oxygenation (MOX). It is a special process of continuous oxygenation and has become an opinion-leading instrument for producing red wines worldwide.

MOX has been developed in response to a need to improve the balance and harmony of red wines, especially those with a high total phenol content, and to be a more economic means for wine producers to store wines in tanks, rather than in barrels. Hence, it aims to imitate the slow oxygen uptake and maturation process occurring during barrel aging when wines are stored in tanks, to promote the polymerization of tannins and anthocyanins via an ethyl bridge, and thus to build body and smoothen tannins. For the formation of the ethyl bridge, acetaldehyde is necessary and produced by oxidation of ethanol (Chapter 7.2) by means of oxygen supplied by MOX. The deliberate addition of acetaldehyde during (Sheridan and Elias 2015) or after (Aleixandre-Tudo et al. 2016) alcoholic fermentation would lead to the same effect (Figure 56) if it were permitted by law.

The original concept of MOX was based on the proposition that it would control the amount of oxygen added to the wine, rather than the more random oxygen ingress that occurs in barrels. During the MOX, a precisely measured oxygen supply ideally produces only as much acetaldehyde as is necessary to form the ethyl bridge during oxidative polymerization. Thus, potential negative effects of over-oxidation by accumulation of dissolved oxygen, intermediate peroxide, and acetaldehyde (Chapters 7.2 and 7.3) during discontinuous macro-oxygenation (Chapters 8.1.3 and 8.2) can be avoided. In sensory terms, MOX is claimed to have the following benefits (Gonzáles-San José et al. 2008, Schmidtke et al. 2011, Parpinello et al. 2012):

- increasing suppleness or roundedness of the wine,
- building optimal structure,
- mitigating green and dry tannins,
- improving body and aroma,
- increasing and stabilizing color,
- preventing or removing reduction flavor,
- reducing herbaceous or vegetal notes,
- making red wines drink well earlier, producing a more aged structure in young wines,
- being less expensive than barrel aging when used in conjunction with oak chips.

As a relatively new oxygenation technique, MOX has been adopted with much enthusiasm by some winemakers, even though all the details of the underlying science are not yet completely understood.

8.3.2. Technical implementation

The technical implementation of MOX is conducted such that pure oxygen in the form of small bubbles is slowly and almost continuously added to the wine in the tank. This is done in such a way that ideally the oxygen dissolves completely and cannot escape as gas bubbles through the wine surface. Two methods are used for this purpose:

Semi-continuous injection of oxygen micro-bubbles using diffusers

The oxygen from the steel bottle is fed into a ceramic diffuser with a nominal pore size between 1 and 10 µm, which ensures the desired small bubble size, maximizes the bubble surface area to volume ratio, and minimizes the bubble rise velocity. This diffuser is suspended from the top of the vessel and positioned near its bottom. The oxygen dosage is performed by use of a variable-capacity dosing chamber that is located between two sequentially operated gas valves. It consists of a tube closed at its end on which a movable clamp is mounted and whose position determines the chamber's capacity. Pressurized oxygen is allowed to enter the dosing chamber by the sequential closing of the downstream and opening of the upstream valves. Once the dosing chamber is full the valve operation alternates so that the gas flows into a second chamber, designated as 'expansion chamber', which is connected by tubing to the diffuser. The volume of gas injected depends on the pressure difference between both chambers, which is maintained constant to ensure a stable gas flow. Under these conditions, the number of injection sequences, i. e. the duration between filling and emptying cycles of the dosing chamber along with the chamber volume thus determines the rate of oxygenation, expressed as mL/L oxygen per month. At high dosage rates, the sequential injection of oxygen from the dosing chamber may be closely spaced and thus appear continuous. However, in reality it is a discontinuous process as the oxygen flow into the wine is interrupted during filling of the dosing chamber (Schmidtke et al. 2011, Nevares and del Alamo-Sanza 2016).

Figure 63: Bubble plume formation during MOX with diffusers.

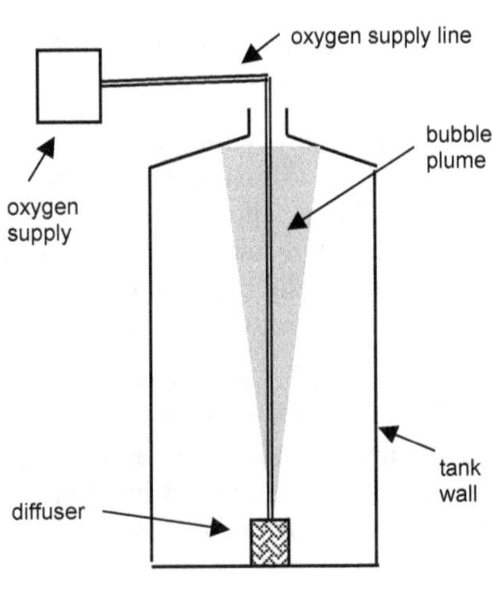

Subsequently, more sophisticated MOX systems have been developed. Instead of volumetrically dosing oxygen, depending on the fine adjustment of the pressure, they do so based on mass, which means in mg/L•month instead of mL/L•month.

The oxygen enters the wine as a stream of fine bubbles creating a bubble plume (Figure 63). Bubble rise velocity has been estimated at around 0.28 m/s, and bubble diameter at 310 to 668 µm at the bubble release point under typical MOX conditions (Schmidtke et al. 2011). The oxygen contained in the bubbles dissolves in the wine as the bubbles rise. At this point the bubbles continuously shrink and finally disappear. Regardless of how the oxygen bubbles are injected, the wine height above the diffuser and the distance to be covered by the bubbles determine the time available for the oxygen to be dissolved in the wine. It is crucial that the dosed amount of oxygen be dissolved during its ascent from the diffuser at the bottom of the tank. Otherwise remaining undissolved oxygen will escape through the wine surface. Insufficient wine-column height, large bubbles, or oxygen application rates that exceed the gas transfer rate from the bubbles into the liquid will result in losses of oxygen and its accumulation in the headspace.

For complete dissolution of the bubbles during their ascent, a minimum gas flow rate, a liquid column of at least 2.5 m (8 ft.) and a diffuser pore diameter of approx. 5 µm are required. The CO_2 content of the wine should be less than 0.5 g/L, because otherwise CO_2 would diffuse into the oxygen bubbles. This would cause the bubbles to rise faster and thus reduce the ascent period available for complete dissolution of the oxygen. Advantages and drawbacks of different dosing devices and diffusers with respect to oxygen dissolution have been described extensively (Pour Nikfardjam 2004, Bartolini et al. 2008).

The bubble plume formed when the gas stream enters the wine is shown in figure 63. It can be explained by a limited lateral oxygen dispersion from the diffuser and results in its uneven distribution in some cases, in particular in large tanks. Differences in DO concentration within the tank cause differences in oxidative reactions and wine composition between specific areas within the wine. Thus, any lack of homogeneous distribution of DO creates a problem of representativeness of sensory or analytical assessments used to monitor the development of the wine during MOX because the results would depend on the sampling point of the tank. This problem can be overcome by mixing the wine with the help of mini submersible pumps or propeller mixers mounted on the tank port. This clearly shows the limitations of bubble-plume devices in large commercial-scale tanks, but it does not arise in small research vessels.

Continuous oxygen diffusion through semi-permeable membranes

The requirement for hydrostatic pressure to retard the ascending movement of bubbles is circumvented by the use of pressurized semi-permeable tubular membranes pervious to oxygen. As these membranes do not rely on the formation of bubble plumes for oxygen mass transfer, they have an additional advantage for the application of MOX to small wine volumes stored in research-scale vessels or barrels.

Plastic tubes allow the oxygen they contain to diffuse so slowly through their walls that no bubbles on the tube surface can be observed. This diffusion is a continuous process while the gas pressure is maintained within the tubes. Thus, in the absence of the micro-bubbles of a 'classic' MOX injection system, there are no longer requirements to ensure their dissolution. Accordingly, the previously discussed considerations regarding the rising height, bubble size and CO_2 disturbances cease to apply.

Typical tubes are made of polydimethylsiloxane or polytetrafluoroethylene, which are hydrophobic materials limiting the formation of bubbles at the membrane-liquid interface. They have a wall thickness of 0.3 to 0.5 mm and a diameter to wall thickness ratio of approximately 10:1. Using such a simple device, the oxygen input can be infinitely adjusted exclusively via the hose length and the differential pressure. There is no temporal switching of the oxygen supply, which can be observed to a certain degree when using dosing devices (Kelly and Wollan 2003).

Theoretically, these tubes can be operated with pure oxygen or air. However, since they are also permeable to carbon dioxide dissolved in the wine, they require renewal of the gases they contain. Otherwise carbon dioxide would accumulate within the tubes, preventing the desired differential oxygen concentration between the wine and the gas inside the tube from being similar to that of the atmosphere. Thus, the oxygen supply rate would rapidly decrease because the tubes would cease functioning under the conditions in which their nominal oxygen permeability was determined. By making both of the tube's ends accessible it is possible to pump air or oxygen through the tube continuously and vent it to the atmosphere to avoid carbon dioxide accumulation. However, in practice, this means that both ends of the tube must protrude from the vessel and be accessible.

Duration and other characteristics of MOX as compared to macro-oxygenation

However the oxygen is dissolved in the wine, the oxygen dosage is performed over several weeks or months in such a way that the oxygen is not supplied faster than it is bound. The oxygen dissolved in wine during MOX should not exceed 0.5 mg/L (Bartolini et al. 2008). A greater accumulation of DO would negate the advantages of the process and lead to over-oxidation. The typical oxygen dosage during MOX is 5 to 80 mg/L•month during one to two weeks when performed before MLF, and 1 to 5 mg/L•month over two to six months when performed post MLF. The optimal dosage must be determined individually for each wine within the given range.

Table 13 summarizes the essential characteristics of micro-oxygenation in contrast to macro-oxygenation, assuming best-worst case scenarios.

Table 13: Differences between micro- and macro-oxygenation.

macro-oxygenation	micro-oxygenation
discontinuous	semi-continuous or continuous
Fast and unique oxygenation in the order of 5 mg/L O_2 in one day *	Slow oxygenation in the order of in 0.1 to 0.7 mg/L O_2 per day over several months*
Temporary accumulation of DO up to several mg/L.	O_2 binding faster than O_2 supply, no (hardly) DO detectable.
Regenerative polymerization follows oxidation with a time delay.	Regenerative polymerization immediately cancels the effect of oxidation.
Oxidizable phenols are consumed.	Oxidizable phenols are regenerated.

* In connection with macro- and micro-oxygenation, different units of oxygen supply measurement are used, which can easily lead to confusion. The following conversions apply:

1 mg/L•month = 0.03 mg/L•day = 0.7 mL/L•month = 0.023 mL/L•day

1 mg O_2 = 1.43 mL O_2

Since a large number of wines often have to be treated at the same time, most dosing and control units have several separately controllable outputs, so that one unit can control several tanks in parallel and supply them with variable amounts of oxygen.

The meaning of temperature in the context of micro-oxygenation

During MOX temperature plays a double role:

1. The solubility of oxygen improves with decreasing temperature, so losses of undissolved supplied oxygen are reduced.
2. The chemical binding of DO (Figure 52) and the kinetics of the resulting secondary reactions decrease.

Below a threshold of 12-13° C (54-55° F), the oxygen supply rate should be very low (<< 1 mg/L/week) and completely stopped at 10° C (50° F). Otherwise an accumulation of DO occurs, which would lead to over-oxidation when the temperature rises again. At temperatures above 20° C (68° F) the oxygen consumption rate increases so much that detrimental consequences for aroma and color can occur. These risks are particularly acute for red wines with low total phenol content, which are more sensitive to oxygen.

The optimum temperature range during MOX is between 15 and 17° C (59-63° F). MOX systems equipped with a temperature probe are able to automatically adjust the oxygen supply rate to the temperature (Bartolini et al. 2008).

Practical hints

Readers are cautioned against sparging wine with uncontrolled amounts of air or pure oxygen by means of a diffuser or simple gassing units such as those used for flow-through carbonation. If dosing is possible at all, such devices can only dose gases in the g/L range, as it is required when refreshing white wine with carbon dioxide. The consequences would be an over-oxidation or even an autocatalytic reaction cascade (Chapter 7.3) with subsequent quality losses. If the control technology of professional micro-oxygenation units is not available, one has to restrict oneself to the classical possibilities of oxygenation summarized in tables 11 and 12, such as pumping-over, for active oxygen supply. Injecting air (20.8 % oxygen) has the additional disadvantage that the oxygen supply cannot be measured because a variable and in practice unknown portion of the oxygen is washed out with the nitrogen contained in air, which is insoluble.

8.3.3. Timing

Phenols are most reactive in the young wine stage (Garrido-Bañuelos et al. 2019). The more advanced the wine is in its development, the more strongly its phenols are polymerized, either by already existing ethyl bridges, or by direct binding of the flavonoids in the absence of oxygen via C4→C8 and C6→C8 links. However, high polymer tannins are less prone to form an ethyl bridge, making it difficult to catch up with MOX in a late stage of development. It would remain almost without effect (Nickolaus and Durner 2016). For this reason, MOX is usually started in young wine before the first SO_2 addition takes place, either directly after primary fermentation or after completion of malolactic fermentation (MLF).

When MOX is performed prior to MLF, oxygen supply is limited to a period of one to two weeks using relatively high dosages of up to 80 mg/L•month to drive the preferred chemical reactions with phenols, because a large portion of the oxygen is consumed by suspended yeast lees (Chapter 6.2) that are still present in high amounts. The primary objective is to promote flavonoid polymerization reactions in which acetaldehyde, an oxidation product, participates in the bridging of flavanols and anthocyanins. From a sensory perspective, a lowering of astringency is expected in the long term. Any excess acetaldehyde built up by MOX is largely metabolized during subsequent MLF.

When MOX is performed post MLF, the oxygen supply is reduced to less than 5 mg/L•month over a longer period of several months. It is used to mimic the oxygen ingress associated with barrel maturation and often applied in conjunction with alternative oak products. In oxidation-sensitive wines such as Pinot Noir, MOX can show better sensory results before MLF (Durner et al. 2010 a), because the still abundant yeast lees participate in the oxygen consumption and thus protects the wine from overoxidation.

Sulfur dioxide acts similarly to suspended yeast lees. If free SO_2 is present during MOX, it will largely impede the formation of acetaldehyde by two reactions:

- It scavenges hydrogen peroxide generated from oxidation of phenols and required for the oxidation of ethanol to acetaldehyde (Chapter 6.4 and 7.2).
- It binds any acetaldehyde produced to form the bisulfite-acetaldehyde adduct. Only when free SO_2 has become completely depleted by H_2O_2 during the course of MOX, is free acetaldehyde able to occur.

As a consequence of the competing action of free SO_2, less secondary products such as polymeric pigments and aging aromas appear (Tao et al. 2007, Gambuti et al. 2015). Depending on the initial free SO_2 level, MOX can thus become almost ineffective. Therefore, SO_2 additions are usually postponed after MOX.

MOX does not provide either the aromas or the ellagitannins that barrel aging provides. Therefore, it is frequently performed in conjunction with the simultaneous addition of oak fragments (Chapter 11). They impart additional wood-derived aroma compounds and release ellagitannins, which facilitate the formation of the ethyl bridge. Alternatively, the addition of commercial ellagitannin preparations (Chapter 9) is possible and actually performed by some winemakers, but not mandatory.

The use of MOX during fermentative maceration requires a much higher oxygen supply than on young wine. Since during primary fermentation most of the added oxygen is consumed by the yeast, it requires a dosage of 1 to 6 mg/L O_2 per day to achieve advantages in taste and color. Thus, such a procedure corresponds more to a macro-oxygenation with the main purpose of improving yeast growth and vitality. The sensory effects of a macro-oxygenation during fermentative maceration are less pronounced than those of micro-oxygenation of the young wine over several months (Durner and Fischer 2009 a, Weiand 2014).

8.3.4. Some results

MOX of young wines requires considerable practical experience in order to adjust the oxygen supply to the individual wine, the contribution of yeast lees to oxygen consumption, and to the differences between grape varieties and years. Analytically, these differences are reflected in the total phenol content, the tannin-anthocyanin ratio and in the occurrence of non-phenolic oxygen acceptors. As a consequence, the results obtained worldwide with MOX are quite variable (Anli and Cavuldak 2012) and cannot be transferred from one wine to another. The lowest common denominator is of analytical nature: In comparison with the untreated reference, MOX decreases anthocyanins, whilst it increases the concentrations of ethyl-bridged flavonoids, pyranoanthocyanins, and nonbleachable pigments. Such effects are as expected but not limited to MOX since they also produced by other techniques of oxygenation.

Many contradictions and little concrete knowledge

On the other hand, the increase of the mean degree of polymerization (mDP) is not significant or not generalizable (Atanasova et al. 2002, Dykes and Kilmartin 2007). This finding challenges the popular theory underlying MOX, according to which it causes a decrease in astringency by increasing mDP of the tannin-anthocyanin complex. In contrast, it corroborates the finding that a decrease of perceived astringency observed during wine aging is rather accompanied by an average size reduction of tannins (Chira et al. 2012), and that astringency is more affected by the subunit composition of procyanidins than by their mDP or concentration (Quijada-Morín et al. 2012). Moreover, several earlier studies performed on pure procyanidin solutions have shown that the more polymerization progresses, the more astringency increases (Arnold et al. 1980, Delcour et al. 1984, Noble 1994, Sarni-Machado et al. 1999 a, Vidal et al. 2004 b). Thus, if there was really an increase of the mDP during MOX, it could not explain a smoothening or better sensory integration of astringency. Monitoring reactions on the colloidal level, including polysaccharides, and changes of the average particle size (McRae et al. 2017) during MOX could possibly provide more useful information than focusing one-sidedly on the phenolic profile and its mDP.

As a consequence of the previously discussed contradictions, the sensory outcome of MOX is equivocal. The proponents of MOX emphasize an improvement of the fruitiness as well as a fuller and rounder palate structure that is considered to be more favorable than in control wines. However, peer-reviewed publications showing results from well-defined, rigorously executed sensory investigation are rare. They suggest that sensory results of MOX strongly depend on wine composition prior to MOX. Furthermore, they highlight the empirical nature of MOX and the lack of scientific understanding.

Originally, the sensory development of wines during MOX was divided into three successive phases designated as structuring, harmonization, and overoxidation with formation of dry tannins. However, terms such as structuring and harmonization are badly defined and have been replaced by more precise sensory terms in the meantime.

In Pinot noir, a systematic decrease of green tannins can be achieved by MOX. However, a softer mouthfeel only occurred in combination with a loss of color and a stronger browning. This was due to the preferred formation of polymers of the type 'flavanol-flavanol', more precisely 'tannin-tannin'. When a color deepening was observed, it was usually accompanied by a harder mouthfeel and associated with the preferred formation of flavanol-anthocyanin polymers (Durner et al. 2010 a). In general, MOX of wine with few anthocyanins favors polymerization of tannins, which easily leads to dryness and browning (Arapitsas et al. 2012).

The ratio of monomeric flavanols (F) and anthocyanins (A) determines which type of compound is formed predominantly. Expressed as a quotient (F:A), it allows an approximate prediction of the changes in color and mouthfeel caused by MOX (Durner et al. 2015). If this quotient exceeds a certain value, colorless flavanols will polymerize between themselves to brown pigments with a softer mouthfeel and less color. Below a certain minimum value, no change in color is observed; the oxygen supply can be increased. This outcome underlines the importance of the F:A ratio (Chapter 7.4.3). Average values are given for a balanced flavanol-anthocyanin ratio, which makes wines suitable for micro-oxygenation. For this purpose flavanols and anthocyanins were determined by HPLC (Durner and Fischer 2009 b, Durner et al. 2010 a), albeit less labor intensive measurements (Chapter 3.2.3) can also serve the cause in commercial winemaking.

Due to their low anthocyanin content, Pinot noir wines easily tend to browning and are not very suitable for MOX (Durner and Fischer 2009 b). However, it is useful to evaluate such wines on an individual basis in order to adapt duration and intensity of MOX to their total phenol content; corresponding data have been presented (Bernath et al. 2003). In that case, the monthly oxygen addition was gradually reduced and adapted to the continuously changing F:A ratio.

For striking a balance, starting MOX with an oxygen dosage of 10-20 mg/L•month is advisable. This rate can be maintained as long as the color intensity increases. After one to two months, it should be lowered to 5 mg/L/month to gradually reduce the DO and acetaldehyde levels that have accumulated in the wine (Nickolaus and Durner 2016). In general, wines with low total phenols contents are little or negatively affected by MOX, whilst wines with highest total phenol levels benefit the most from it in terms of color intensity and other chromatic attributes (Cano-López et al. 2008 a), although they need it the least. As outlined in chapter 8.3.5, MOX makes absolutely no sense in wines with less than 2,000 mg/L total phenols, and even at levels from 2,000 to 2,500 mg/L its sense is debatable. It is highly dangerous for light wines, where it can oxidize them rather than build up structure or any other feature that might be considered beneficial.

Wines obtained by thermovinification of strongly colored varieties do not respond to MOX, or at least not in a positive way, due to their low tannin content (Durner et al. 2009). This is in line with the role of the tannin-anthocyanin ratio outlined in chapter

7.4.3 and table 8. When this ratio is low, chemical aging and storage do not improve quality.

However, MOX of red wines obtained by thermovinification can improve quality provided that total phenols and the F:A ratio are within the optimal range. This is particularly true when less than 5 mg/L O_2 are supplied per month after addition of commercial tannin preparations. On the other hand, it is questionable whether MOX really makes sense to supply these low amounts of oxygen when there are other means available to do so.

With regard to wine behavior during barrel aging, previous MOX has shown variable results. In one study focusing upon aroma compounds, MOX slowed down the extraction of some of the volatile compounds from wood without causing significant changes in the varietal and fermentation volatile compounds (Ortega-Heras et al. 2008). In another study, a clear impact of wood aromas and a drastically lower astringency was observed (Llaudy et al. 2006). Since the effects in the finished product depend on many variables, differences of this kind must always be expected.

The application of MOX in the presence of oak alternatives is addressed in section 11.3.

At its current stage of development, MOX is surrounded by a lot of scientific uncertainty, mostly based on trial and error rather than on precise knowledge, and frequently shrouded in secrecy. Due to a lack of objective measurements, it has not yet come close to being a scientific method. There are both short-term and long-term results that cannot yet be exactly predicted. At best, it is undisputed that it improves high-phenol red wines that would otherwise be stored in stainless steel tanks without any oxygen uptake.

The sensory results obtained by MOX can be better interpreted if one takes into account other compositional changes, which occur concurrently and which are not related to phenol chemistry and oxygen. These changes comprise losses of carbon dioxide and total acidity. The effect of CO_2 on the palate can be easily understood when it is removed from a sample by intensive shaking. Total acidity, associated with pH, is known to strongly modulate perceived astringency (Chapter 5.3.4). It is lowered to a variable extent due to potassium bitartrate precipitations occurring in young wines during MOX. Consideration of these non-phenolic aspects of the MOX process has so far been recklessly neglected in practice and research.

8.3.5. Analytical prerequisites

The validity of the tannin-anthocyanin ratio, however defined and measured in analytical terms, has been confirmed in the previous section. Successful MOX requires a tannin-anthocyanin ratio (in mg/L) between 1:1 and 1:3, a minimum anthocyanin concentration of 100 mg/L, and a minimum tannin concentration of 150 mg/L as determined by the Harbertson-Adams assay (Nickolaus and Durner 2016). Care should be taken to not confound tannins measured by the Harbertson-Adams assay (Harbertson et al. 2003) with those obtained by other assays or total phenols. It is always important to clearly specify the analytical methods from whose results any ratio is derived.

If insufficient anthocyanins are present, flavanols and tannins react preferentially among themselves to form polymers that display a dry astringency in most cases. If too little tannin is present, anthocyanins react preferentially with each other to form colorless or brown oxidation products, which can eventually lead to browning of the wine (Table 8).

If none of the aforementioned analytical tools is available, as is usually the case in quite a few wineries, it should at least be ensured that the wine contains a minimum of 2,500 mg/l total phenols according to the Folin-Ciocalteu method, corresponding to 40-50 of the UV index (Chapter 3.1) before MOX is performed. These values also constitute minimum levels required for barrel aging. For wines with a lower total phenol content, passive oxygen uptake occurring during barrel aging and standard winery operations and, if necessary, one or two splash-transfers (Chapter 8.1.3 and 8.2) is usually sufficient. Many wines grown under cool-climate conditions fall into this category. This explains the limited dissemination of MOX under these conditions.

8.3.6. Process control

Process control during MOX implies the monitoring and controlling of oxygen-induced changes of critical sensory and analytical parameters. In this context, it is vital to note that sample bias can arise when monitoring wines in large non-homogenized tanks. Samples collected from sample valves located on the side or near the bottom, or from the top of the tank frequently do not represent the wine that has been exposed to oxygen. Therefore, previous mixing of the wine is advisable.

One of the options of MOX monitoring would be that of continuous DO measurement to adjust O_2 supply to DO consumption and thus avoid DO build-up. If DO exceeds 0.3 mg/L, the oxygen supply must be reduced. Monitoring DO during MOX is perfectly possible using modern DO sensors placed at various measuring points within the tank. However, in practice, analytical process control during MOX is hardly feasible for average size wineries. Therefore, daily sensory evaluation of MOX wines is the most critical aspect for decision making in the management of oxygen additions.

Key sensory attributes affected by MOX are wine color, the concentration of free acetaldehyde, perceived astringency, and fruitiness. Astringency and similar mouthfeel characters are the most commonly used attributes for monitoring MOX, but confusion abounds among winemakers regarding the astringent properties of phenols. The consistent use of astringent sub-quality terms such as green, harsh, silky, rough, grainy, etc. is difficult to achieve; frequently such terms are used interchangeably (Chapter 5).

Monitoring acetaldehyde by smell is another possibility. The appearance of the distinctive smell of acetaldehyde during MOX is quite normal, but it indicates over-oxidation when it becomes too strong. It disappears by itself after the MOX is finished and when the acetaldehyde is bound to tannins. This requires a waiting period of about one to two months (Figure 56). Only after that time the first SO_2 addition should take place. If SO_2 is added earlier, acetaldehyde is bound to it and removed from the desired reaction with the phenols. This inevitably also increases bound SO_2 and, eventually, total SO_2 levels, which are legally limited.

Much confusion about acetaldehyde smell thresholds has been created by the literature because due attention is not always made to differentiate between free and total acetaldehyde, the difference being attributable to SO_2-bound acetaldehyde. All acetaldehyde fractions can be measured, but only free acetaldehyde that is not bound to SO_2 is volatile and accessible to smell. Trained winemakers are able to detect 1 mg/L of free acetaldehyde, though there might appear masking effects in full-bodied wines. No free acetaldehyde can be smelt in the presence of free SO_2 as one excludes the other.

CO_2 in young wines might disturb on the palate but is not necessarily a quality criterion. A large part of it is removed by sparging at the beginning of MOX in the form of O_2-CO_2 mixed gas bubbles that escape. However, as long as it is present in the wine, it reduces the effect of the oxygen supply on the tannin (Durner and Fischer 2009 c). Sensory changes on the palate due to CO_2 losses should not be mistaken for a qualitative change of the tannin, although such a fallacy often occurs in practice.

The influence of MOX on red wine aroma is in the broadest sense comparable to that which occurs during storage in older barrels (Durner and Fischer 2009 c). But only when free SO_2 is present it is possible to definitively assess the wine's aroma and the general outcome of MOX. Over-oxidation leads to strong changes in the aroma by irreversible formation of higher aldehydes such as methional, phenylacetaldehyde, furfural 2-methylpropanal, 2-nonenal and 2-octenal. The odor produced by these molecules is reminiscent of honey, cooked potatoes, cooked vegetables, curry, etc. (Kilmartin 2009, Balboa-Lagunero et al. 2011). Chemically and sensorially, it resembles the aroma pattern of white wines spoiled by oxidative aging (Schneider 2019).

8.3.7. Comparison with storage in HDPE tanks

Historically, high density polyethylene (HDPE) tanks have been mainly used for storing wine surpluses, but it was soon discovered that atmospheric oxygen enters these vessels through their walls. In current times, there are many vessels on the market made of HDPE. They differ in shape, size, and oxygen transmission rate (OTR). The OTR can be adjusted to match the level of oxygen transfer on par with barrels when used at normal cellar temperature, or to lower levels using greater wall thickness.

Comparable to oak wood containers, storage in HDPE containers represents a passive oxygen incorporation. Their oxygen uptake works under a process of diffusion governed by Fick's law and the oxygen partial pressure gradient from the outside of the vessel into the liquid. This mechanism is different from the dissolution process observed with active MOX, in which differences in oxygen barometric pressure control the process.

HDPE tanks commercialized as maturation tanks are claimed to replicate the oxygen permeation rate of wooden barrels of 225 L standard size. Thus they are regarded as potential inexpensive alternatives to barrel aging when the wine stored in them is supplied with oak alternatives (Chapter 11.3). However, their practical use is still very limited, and specific research results are scarce. One study on HDPE vessels of 140 L reported an oxygen diffusion of 22 mg/L per year, which is within the commonly accepted range of values for new barrels. As with other vessels used in winemaking, an asymmetrical DO distribution was observed within these tanks, with regions near the lid and the inner walls of the tanks having more DO than the liquid in the center of the tanks (del Alamo-Sanza et al. 2015). Furthermore, they better preserve carbon dioxide than wooden barrels, which is not desired in red wines (Durner and Fischer 2009 c).

In contrast to active MOX but similarly to barrel aging, the oxygen uptake rate in HDPE vessels is constant over time and cannot be adjusted to the wine's specific DO consumption rate. Moreover, it is often unknown when such tanks are used for the first time. The easiest way to measure it is to store an aqueous solution of ascorbic acid (100 to 200 mg/L) in them for a certain period of time under avoidance of any headspace. Since ascorbic acid is the only oxygen acceptor under these conditions, its decrease can be stoichiometrically converted into oxygen uptake, both expressed as mg/L.

8.3.8. Comparison with periodical macro-oxygenation

There is no doubt that most wines are rated higher after a properly performed MOX when compared to the standard without any oxygenation. This is the basis for the worldwide prestige of micro-oxygenation.

A key premise of MOX is that a continuous low-level oxygen addition should be applied in order to produce the benefits reported, and to avoid oxygen buildup and uncontrolled oxidation. This demand is rarely challenged by the proponents of MOX. On the other hand, the actual DO concentrations during MOX have been reported to easily exceed 1 mg/L DO in industrial scale winemaking regardless of the oxygen delivery system used, thus clearly demonstrating the difficulty of adjusting oxygen supply to actual DO consumption rate (Laurie et al. 2008, Durner et al. 2010 b, Nevares et al. 2010). These levels are not very different from those that can also be achieved by certain means of traditional macro-oxygenation. Nevertheless, only rarely has the question been raised as to whether MOX is actually superior to these traditional approaches under the condition that an identical amount of oxygen is supplied. Traditional macro-oxygenation is understood here to mean one or several pump-overs, filling the vessel from the top (Tables 11 and 12).

At least for wines of average phenolic composition this question was answered by an assay on industrial scale performed on Carménère at 16-18° C. In that assay, the continuous supply of 6.7 mg/L O_2 per month using MOX over four months was compared with an aeration treatment that avoided the use of MOX. The aeration involved a small volume of wine that was splash-racked into a tub comprising a screen that helped aerating the wine. Once the wine in the tub reached ~7 mg/L DO, it was reintroduced into the original tank. This procedure was repeated 12 times in order to incorporate the same nominal amount of oxygen added with conventional MOX every week. Interestingly, the aeration treatment yielded results equivalent to the MOX treatment, suggesting that well-managed discontinued aeration treatments could be used as an alternative to MOX (Laurie et al. 2014).

In a similar approach (Petrozziello et al. 2018) on pilot scale mimicking oxygen exposure levels similar to those occurring during common cellar handling, four Nebbiolo wines were supplied with various levels of oxygen each during the first month post MLF. Total phenols ranged from 2,833 to 3,635 mg/L, anthocyanins from 118 to 179 mg/L, and initial free SO_2 was adjusted to 25 mg/L. The wines were enriched with oxygen by stirring with air until a concentration of 7 mg/L DO was achieved, according to a macro-oxygenation process. Then the wines were poured in 2-L glass bottles, with DO constantly monitored by non-invasive luminescence measurements. For each individual aliquot, when DO reached a value lower than 0.5 mg/L, the oxygenation was either stopped, or another cycle of DO enrichment and consumption was initiated up to a maximum of four cycles. Thus four levels of oxygenation were obtained, corresponding to 7, 14, 21, and 28 mg/L of consumed oxygen respectively. Afterwards, wines were stored at 20° C in the absence of oxygen and analyzed after 30 and 300 days.

Increasing amounts of oxygen favorably influenced the evolution of color, albeit differences in color parameters between different doses of oxygen were modest. This is explained by the high tannin-anthocyanin ratio of Nebbiolo wines. In contrast, free acetaldehyde measured 60 days after oxygenation increased with increasing doses of oxygen, but decreased during further storage due to consumption by polymerizing phenols. The concentration of higher aldehydes such as methional and phenylacetaldehyde associated with oxidized aroma notes did not markedly change with increasing oxygen supply, but increased during storage under anoxic conditions. Overall, the four oxygenation cycles did not detract from quality and showed results similar to those of MOX.

In both studies discussed above, oxygen supply was performed by single or periodic macro-oxygenations and in amounts similar to or exceeding the amounts of oxygen provided by MOX or splash-rackings. Moreover, storage in tanks submitted to several splash-rackings in monthly or longer intervals has been shown to produce an effect comparable to storage in old barrels (Pontallier and Ribéreau-Gayon 1983, Schneider 2001). Considering these results, it is not without reason that periodic pump-overs continue to be maintained as a traditional, widespread, and successful procedure of oxygenation that does not require any investment. However, whilst it is well known that each pump-over incorporates 1 to 3 mg/L O_2 with fairly predictable consequences, such an approach has never deserved especial attention in enological research. Quite the contrary, the euphoria that MOX has generated and the endorsement it has received since its introduction in the 1990's has caused the traditional options of oxygenation to be largely underestimated in the academic world, even though these options can be easily adapted to large-scale wine volumes. The essential advantage of MOX is actually the reduced workload when treating very large wine volumes.

The macro-oxygenation trials previously described open the way to experimentally determine the oxygen requirements of individual wines at a given stage of development. The following section explains how this is done.

8.4. Experimental determination of oxygen requirements

Chapters 6 and 7 have shown that oxygen requirements are highly variable and cannot be derived from a single set of analytical data. The very nature of MOX and any other means of oxygenation, which is the addition of sufficient oxygen to produce desired changes in wines' sensory properties, implies that it is essentially impossible to specify general guidelines and recipes for oxygen dosing, as each wine will differ in the amount of chemical change required for the desired sensory outcome. Things are further complicated by the fact that quality is a very personal matter. Therefore, the experimental assessment of the oxygen demand provides much more useful information. It takes into account the type, the matrix, and the current state of the wine, because all oxygen acceptors (Chapter 6) – phenols, yeast lees, and possibly SO_2 – are included in the result.

In the past there have been attempts to assess oxygen requirements by use of a manometric method using the so-called Warburg's apparatus. This is a device for measuring the pressure of a gas at constant volume and constant temperature so that the pressure is a measure of the quantity of gas, whilst changes in pressure reflect the production or absorption of gas. It is hermetically connected to a manometer. When it is placed on a magnetic stirrer and filled with a defined volume of wine, the oxygen consumed by the wine creates a negative pressure in the headspace, which is monitored and converted into mg/L O_2 the wine consumes. The result is an oxygen consumption curve as given in figure 64.

Increasing wine temperature or pH accelerates the oxygen consumption rate, makes the curve steeper, and shortens the measurement. When the curve flattens, the wine's reductive strength is depleted. The oxygen consumed until that point is assumed to indicate its effective oxygen demand.

Figure 64: Oxygen consumption of red wine at 20° C in a Warburg apparatus.

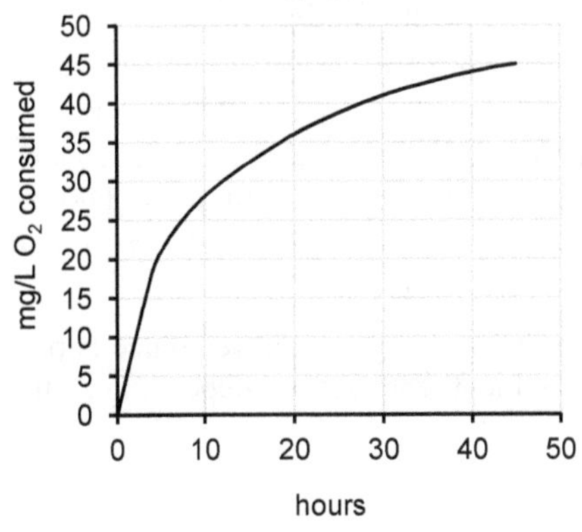

This method, as fascinating as it may be and as much as it has yielded information in past enological research, has two serious disadvantages: Firstly, it is seriously disturbed by CO_2 which develops, releases, and distorts the pressure values if very elaborate countermeasures are not taken. Secondly, the result obtained does not give any information about the sensory outcome to expect after the oxygen has been added to the wine in practice.

Oxygenation bottle test

To overcome these drawbacks, a simple to apply method of experimentally approaching the optimal amount of oxygen has been developed and proven to be useful. One might call it 'oxygenation bottle test'. It comprises the following steps:

1. Four screw cap bottles of 750 mL nominal volume (rim-full volume = 785 mL) are filled with wine to the rim, using a sampling hose whose end is led to the bottom and constantly left in the wine during filling in such a way that the wine does not splash through the air. The aim is to avoid any oxygen absorption during the filling process. In addition, the bottles can be sparged with nitrogen for about two minutes before filling.
2. A bottle filled to the brim is immediately sealed with a screw cap. It serves as a reference for subsequent evaluation.
3. Different volumes of wine are removed from the remaining bottles with a pipette before they are also sealed with screw caps. The headspace thus created contains air, which consists of 20.8 % oxygen (density = 1.43 mg/mL). The headspace volumes shown in table 14 with the resulting oxygen dosages, converted to mg/L O_2, have proven to be useful.

Table 14: Oxygen supply from the headspace of 750 mL-bottles.

headspace volume (mL)	25	50	100
mL O_2 in bottle	5.2	10.4	20.8
mg O_2 in bottle	7.44	14.97	29.74
mg/L O_2	9.8	20.4	43.4
time required for O_2 consumption at ambient temperature until tasting	1 week	2 weeks	3 weeks

4. To accelerate the dissolution of headspace oxygen, the closed bottles are shaken daily. They are stored laid down at 15 to 20° C (59 to 68° F). Higher temperatures are only suitable for filtered red wines because of the microbial risks involved, while low temperatures slow down the DO consumption (Figure 53).
5. A more sophisticated and safer way to overcome problems with microbial spoilage during oxygen exposure is the addition of one or two drops of dimethyl dicarbonate (DMDC, maximum 200 mg/L) as a preservative. Under these conditions, a short-term storage at up to 30° C is feasible to accelerate oxygen consumption. DMDC has the reputation for dissolving poorly (Section 12.4.3), but it can be dissolved in a very small wine volume vigorously stirred on a magnetic stirrer.
6. The sensory evaluation in comparison with the non-oxygenated reference is performed after one to three weeks' storage at ambient temperature, depending on the

headspace volume[54]. The waiting period can be halved for young wines still containing yeast lees.

7. The smell of free acetaldehyde disturbs the olfactory evaluation. In this case the bottles are supplied with approx. 50 mg SO_2. This can be done by adding 3.75 mL of a 2 % solution of potassium disulfite (~ 10 mg/mL SO_2) in water. The SO_2 needs about five minutes to bind the free acetaldehyde.
8. If this test is performed at a temperature higher than the cellar storage temperature, some wines may produce CO_2 bubbles when a remainder of sugar or malic acid is microbiologically degraded. Since these bubbles significantly interfere with the gustative evaluation, the bottles should be shaken vigorously to expel the CO_2 before tasting.

Figure 65 shows sensory results obtained by this test using the example of a largely clarified (residual turbidity < 5 NTU) high-phenol red wine five months post-MLF and stored in used barrels. The palate changes were more pronounced than the olfactory changes. Increasing amounts of oxygen strongly decreased astringency but increased bitterness and the fullness/volume attribute. These results are in contrast to those obtained on low-phenol red wines shown in figures 57, 58, and 59 and supplied with similar amounts of oxygen, to which they reacted in a more negative way.

Figure 65: Sensory results of the bottle oxygenation test performed on a high-phenol red wine (Merlot) five months after MLF and stored in old barrels.

Total phenols = 3409 mg/L GAE, anthocyanins = 389 mg/L, free SO_2 = 0 mg/L.

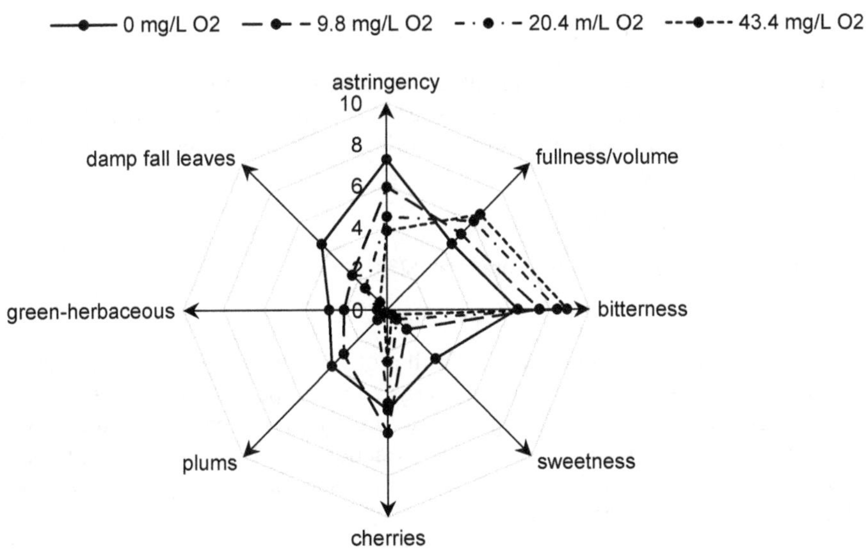

54 Most filtered red wines need 10 days to completely consume one saturation concentration of DO (Ferreira et al 2015), but about 90 % of it is consumed within one week due to the rapid consumption rate in the initial high concentration range (Chapter 6.1).

In straightforward experiments of this kind, one might rapidly find out that many recommendations and doctrines regarding optimal oxygen supply to red wines do not apply to one's own wines. They cannot be generalized to any conditions. More important is the awareness that oxygen cannot solve all problems, is only a part of the whole, and that the factors of time and temperature are equally important or even more important. It can easily happen that the optimum oxygen uptake has already been reached by the time the disturbing carbon dioxide has been completely removed.

With regard to color, storage trials at 3 and 25° C, each with and without oxygen (12 to 24 mg/L O_2), have clearly shown that temperature has the most influence on progressive change of color, and that aging reactions were much increased at the higher storage temperature. The latter is the major factor leading to the conversion of anthocyanins in stable polymeric pigments. They are essentially the same in the presence or absence of oxygen, though oxygen exposure does increase the overall polymerization rate. On the other hand, in many wines the early increase of color intensity under aerobic conditions during the first months post-fermentation is followed by its gradual decline during further storage. Thus, the oxygen can also exert a destructive effect on color in the long term, as least in the absence of ellagitannins as provided from wood during barrel aging (Somers and Evans 1986).

Since color parameters can be easily measured by instrumental analysis, most research in this field overstates their importance without giving due importance to the olfactory and gustatory consequences of oxygenation, which are much more difficult to evaluate. The color alone is of little use when the wine is not drinkable. Therefore, bottle oxygenation tests and their careful sensory evaluation as described in this section is of outstanding importance.

Accelerated oxidation tests

Without any doubt, there is considerable interest in shortening the oxygenation bottle test, and first attempts of this kind have already been made. After atmospheric oxygen saturation by shaking, red wines were submitted to three different accelerated aging protocols comprising:

- heat test by storage at 60° C,

- addition of H_2O_2 followed by storage at 22° C,

- addition of laccase[55] followed by storage at 22° C.

Compared to the control stored at 22° C, all protocols reduced DO consumption from several days to less than three hours. Untargeted mass spectroscopy revealed specific oxidation or aging markers with significant differences between test protocols. Color shift measurements suggested that forced oxidation with H_2O_2 was closest to natural oxidation (Deshaies et al. 2020). However, no sensory data or correlations with long-term oxygenation have been presented.

Undoubtedly, such accelerated oxidation tests have considerable practical interest and deserve further investigation. At present, however, there is no practical experience

[55] Laccase is a polyphenol oxidase used in this context to enzymatically accelerate phenol oxidation.

with the transferability of data obtained in this way to commercial winemaking conditions. Therefore, their application cannot yet be recommended.

8.5. Effects of oxygenation in the time course of tasting

Both consumers and many producers tend to believe that aeration makes red wines softer and more harmonious on the palate in a relatively short period of time. It is even a widespread custom to decant great red wines before drinking them or at least to swivel them in large-volume glasses. The aim of this procedure is the aforementioned oxygenation.

From the measurements of passive oxygen uptake via the surface (Chapter 8.1.1) it is clear that wine absorbs approx. 500 mg O_2/m^2 per day through its stagnant surface at room temperature. For a common tasting glass with a diameter of approx. 8 cm and a liquid surface of 50 cm^2 after filling with 50 mL wine, this results in an oxygen uptake of 5 mg/L within 15 minutes according to some basic calculations. If the wine is decanted or swiveled in the glass, it is saturated with oxygen in even less time thanks to the agitated surface.

On the other hand, chapter 6.1 and figure 35 show that an average wine can under no circumstances bind more than 0.5 to 1.0 mg/L DO per day. Therefore the wine contained in our hypothetical tasting glass will need several days to completely bind the oxygen it has taken up. If this time is allowed, there might be far-reaching changes in the mouthfeel, because the quality of the tannin changes through reaction with the oxygen.

However, tannin is present in red wines in a concentration range of usually more than 1000 mg/L. Hence, it must react with a correspondingly high amount of oxygen of approx. 10 mg/L O_2 to obtain qualitative changes on the palate that are measurable (Chapter 7.4). Before this moment occurs, however, the wine is drunk.

Within a time window of only a few hours, a wine will have bound only a small fraction of the oxygen dissolved in it, let's say 10 to 100 µg/L. This amount is sufficient to change the olfactory characteristics drastically by chemical reactions with volatile aroma compounds, whose concentrations are measured in µg/L and are 4-5 orders of magnitude lower than those of tannin. However, this amount of bound oxygen is definitely too small to impact the mouthfeel provided by the tannin it reacts with in a significant way.

The conclusion is simple: Aeration before consumption may open up the aroma assessed by smell, but does not soften the tannin of harsh red wines within the short time available. Opposite statements and sensations are to be explained in some cases by the release of carbon dioxide, which distracts on the palate and which is in no direct connection with the oxygen uptake. In most cases, however, suggestion and auto-suggestion might play an even greater role.

9. Ellagitannin additions post-fermentation

Ellagitannins are extracted from vegetal materials other than grapes. They differ largely from grape-derived tannins in that they bind oxygen faster and thus protect wine from the detrimental effects of excessive oxygen uptake. Their action as a redox buffer is accompanied by the promotion of the formation of stable polymeric pigments, more color intensity, and stability. Therefore, they continue to be recommended for addition to musts, where their effect is not always clear. For the same purpose, they are also sometimes added post-fermentation, when their effect tends to become more obvious. The increase in color intensity they provide is particularly pronounced in slightly colored wines, in which it goes along with a marked increase in astringency, which such wines are not always able to integrate. Indeed, the various commercial preparations of ellagitannins differ primarily in their astringency, which can be used to deliberately enhance this sensory attribute. The relative increase of perceived astringency is less pronounced in high-phenol red wines with strong color, which in turn do not require such additives to enhance their color.

9.1. Use and effects of post-fermentation additions

The effects of tannin additions to must are discussed in chapter 4.1.15. They can be summarized as an increase in color and astringency in some cases, but they are hardly reproducible and should not be overestimated in wines obtained from Vitis vinifera varieties. In the following results of post-fermentation ellagitannin additions will be focused.

Distinguishing different commercial tannins

The commercially available tannins are hydrolysable tannins extracted from sources other than grapes, grape-derived tannins obtained from winemaking residues, and mixtures of these two. Within each category, they show considerable variability in composition, average molecular sizes, and degree of purity. Not all of them show agreement between product composition and their designated material of origin (Li et al. 2018). According to legal requirements, they should be odorless and not impact wine aroma.

Hydrolysable tannins comprise essentially ellagitannins (Chapter 2.3.2). Both terms are often arbitrarily exchanged with each other. They are the preferred ones in winemaking because they are less expensive and have characteristics that differ from those of grape-derived tannins with flavonoid structure (Chapter 2.2). These characteristics first became known in wines aged in new oak barrels. The ellagitannins extracted from the oak alter the wine in terms of taste and chromatic characteristics. The simultaneous olfactory changes are not due to ellagitannins, but to other volatile components of the wood. Ellagitannins in their pure form are not volatile and therefore odorless, though some of their commercial preparations display odor-active impurities when they are obtained from toasted wood.

Catalytic effects of ellagitannins

Ellagitannins have a higher density of phenolic hydroxyl groups on their molecule skeleton than grape-derived tannins of flavonoid structure. This property confers them with a large array of interacting effects (Vivas and Glories 1996, Lempéreur et al. 2002), which can be summarized as follows:

- They cause an immediate color increase after their addition to young wine due to their own color and by enhancing the co-pigmentation level (Chapter 2.2.4) of anthocyanins. This reaction does not require the availability of oxygen.
- Under conditions of oxygen uptake, they elicit a further color increase, which is greater than it would have been without their presence.
- They increase the redox potential without requiring dissolved oxygen. Thus, they render the wine more oxidizable and favor oxidation processes.
- They bind with free oxygen radicals, thus protecting wine compounds from the damaging effect of oxidation by these radicals.
- They have the ability to bind oxygen faster than grape-derived flavonoid phenols and to release it again in numerous reactions. The oxygen withdrawn in this way is no longer available for direct oxidation of tannins and anthocyanins.
- Their rapid oxygen consumption promotes the formation of intermediate peroxide and acetaldehyde, which facilitates the polymerization of tannins and anthocyanins by ethyl bridging (Chapters 7.1 and 7.2). The polymerization occurring in young wines is claimed to reduce astringency, to favor color deepening and stabilization, and to limit the oxidative degradation of grape-derived phenolics.
- Ultimately, they allegedly intervene in a regulating way in oxidation phenomena, favoring the positive effects of oxidations, i.e. an improvement of structure and color, while limiting the damaging impact of oxidation. In a nutshell, the role of ellagitannins is essentially a catalytic one.

For the reasons elucidated above it is not surprising that micro-oxygenation is often performed in conjunction with the addition of oak alternatives (Chapter 11). The oak provides the ellagitannins, which direct oxidation and polymerization of the original grape tannin in a different direction towards allegedly less aggressive molecule structures.

Mouthfeel effects of grape tannin additions

Similarly to ellagitannins, grape-derived tannins are also commercially available as numerous preparations obtained from either grape skins or seeds, or a mix of both.

The enormous sensory differences between skin and seed tannins are outlined in section 4.1.7. The primary purpose of adding grape tannins is an improvement of body and structure in wines whose initial tannin levels are considered too low due to insufficient extraction or poor fruit, or to correct for a low tannin-anthocyanin ratio. Due to their high price, grape tannins find little use, though they are very interesting for a phenolic reprofiling of wines.

Ellagitannins protect low phenol wines from oxidative damage

It is to be expected that the post-fermentation addition of any tannin will produce more evident effects than an addition to must because the influence of the yeast is less or completely absent. The chromatic effect of an ellagitannin addition to a filtered low-phenol red wine is shown in figure 66. Absorbance measurements at 420, 520, and 620 nm represent brown, red, and blue color, respectively, whilst their sum indicates color intensity (Chapter 3.3). Oxygen as an additional variable was also added at a rate of 8 mg/L. The evaluation of the treatment effects was performed two months later and after adjusting free SO_2 to a comparable level of 40 mg/L in all lots.

Figure 66: Impact of ellagitannin (100 mg/L) and oxygen (8 mg/L) additions to a filtered low-phenol red wine (total phenols = 1080 mg/L GAE) on color intensity and color spectrum.

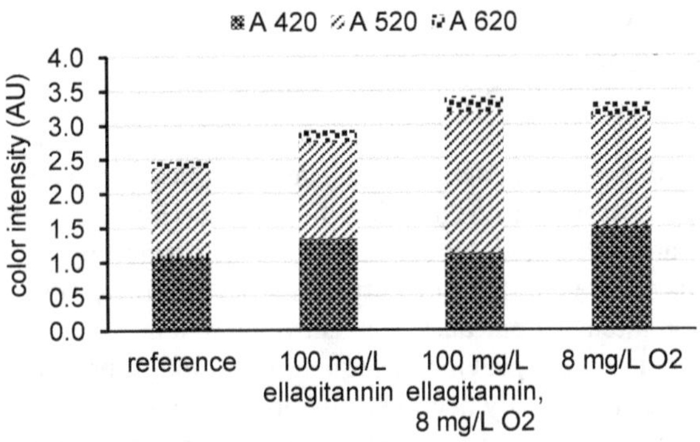

The sole addition of ellagitannin in the absence of oxygen led to a higher color intensity due to an intensification of the red (A 520) and brown (A 420) color component to a comparable extent. In conjunction with the additional oxygen uptake, a further color intensification with emphasis on the red (A 520) and blue (A 620) color component occurred. However, the same oxygen uptake in the absence of ellagitannin resulted in less color intensity with increased browning (A 420). This behavior corresponds largely to theoretical expectations. It can be visually traced in lightly colored wines, but barely in strongly colored ones.

Color is only one of several quality criteria. A strong color cannot compensate for deficiencies in taste and smell. Figure 67 concerning the same wine as before shows how the presence of the ellagitannins affected the percentage changes in olfactory, taste and pertinent analytical parameters caused by the oxygen consumption. The average of all three variants was taken as 100 %.

Figure 67: Impact of ellagitannin (100 mg/L) on the sensory outcome of oxygen (8 mg/L) consumption in a low-phenol red wine (total phenols = 1080 mg/L GAE).

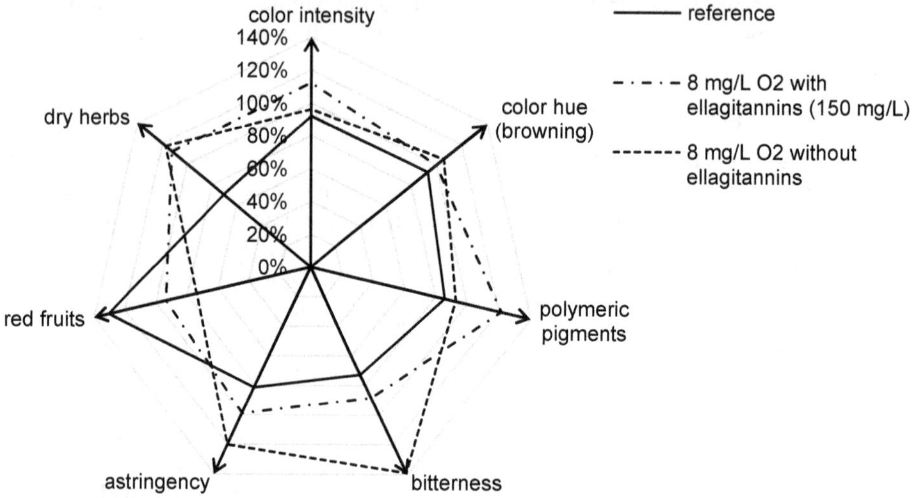

The untreated reference stored under oxygen exclusion displayed highest intensities of the much sought varietal aroma described as 'red fruits', but also the lowest levels for color, intensity, astringency, and bitterness. Under conditions of oxygen uptake, the lot provided with ellagitannins showed higher intensity rating of 'red fruits', more polymeric pigments, higher color intensity but also less astringency and bitterness than the variant that was oxygenated without previous ellagitannin supply. It becomes obvious that any advantages of ellagitannin only become fully obvious in conjunction with oxygen. On the other hand, the totality of the data shows that the oxygen uptake of 8 mg/L O_2 was too high for this low-phenol red wine. It leads to a loss of quality in smell and taste. However, these quality losses were less after the addition of ellagitannin.

The presence of ellagitannin actually protects thin red wines with low total phenol and anthocyanin levels from over-oxidation, while at the same time intensifying their color. However, the addition of such tannins can also disfigure the palate of these wines due to a badly integrated and lasting astringency, depending on the kind of tannin preparation used.

Impact of the wine matrix

It is obvious that the effects of ellagitannin are less evident in high-phenol red wines to the extent that they can become hardly detectable (Schneider 2004). This applies in the broadest sense to all treatments impacting the wine redox balance such as SO_2 additions, filtration, or the addition of grape-derived tannins. This is illustrated in figure 68 using the example of 'color intensity' as a single parameter of two young red wine blends (A and B), which differ significantly in total phenol and anthocyanin content. It displays the percentage changes of color intensity (A 420 + A 520 + A 620) regardless of sign induced by the following treatments:

- pad filtration causing color decrease,
- pad filtration and addition of 50 mg/L SO_2 causing color decrease,
- pad filtration and de-novo addition of 3 % yeast lees causing color decrease,
- pad filtration and addition of 50 mg/L ellagitannins causing color increase,
- pad filtration and addition of 500 mg/L pure catechin mimicking longer skin contact or the addition of grape tannin causing color increase.

Figure 68: Percentage change of color intensity (A 420 + A 520 + A 620) upon various treatments of two young red wines.

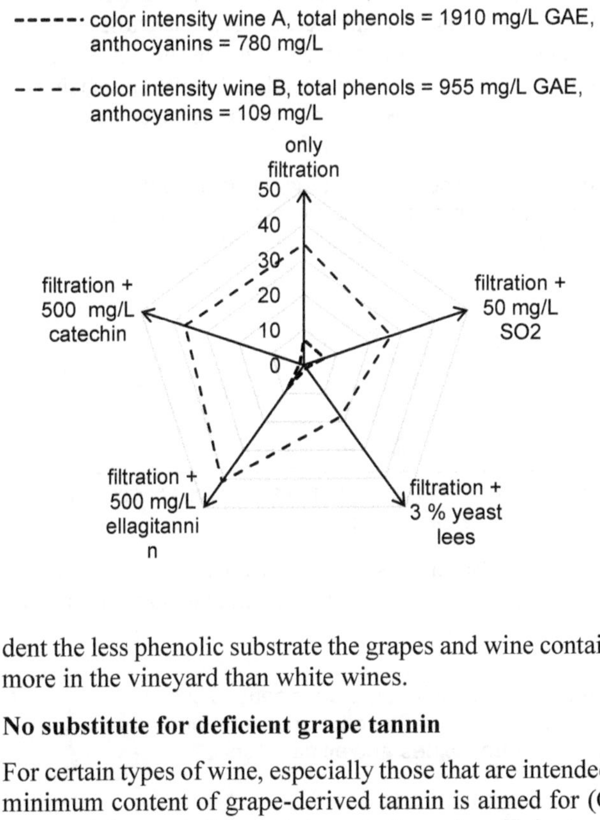

Compared with the low-phenol wine B, the effects of all treatments are insignificant in the high-phenol red wine A with its stronger matrix background. Section 4.1.15 states that ellagitannin additions to must show only small and hardly reproducible effects compared to the effects of other variables such as maceration length, temperature, bleeding, and cap management. In wine, the effects of ellagitannin additions are close to zero when the wine has been obtained from grapes of high phenolic maturity and when the phenolic potential of the fruit has been exhaustively extracted. They become more evident the less phenolic substrate the grapes and wine contain. Red wines are made even more in the vineyard than white wines.

No substitute for deficient grape tannin

For certain types of wine, especially those that are intended to be barrel-aged, a certain minimum content of grape-derived tannin is aimed for (Chapter 3.1.4). If this is not achieved due to deficient fruit maturity or insufficient extraction during maceration, a lack of tannin can hardly be compensated by adding commercial ellagitannins. A simple comparison illustrates the orders of magnitude: With an average tannin content of 2,000 mg/L, measured as total phenols, a tannin addition of 100 mg/L, which is common in practice, only leads to a modest increase of 5 %. This applies to both ellagitannins and grape tannins. However, with a considerable expenditure of material and costs, the addition of grape tannins can work wonders.

9.2. Sensory differences between ellagitannin preparations

Ellagitannins differ in their botanical origin and chemical constitution. Thus there are ellagitannins in the actual sense and gallotannins (Chapter 2.3.2). As tannins, they are inevitably distinguished on the palate by their astringency. It is the primary sensory expression of all tannins. In terms of concentration units, the astringency of ellagitannins is stronger than that of grape tannins. Figure 69 shows how much it varies between 28 different commercial ellagitannin preparations. The data were determined two months after their addition to filtered wines. Panelists had been previously trained on astringency intensity ratings using aqueous solutions of alum. They were asked to score astringency as an over-all quality without specifying astringent subqualities, adhering to a time interval of five minutes between samples (Chapter 5.2).

An average astringency increase of 49 % (25 to 77 %) at a comparable and frequently used addition of 100 mg/L makes astringency a central criterion in the choice of such products. In some wines and when the addition is performed in the short term before bottling, it can lead to wines being distorted in taste and rendered indistinguishable by a one-sided astringency, which can be more decisive for consumer acceptance than the improvement of color parameters.

Figure 69: Percentage increase of astringency caused by the addition of 100 mg/L ellagitannin of 28 commercial preparations. Averages over three wines and all panelists.

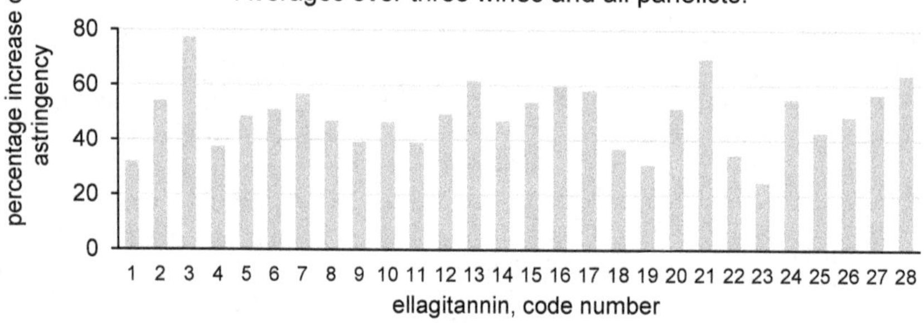

By analogy to astringency, bitterness also increased by 26 % on average (6 to 42 %). Panelists had been trained on bitterness intensity ratings using aqueous solutions of quinine chloride. Figure 70 gives the results.

Figure 70: Percentage increase of bitterness caused by the addition of 100 mg/L ellagitannin of 28 commerical preparations. Averages over three wines and all panelists.

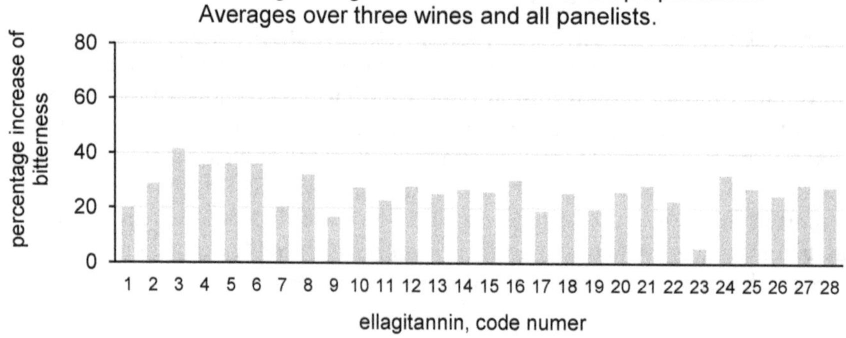

Under the same conditions, an increase in the 'fullness, volume' term occurred within a range of 0 to 46 % and an average of 13 %. This sensory term was defined and trained as the tactile sensation caused by the addition of 20 g/L glycerol to an average dry white wine. Interestingly, whilst most winemakers seek an improvement of what is commonly described as fullness, body, or volume on the palate when they work with legal additives, this is the gustative attribute that is least enhanced by ellagitannin additions. To the contrary, the primary gustative change they trigger is an increase of astringency, which is not related to, or even opposed to, the 'fullness, volume' term.

Influence of wine matrix and aging

The sensory characteristics of commercial tannins become more prominent after addition to clarified wine than after addition to cloudy young wine or even during fermentation, where they are partially absorbed by yeast lees (del Barrio-Galán et al. 2015). Furthermore, during the aging process, the ellagitannins undergo degradation by acid hydrolysis and oxidation. In both cases a reduction of the initially dominant astringency occurs, but this process requires time and cannot be relied on when the wine is to be marketed in the short term.

Ellagitannin additions post-fermentation in a nutshell

According to the recommendations of product promotional material, ellagitannins are mostly used for the purpose of color stabilization, despite color-stable red wines having been achieved long before these tannins were introduced in the market. Of greater importance is a certain color intensification in slightly colored red wines.

Under conditions of high oxygen exposure of low-phenol red wines, fruity varietal aromas are better preserved in the presence of ellagitannins. The additional astringency they provide immediately after their addition slowly decreases over a period of many months. This also happens when ellagitannins are extracted from oak during barrel aging (Chapter 10.3). However, a one-sided and poorly integrated astringency in ordinary red wines most often does not disappear during wine maturation, albeit many people might hope so.

Astringency is the primary criterion that most distinguishes the commercial tannins from each other. The changes it causes on the palate are more important than color and aroma aspects. It can have dramatic consequences for the overall quality and consumer acceptance if it is not recognized in time or if it is not reduced by fining or otherwise (Chapter 12.1). The gain in quality through color intensification is thus nullified. Therefore, it is risky to use tannins without appropriate preliminary bench testing, in particular in low-phenol wines.

In high-phenol red wines, both the negative and positive effects of ellagitannin additions are measurable by analytical means, whilst their sensory outcome tends towards zero. Therefore, such additives have no practical interest above a certain quality level.

10. Barrel aging

Barrel aging is the high art of redox management. Whilst various wood species such as chestnut, acacia, or cherry have been used for wine storage and aging, oak is the most prominent of them in modern times. Apart from a short introduction into oak characteristics, differences, aromas, and defects, this chapter focuses upon oxygen ingress into the barrels, the ellagitannins they release, and the way of keeping an appropriate balance between reactions of oxidation and reduction to optimize the sensory outcome. It is an example of practical application and the synthesis of the theoretical aspects discussed in the previous chapters. As a matter of fact, it is impossible to establish a standard protocol for barrel aging since it is subject to a large number of possibilities and variables. They include the botanical and geographical origin of the wood used for manufacturing the barrels, their age, the phenolic strength of the wine, the sensory objectives sought, the storage conditions, and the care taken of the wine during the aging process. The enological tools that are available to drive the wine's development in the best way comprise oxygenation by splash rackings, SO_2 additions, yeast lees stirring, temperature settings, and just doing nothing. Whilst none of these is compelling, if any of them are undertaken, its outcome will be strongly influenced by wine composition and barrel age.

10.1. Barrel aging between maturation and aromatization

Wooden barrels have been used since antiquity to store wine. What we now know as a standard 225 L barrel differs from other barrels basically by its limited volume. This results in a relatively large internal surface area per unit volume and, associated with this, an intensive mass transfer at the phase boundaries of the wine, wood, and surrounding atmosphere. The relationship between barrel size and internal surface area is shown in Figure 62. There is a small, but significant difference between common barrels of 225 L and those containing 300 or 500 L, which explains why these larger ones are sometimes used to reduce the impact of wood during barrel aging.

The use of new oak barrels as a stylistic tool to deliberately and sometimes profoundly alter the sensory characteristics of wine has only become known since the 1970's. Currently barrels are a powerful symbol of artisanality for both winemakers and consumers. The reason is primarily that oak is a wood species with an incomparably high aromatic potential, which is made accessible for the production of barrels by drying and toasting. In contrast, other woods used for barrel production such as chestnut, acacia, or cherry are relatively neutral in aroma.

The sensory contribution of oak to wine flavor can range from barely perceptible to dominant. The aim is always to achieve a combination of oak and wine that is subjectively perceived as optimal. However, there is no binding definition of what is optimal. Novice winemakers look at oak first and foremost as a flavoring agent. This approach is very short-sighted, because decisions based solely on providing aroma or color embellishment result in unbalanced wines. On the other hand, barrel aging can also be performed as a way to mature the wine and to add texture and suppleness without adding flavors. If presenting the authentic flavor of origin or variety or other natural aspects of wine is the goal, oak extractives that new barrels provide should not play more than a supporting or subliminal role. In this case, the predominant use of old barrels or old large casks can be the better solution. For some winemakers and for certain red wines, there is even the question of whether oak plays a fundamental role in winemaking at all.

In any case, in order to achieve the desired sensory goal, successful barrel aging requires more than just purchasing expensive barrels from a renowned origin and a reputed cooperage. Instead, three interacting factors must be taken into account:

- type of oak and age of the barrel,
- type of wine,
- redox management during aging.

The oak directly or indirectly provides the wine with substances that can be divided into three groups explained in the following chapters.

10.2. Oxygen uptake during barrel aging

The importance of oxygen ingress into wine is well known because oxygen modulates the aging process in wine barrels. Indeed, barrel aging is the traditional way of micro-oxygenation.

The oxygen uptake in barrels occurs through the inherent porosity of the wood, through the joints between the staves, via the ullage around the bung when barrels are not hermetically sealed, and upon periodic topping. It is strongly increased when a vacuum is formed inside the barrels through absorption of wine by the wood and its subsequent evaporation. This happens at the earliest one month after filling the barrels without any kind of topping, sealing them hermetically with silicone bungs, or storing them turned on their side to keep wooden bungs wet (Peterson 1976).

The annual rate of oxygen entry into barrels is the amount of oxygen that enters the barrel full of wine in one year, typically expressed as mg/L per year. For new barrels, highly variable oxygen uptake rates (OTR) have been reported ranging from 10 to 40 mg/L during the first year. European wood of several members of the Quercus genus is more permeable to oxygen than American wood of the Quercus alba variety, and fine-grain oak is more permeable than coarse-grain oak (Vivas 1999, Kelly and Wollan 2003, Vivas et al. 2003 b, Nevares et al. 2014, del Alamo-Sanza et al. 2017. As the number of wines aged in a barrel increases, the wood pores and stave joints become increasingly clogged with microorganisms, bitartrate, and other precipitated wine material, so that the oxygen uptake decreases over time to considerably less than 10 mg/L O_2 per year.

Oxygen permeability of new barrels is dynamic

However, OTR data referring to new barrels can be misleading because they show a dynamic behavior, which depends on the wood moisture and the progression of the moisture front from inside the barrel to the outside. In new barrels, the wood pores still contain air, through which oxygen diffuses much faster than through the wine subsequently filling the pores when the wood becomes wine-saturated after some period of wine storage. Thus, under conditions of continuous OTR monitoring, a marked decrease was reported, from 44 mg/L•year at day 1 to 22 mg/L•year at day 10, to 16 mg/L•year at day 30, and finally stabilizing at 7.3 mg/L•year after one and a half months. These results allowed estimation of the real value of the annual OTR in four barrels, which was 11.3 mg/L•year. As a result of the evolution of the OTR, after two months of aging, a new barrel has already supplied nearly 40 % of all the oxygen it will provide to the wine in a year. Ultimately, the impact of wood moisture on oxygen diffusion is so dramatic that moisture seems to be the key factor leading to very low OTR conditions in wine barrels just one to two months after filling (del Alamo-Sanza and Nevares 2018). Additionally, both the assembly of the staves and the origin of the wood play a role, with French wood permeating slightly more oxygen than American oak (Nevares and del Alamo-Sanza 2016).

The previously discussed studies used model wine, renewing and degassing it each of the 15 times over a year the oxygen entry in the barrels was measured. They revealed that older studies largely overestimated the annual OTR of barrels by extrapolating the rate of oxygen entry into barrels during a shorter period to an entire year to establish a misleadingly high assumed annual rate (del Alamo-Sanza and Nevares 2017). Moreover, they explain why even prolonged barrel aging may not provide enough oxygen to some kinds of red wine, thus suggesting additional measures of active oxygen supply by splash-rackings or other means.

In comparison with the oxygen uptake through the stave joints, the oxygen transfer through the intact wood is less important. Furthermore, it loses up to two thirds of its initial OTR during the toasting process (Nevares and del Alamo-Sanza 2015). Hence, lightly toasted wood can be expected to provide more oxygen than strongly toasted oak. In any case, an important, or perhaps the main, oxygen transfer takes place between oak stave gaps due to the fragile contact between adjacent staves (Qiu et al. 2018). In recent times, progress has been made in manufacturing customized barrels with desired OTR levels by using image analysis of the staves in order to classify wood based on its anatomical characteristics besides grain and stave handling (Prat-García et al. 2020).

Apart from entry through wood pores and stave joints, the third route of oxygen entry into barrels is the bung and the headspace. Both are interrelated, but more associated with winery practices than with wood specifications. With regard to the bung, it has lost most of its former, direct importance for oxygen entry due to the widespread use of silicone bungs that allow a tight seal. However, the gas-tight fit of silicone bungs facilitates the generation of a negative pressure in the barrel when wine evaporates through the wood, thus causing a headspace. As a result of the depressurization effect, the use of silicone bungs exerts an indirect impact on oxygen uptake.

Barrel depressurization impacts oxygen uptake

In this context, it is interesting to note that whilst silicone bungs reduce losses of wine volume by approximately one third when compared to wooden bungs or open barrels, they strongly enhance oxygen entry into the barrel. This might appear counter-intuitive, but the reason can be found in the internal depressurization they facilitate. It creates a pressure difference between inside and outside of the barrel. This pressure difference favors the entry of a greater amount of oxygen by diffusion through the wood, and an even greater amount in the case of silicone bungs when a headspace is formed with the bung being vertical.

The headspace gas is assumed to at least partially result from degassing of the wine when barrels work with a partial vacuum to form a headspace. When the pressure in the headspace is lower than atmospheric pressure, the solubility of oxygen and carbon dioxide is lowered, which causes its degassing from the wine, given that it tends to equilibrate, according to Henry's law, with the gas of the headspace with which it is in contact. Under these conditions, the degassing process of wine produces a gas mixture primarily of CO_2 and O_2 that accumulates in the headspace. The oxygen content of the headspace gas was shown to be only 5-9 % O_2, compared to almost 21 % of O_2 contained in atmospheric air (Moutounet et al. 1998).

According to Henry's law, degassing of only a partial fraction of the wine volume in a barrel would suffice to reach this percentage of oxygen measured in the headspace (Nevares and del Alamo-Sanza 2016). It exerts a significant influence on the entry and consumption balance of oxygen, determining the composition of the headspace. Under more practical aspects, it will be enough to enable the formation of volatile acidity by Acetobacter bacteria and explains why regular topping of barrels is common practice in most wineries.

When a barrel with reduced internal pressure is opened for topping, one can observe that it sucks air from the outside and thus causes an oxygen entry. When the barrel is immediately refilled with a few liters or half a gallon of wine, the DO level of the wine contained in the barrel does not increase to a significant extent. Hence, in depressurized barrels, headspace oxygen and toppings only take a subordinate position in the routes of oxygen uptake during barrel aging. However, this picture will change completely when barrels are frequently opened for sampling in a way that no negative pressure can be formed. In such a case, oxygen uptake from the headspace and regular topping become much more important.

In chemical terms, oxygen uptake during barrel aging triggers a moderate oxidation, the kinetics of which corresponds approximately to that of micro-oxygenation (Chapter 8.3). It leads to an improvement of tannin quality via regenerative polymerization the oxygen promotes (Chapter 7.2), and to a modification and complexation of the wine's aroma.

10.3. Ellagitannins released from oak

Ellagitannins account for around 10 % of the dry weight of the oak heartwood. Their extraction is facilitated by the hydroalcoholic nature of wine, but due to their reactivity, their levels in barrel-aged wines are much lower than might be expected. Therefore, their contribution to the total phenol content is marginal and hardly measurable by total phenol assays (Chapter 3.1). On the other hand, they are highly astringent. Therefore, their extraction from the wood of new barrels leads to a strong increase in astringency during the first three to four months of storage. Their concentration decreases during further storage due to oxidation and acidic hydrolysis, going along with a corresponding decrease in astringency. In this process, only 20 % of the decrease can be attributed to oxygen, with the remainder being governed by pH and time (García-Estévez et al. 2016). Hence, measures of oxygenation can hardly contribute to a significant decrease of perceived astringency caused by ellagitannins. After one year of aging in new barrels, between 1 and 31 mg/L total ellagitannins were measured, depending of the botanical origin of the oak and its toasting level (Navarro et al. 2016).

Ellagitannins protect the grape-derived tannin and anthocyanins from the detrimental effects of excessive oxidation and at the same time promote their polymerization (Chapter 9.1). In this process, the original flavor harshness of young red wines is reduced in favor of a softer and more complex mouthfeel. The combination of micro-oxygenation and its modulation by ellagitannins makes barrels an ideal container for the slow aging of red wines.

The contents of extractable ellagitannins contribute significantly to the differentiation of oak according to its origin and genus. American oak of Quercus alba (white oak) has two to three times less ellagitannins than the European oak species of the genera Quercus petraea (*syn.* sessile oak) or even Quercus robur (*syn.* pedunculate oak) (Chatonnet and Dubourdieu 1998). These differences are primarily reflected in the mouthfeel of the wines. Quercus robur, also known as Limousin oak, has highest levels of ellagitannins but low concentrations of volatile aroma compounds looked for in wines. It is preferred for the aging of spirits.

The basic distinction of oak for wine aging is between American oak and Q. petraea. The letter is frequently referred to as French oak, albeit an appreciable amount of it originates in Central and European countries. Wines aged in American oak taste softer, but are also considered to have less body and structure. Q. petraea contributes more to structure, but to a lesser extent than Q. robur as it releases intermediate levels of ellagitannins. The differences in ellagitannins released from European and American oak have also consequences for the redox management during barrel aging.

Polysaccharides are an additional but poorly investigated oak compound released during barrel aging. They are loosely associated with ellagitannins, whose astringency they mitigate to a certain extent, and comprise essentially cellulose, hemicellulose, and lignin. Their flavor effect consists in a certain contribution to mouthfeel and the masking of aggressive tannins (Chapter 5.3.3). More important, however, is their role as precursors of aroma-active volatile compounds formed during seasoning and toasting of the wood.

10.4. Volatile aroma compounds released from oak

Whilst wood species other than oak only show a limited impact on wine aroma, oak is able to substantially modify it or completely mask it by the volatile compounds it releases. Indeed, an unmanageable diversity of different substances responsible for oak aroma has been identified. They originate from virtually all substance classes of organic chemistry (Chatonnet and Dubourdieu 1998, Chatonnet 1999, Pérez-Coello et al. 1999). Some of them are naturally present in the original wood, but the major part of them is produced during the toasting process. Their concentrations in the wood or wine allow differentiation between oak species and origins to a certain degree. There are major differences between French and American oak, which have been extensively described in the technical literature and summarized by Towey and Waterhouse (1996 a). In a nutshell, whilst American white oak releases less ellagitannins, it imparts more aroma intensity, sweeter notes, vanilla, and coconut than sessile oak from Europe, which instead provides more aroma complexity with spicier notes.

With some experience, such differences between European and American oak are clearly traceable in wine by tasting, though there might be exceptions since wood composition is highly variable and displays differences from tree to tree between trees of the same species and location. These differences can manifest themselves in barrel-to-barrel variations (Towey and Waterhouse 1996 b). Such compositional differences between individual trees, due to forestry conditions and anatomical localization of the wood, can have a greater impact than those differences due to the botanical or geographical origin of the oak. It is one of the main tasks of cooperage to minimize them. However, they always remain to some extent.

The sensory evaluation of barrels' quality on the basis of the wine stored in them requires that the enological variables, essentially the wine and the way it is aged, are identical in all barrels. This particularly concerns the prior adjustment of free SO_2 to a comparable level. Otherwise, there is a risk that the 'wine' factor will have a greater influence on the evaluation than the 'oak' factor. Only when these conditions are met, can barrels of different origins or cooperages be evaluated in sensory terms. Furthermore, due to the barrel-to-barrel variations, objective data can only be obtained by comparing multiple barrels of the same wine. In this way, the standard deviation around the mean value is accounted for.

Seasoning and toasting as performed by the cooperage has an overwhelming influence on aroma quality. In particular, the heat effect during toasting produces the volatile aroma compounds sought in barrel-aged wines. However, amongst their diversity only a few of them have a significant impact on the sensory characteristics of the wine. Table 15 shows the most important of them, including their origin and their olfactory description. They are considered as indicator compounds because their concentrations in wine can give indications concerning the quality of the oak and the toasting profile used. Hence, their analytical measurement is gaining in importance to test the performance of barrels, to analyze their chemical imprint, and to compare barrels from different cooperages, toasting levels, or origins.

Table 15: Properties of characteristic aroma compounds released from oak into wine.
↑ = increase, ↓ = decrease, ∩ = increase followed by decrease at high toasting levels

compounds	origin in the wood	smell attributes	behavior upon toasting
ß-methyl-γ-octalactones occurring as a cis and a trans isomer	lipids	cis: coconut, herbaceous trans: coconut, spicy	↓↓↓
guaiacol, 4-methylguaiacol	lignin	smoke, char, ashes	↑↑↑
vanillin	lignin	vanilla	↑↑∩
furfural, 5-methylfurfural	polysaccharides	caramel, butterscotch, grilled almonds	↑↑↑
eugenol, iso-eugenol	lignin	cloves, spicy	↑↑∩
maltol	polysaccharides	caramel	↑↑↑
(E)-2-nonenal	lipids	pencil shavings, sawdust	↓↓↓

These compounds strongly contribute to the aroma of barrel-aged wines as they have developed since the 1970's, but to a widely varying degree. Their amounts and proportions depend on the botanical and geographical origin of the wood, the seasoning techniques, the intensity of toasting, and the number of times the barrel was used previously. In old barrels that have already been used for several wines, their contribution is marginal or not perceptible. In untoasted or poorly toasted oak, the aroma suggesting coconut or sawdust can be dominant. The most decisive contribution to the formation of the desired oak aroma compounds stems from toasting.

It is noteworthy that the toasting process also produces some acetic acid by hydrolysis of polysaccharides. It leads to a non-negligible increase of volatile acidity in the range of 0.1 to 0.2 g/L in wines aged in new barrels.

The importance of wood drying by seasoning

In order to shape the barrels, the oak staves must be dried to some 14-18 % residual moisture. This can be achieved by natural seasoning or forced drying. The drying process also contributes to what the wood releases into the wine. Naturally seasoned oak has become a standard in barrel manufacturing because it releases lower levels of ellagitannins and bitter tasting coumarins than artificially dried oak. The lower presence of ellagitannins is explained by rainfalls repeatedly washing the staves, thus partially solubilizing and removing them. Moreover, natural seasoning results in higher levels of aromatic volatiles that are sought in barrel-aged wine. In the broadest sense, artificial drying provides fewer aromatic and more bitter and astringent tasting substances.

The importance of toasting

As discussed above, a large number of compounds that are transferred from the oak into the wine are either produced or degraded during the toasting of the staves. This makes the toasting level a key criterion in quality management of the cooperages and in the selection of barrels. Every cooperage has its own 'house style'.

Three main toasting intensities are distinguished: light, medium, and strong. Sometimes further differentiations are made, for example between medium and medium$^+$. Table 16 shows the sensory consequences of the different levels of toasting as perceived in wine. However, these criteria are poorly standardized and differ from one cooperage to another as each one has its own toasting protocols. Ultimately, seasoning and toasting protocols of the cooperage have as much influence on the sensory outcome as the botanical and geographic origin of the wood.

Table 16: Sensory impact of different oak toasting levels on wine.

toasting intensity	untoasted	light toasting	medium toasting	strong toasting
smell	raw wood, coconut, pencil shave	less coconut but celery, walnuts, and pastry.	cinnamon, vanilla, cloves, and spice	Smoke, caramel, coffee, toasted bread
taste	intensive and lingering astringency	less astringent than without toasting	further reduction of astringency	lack of astringency
time-temperature conditions of toasting	- - -	~10 minutes at 120-180° C	20-30 minutes at 180-200° C	30-60 minutes at 200-230° C

The higher the toasting level, the more ellagitannins are degraded, which in turn confers less astringency to the wine. In contrast, the concentration of volatile aroma compounds does not follow the toasting level linearly. By varying the temperature and duration of toasting, it is possible to respond very individually to the specific wood, the type of wine, and customer requirements. For wine aging, medium toasted barrels are generally preferred because they confer the best sensory balance to the wine.

Sensory features of bad oak

Of course, not all barrels are good, and among the aromatic and other compounds oak releases into the wine (Table 15), there are also some less desirable ones. The presence of their sensory attributes in barrel-aged wines usually indicates poor wood quality or insufficient toasting. They must not be confounded with deficient fruit quality, microbial faults, bad cellar techniques, or redox-related issues. Therefore the most common sensory characteristics of wine aged in bad oak are summarized in table 17.

Table 17: Sensory properties of bad oak as perceived in wine.	
Sensory feature	Causes
Smell reminiscent of pencil shavings and sawdust.	Elevated levels of (E) –2-nonenal caused by insufficient toasting or inappropriate wood.
Smell reminiscent of green wood and green walnut shell.	Bad oak, sometimes aggravated by insufficient toasting.
Aggressive, lingering astringency.	Elevated levels of ellagitannins caused by deficiencies in wood quality, season, or toasting.
Complete absence of oak aroma in the 1st and 2nd wine.	Unusable wood.

Lightly toasted barrels are occasionally promoted claiming that they better preserve the original wine varietal aroma by masking it less by oak aroma compounds. This is not true, because even lightly toasted oak releases a distinct aroma. However, such oak is not reminiscent of vanilla, cloves, or cinnamon, but of walnuts, celery, or even sawdust (Tables 16 and 17). When the intention is to better preserve the original wine aroma, using less new barrels and blending back with a portion of the wine stored in stainless steel or old barrels is a better solution than aging in poorly toasted barrels.

10.5. The influence of the barrel age

As is commonly known, all volatile compounds and ellagitannins as well as their sensory impact on the wine decrease as a consequence of their progressive depletion when barrels are used repeatedly for successive batches. Assuming that wines spend 12 months on average in barrels, the sensory impact of oak is reduced by approximately half with each wine, but with the most drastic decrease occurring during the first year of use. After the third wine, the aroma and tannin potential of the oak is largely exhausted (Towey et al. 1996 a). Figure 71 shows the decrease of some characteristic substances over time.

Figure 71: Decrease of substance input (% of original material) as affected by the useful life of barrels.
(Adapted from Vivas 1998)

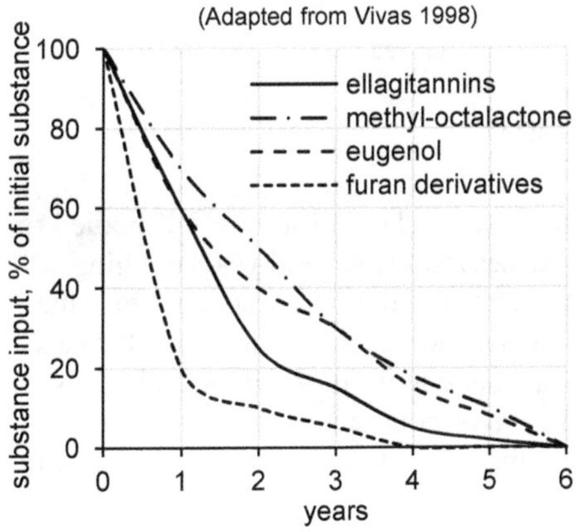

After depletion of oak compounds, the function of the barrel is limited to oxygen supply, albeit the oxygen transmission rate is also gradually diminished due to an increasing clogging of the wood pores by precipitation of bitartrate, high-molecular tannins, and other colloidal material (del Alamo-Sanza et al. 2017. The result is a flavor maturation without aromatization. Thus, even when older or larger barrels, which do not have a direct flavor impact, are used, their ability to allow small amounts of oxygen to be taken up is still important in the development of red wines. This essential function old barrels provide should not be underestimated.

Not everyone likes the flavor imprint that new-oak barrels stamp on a wine. Indeed, the exclusive use of new oak can be detrimental when oak elements in the wine become too strong and mask grape-derived sensory features such as fruity varietal and terroir characteristics. The radar plot of figure 72 gives an example of a wine whose varietal characteristics have become completely masked after nine months of aging in new oak, whilst another portion of the same wine aged in once used barrels turned out to show more complexity and more balance between fruit and oak attributes. As to be expected, the lot stored in stainless steel was rated highest for fruits, lowest for volume/fullness, and without any conspicuous oak aroma descriptors. In this example, a blend of the wine aged in new barrels and in stainless steel would also have been conceivable.

Figure 72: Impact of barrel age: Sensory comparison of Pinot noir (total phenols = 2280 mg/L GAE) aged nine months in new oak (1st wine), once used oak (2nd wine) and stainless steel tank. Oak = american oak, toasting = m+, intensity scoring range 0-5.

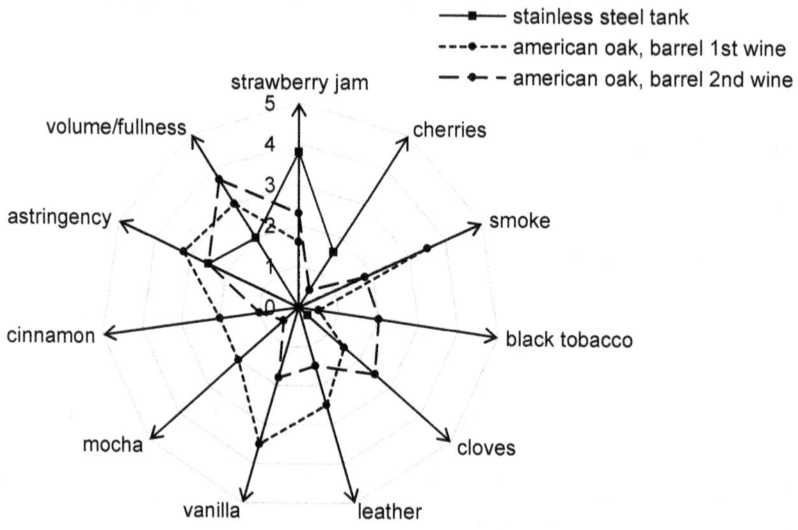

Indeed, exclusively new barrels are only rarely used and only for very full-bodied red wines with high total phenolic content. Such wines require several years of aging after bottling to achieve sensory integration of the oak. In common practice, red wine is aged for one to two years in barrels of the first, second and third wine, one third each, which are blended before bottling. This means that one third of the oldest barrels are replaced by new ones each year to maintain flavor consistency over the years. The actual turnover rate will depend basically on the characteristics of the wine and the marketing goals.

There is no law stipulating that red wines must necessarily taste of oak. Since oak has become ubiquitous in almost all red wines of the world, it is sometimes also perceived as banal and boring. For this reason, efforts have recently been made in some wineries and growing areas to produce high-quality red wines completely without oak, thus setting themselves apart from the mainstream. The goal is to focus more on the primary characteristics of variety and terroir.

10.6. Redox management during barrel aging

Skillful aging in barrels involves knowledge from the high school of redox management, because all elements of oxidation and reduction interact. They comprise oxygen uptake and its consumption by phenolics, yeast lees, and SO_2 (Chapter 6). In contrast, only reactions of reduction take place during aging in topped stainless steel tanks.

The oxygen supplied by the barrel can be optimal, insufficient, or excessive, depending on the individual wine and its current state of development. In practice, this means that barrel aging is often too oxidative or too reductive. This results in certain requirements for the redox management.

In this context, redox management means the maintenance of a balance between oxidation and reduction processes that is suitable for the wine. For this purpose, various technical tools are available, the use of which challenges the enologist's skills. They include the stirring of the yeast lees and SO_2 additions to drive the wine's development in the reductive direction (Chapters 6.3 and 6.4), and oxygenation by splash-rackings (Chapter 8.2) to drive it in the oxidative direction.

Oxygen promotes oak aroma

Wines aged in new barrels develop a more intense oak aroma than those reductively aged in stainless steel tanks with an equivalent amount of oak fragments of comparable wood quality (Wilker and Gallander 1988). Reactions of oxidation obviously promote the olfactory development of certain oak-derived aroma compounds. This can also be observed when a barrel-aged red wine is aerated by decanting (Chapter 8.5). It is supposed that a dynamic equilibrium between the alcoholic, aldehydic, and acidic state of some oak aroma compounds plays a role in this process. Vanillin as an aldehyde is an example. It can be oxidized to vanillic acid with a much higher odor intensity. A biological reduction of vanillin associated with yeast activity has been observed during alcoholic fermentation and aging on the lees in barrels. After racking, the vanillin accumulated again (Spillman et al. 1999).

As a consequence of the redox sensibility of oak-derived aromatics, aging of the same wine in the same oak gives different sensory results, depending on the amount of yeast lees, free SO_2, and oxygen taken up upon rackings. These variables have a decisive influence on the balance between oxidation and reduction. To a certain extent, the intensity of oak aroma follows the time course of the redox potential, measured in mV (Vivas 1999). This refers to both orthonasal and retronasal olfaction.

During barrel aging, only a fraction of the absorbed oxygen is available for aroma development. The more oxygen acceptors such as yeast lees, tannins, anthocyanins and SO_2 compete for the available oxygen, the smaller this fraction is. The higher the total phenolic content of the wines, the more SO_2 loses its importance as a reducing agent (Chapter 7.4).

Many wineries attach a biased importance to the origin of the oak and the cooperage that has manufactured the barrels, whilst they handle somewhat carelessly the enological parameters we have just discussed. A largely clarified wine develops oak aroma more easily and more quickly in barrels than a cloudy one with high levels of suspended yeast lees. On the palate, however, the clarified wine will show less fullness and more aggressive tannins because there tend to be less polysaccharides able to mask them (Chapter 5.3.3).

A closed and tight aroma due to oxygen deficiency or too much free SO_2 can be improved by aeration. However, too much oxygenation has the opposite effect if undesirable reaction products are formed by irreversible over-oxidation or autocatalysis (Chapter 7.3). Periodic pumping-overs, leading to an oxygen uptake of approx. 2 mg/L O_2 each, are quite able to promote the aroma development of high-phenols red wines aged in barrels.

Discussing these possibilities is not meant to be an automatic call to actionism. Frequently no action must be taken at all. When this happens, any non-interventionist winemaker will be happy. Nevertheless, it is important to understand the interrelationships and interpret the sensory impressions correctly when sampling through barrels.

Unfortunately, there are no reliable instrumental measures for conducting an appropriate redox balance measurement, although such methods based on the use of specific electrodes have been touted for decades. Whilst they have been supposed to indicate the wine's level of oxidation or reduction, they rather indicate the instantaneous oxygen concentration and are probably caused by the oxidation of ethanol coupled to the reduction of oxygen. Therefore, the so-called redox potential is of little value for wine (Danilewicz 2012). This in turn makes periodic tastings particularly important[56].

Importance of temperature during barrel aging

Whenever pump-overs are performed, the wine temperature should be below 15° C (59° F). Higher temperatures considerably accelerate the oxygen consumption rate (Figure 53) in a way that regenerative polymerization of phenols lags behind their oxidation, with the risk of over-oxidation as the consequence. Thus, depending on temperature, a given amount of oxygen supplied can have different sensory outcomes.

In principle, high storage temperatures accelerate barrel aging, but also lead to more wine losses by evaporation and, in particular, to an increasing risk of undesirable microbial activities triggering the formation of VA or Brett flavor.

[56] Cyclic voltammetry as an emerging technique has proved very useful in determining the reduction potential of wine constituents and estimating the concentration of the most readily reduced or oxidized phenols, but the experience with the practical application under real-world conditions is still missing.

Storage temperatures above 17° C (63° F) require careful sensory and, if necessary, analytical monitoring of the wines. Measuring VA from time to time can be extremely helpful when you are not sure of your sensory judgment. At a storage temperature of less than 10-12° C (50-54° F), almost no aging takes place. This is a typical problem during winter in cool-climate growing regions. Under these conditions, a clear leap forward in the development of the wine can only be observed after the cellar has been warmed up in spring.

It is not without reason that traditional barrel aging in the Bordeaux area consists of transferring the unclarified wines into the barrels right after MLF and subjecting them to several rackings during the following cool season. Thus, by gradually decreasing yeast lees as a reducing agent and periodically allowing for some oxygen uptake, the redox potential is repeatedly corrected towards the more oxidative side.

In conjunction with maintaining a free SO_2 level of 20-30 mg/L, this procedure is referred to as 'oxidation under reductive conditions' (Chapter 7.4.1). After six months and two rackings, the barrels are turned with the bung at a 45° angle. This creates a vacuum in the headspace, which reduces oxygen uptake during further storage. Although the above practice cannot claim universal validity, it nevertheless shows the importance to be attached to redox management.

No microbial safety through filtration

Sometimes, especially when novices are going through a learning and orientation phase, the attempt is made to age filtered wines in barrels. One of the reasons for doing so is the search for microbial safety. In doing so, they ignore the fact that filtered wines in barrels have a greater tendency to produce VA and other microbial faults than unfiltered wines. The primary explanation lies in the yeast's ability to consume excess oxygen, which is thus no longer available for the proliferation of aerobically growing microorganisms such as Acetobacter. Practical experience shows that such microorganisms are present at the latest after a few samples have been taken for tasting, even in sterile-filtered wines. In barrels, contamination is unavoidable and sterility impossible.

10.7. Influence of the wine

Without any doubt and even above other things, redox management during barrel aging depends on the phenolic profile of the individual wine. Most deficiencies in fruit quality and phenol extraction during maceration intensify during aging. Regardless of initial must Brix readings, meager wines with low total phenol levels become even more meager. Whilst a satisfactory tannin level is not the only criterion for successful barrel aging, it is a very important one. Wines of the Pinot noir type should have at least 2,500 mg/L GAE total phenols, more color-intensive ones such as Cabernet Sauvignon, Syrah or Merlot even 3,000 mg/L GAE.

Wines obtained by thermovinification without any subsequent maceration period are not suitable for barrel aging due to their lack of tannins, despite their high anthocyanin levels possibly evoking a strong volume on the palate whilst they are still young. This sensation will disappear during aging.

On the other hand, satisfactory anthocyanin levels are also crucial for successful barrel aging. They combine with colorless tannins forming polymeric pigments, thus softening the tannin (Chapters 2.2.5 and 5.3.2) and contributing to flavor maturation. Therefore, barrel aging of slightly colored varieties such as Pinot noir is a major challenge as the contribution of anthocyanins to tannin softening must be found elsewhere and replaced, for example by mannoproteins released from yeast lees (Chapter 6.3). Depending on the initial tannin-anthocyanin ratio, mouthfeel and aroma profile evolve in different ways during barrel aging.

Figure 73 shows by the example of two completely different red wines how the type of wine differentiates the sensory expression of oak during aging in comparable barrels. In this example, the Pinot noir represents the kind of a weakly colored wine with a low initial anthocyanin level, whilst this level was several times higher in the Merlot. As a result, the Pinot noir was rated higher in astringency and lower in the volume/fullness attribute. The evaluation by profile analysis was conducted after nine months of barrel aging under comparable conditions of redox management in new French oak manufactured by the same cooperage.

Figure 73: Impact of the phenolic profile of red wine on the intensity and quality of oak aroma imparted by 9 months' barrel aging in new French oak, medium toasting.

Intensity scoring range 0-5, means of three barrels.

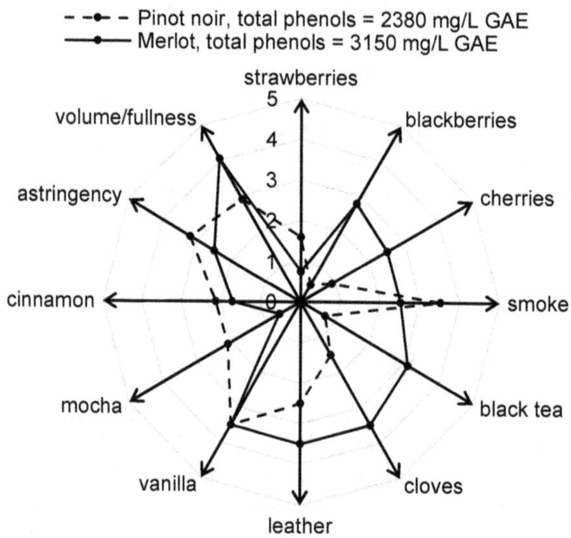

The sensation of fullness imparted by anthocyanins in young wines can be deceptive. It gets lost when anthocyanins are degraded by oxidation and polymerization. This process is accelerated under the semi-oxidative aging conditions in barrels. If there are not enough tannins available to take over the role that anthocyanins play in providing fullness in young wine, the wine will be meager and hollow after barrel aging.

Such wines are better suited to more reductive aging in tanks. Chapter 3.1.4 and table 1 give an overview of how the tannin content can be roughly estimated from the total phenol content if no more specific analytical methods are available.

The sensitivity of anthocyanins to oxidation means that they are rapidly destroyed under conditions of strong oxygenation before they can be incorporated into polymeric pigments by way of regenerative polymerization (Chapters 7.2). For this reason, in wines with low initial anthocyanin contents, it is useful to make efforts to counteract the premature degradation of anthocyanins by maintaining generally more reductive conditions during barrel aging. This can be achieved by abundant yeast lees and periodic bâtonnage, or by maintaining moderate levels (10-30 mg/L) of free SO_2. Yeast lees are considered to absorb anthocyanins, but this argument loses validity as their absorption capacity is already exhausted during fermentation. An exception arises when external yeast less from white wine or industrially produced yeast biomass is added for that purpose (Chapter 5.3.3).

Conversely, wines with high levels of anthocyanins allow us to temporarily dispense with free SO_2 and to even provide additional oxygen by pumping-over. For some of these wines it can even be recommended, depending on the passive oxygen uptake that can be expected in the barrels and taking into account that new barrels show higher oxygen permeation rates than old barrels.

It is not a good idea to fill young red wines into barrels, leave them to themselves and wait for a miracle. Such an approach can work if you have enough experience with your own wines or the type of wine in question, but it does not have to. The effects of the wine type and the various enological practices interact with the wood; they may even override the effects of the wood. The quality of the oak is not necessarily responsible for quality deficiencies in the finished wine.

Practical hints: The most common mistakes during barrel aging

The most frequent mistakes during barrel aging can be summarized in a few key words:

- Under hot-climate conditions, the use of wines too high in alcohol content. The burning sensation caused by alcohol can add to the astringency of the tannin and make the wines less enjoyable in the aftertaste.
- Under cool-climate conditions, the use of too meager wines with insufficient tannin content. They become even thinner during barrel aging.
- Excessive use of new barrels leading to a one-dimensional oak aroma also known as carpentry flavor, a complete loss of fruity varietal aroma by masking effects, and an aggressive astringency on the palate.
- Too short a storage time in new barrels, as a result of which the astringent ellagitannins taken up from new oak have not enough time to break down under semi-oxidative conditions in the barrels.
- Increased levels of volatile acidity and/or ethyl acetate due to acetic bacteria as a result of insufficient topping or excessively high storage temperature.

- Too little yeast lees, which might result in aggressive tannins and deficiencies in fullness on the palate.
- Over-oxidation and subsequent losses of body due to tannin precipitation.
- Reluctance to dispose of barrels which have been identified as contaminated by Brettanomyces.
- Inadequate preservation of empty barrels: When barrels lie empty between fillings with wine, storage with a water solution containing 3 g/L of citric acid and 200 mg/L SO_2 is preferable to repeated SO_2 gassing. The water prevents the wood from drying out. The low pH of approx. 3.3 brought about by the citric acid ensures microbial activity of SO_2 comparable to that in real wine. Thus, microbial activity is reliably inhibited.

11. The use and effects of oak alternatives

Oak is directly involved in chemical reactions occurring in wine that can have positive effects on its flavor and structure. It is considered vital in the production of many of present-day red wines. However, oak barrels are expensive and short-lived, and their use relies on limited resources with serious ecological implications. Therefore, the addition of oak fragments to wines stored in tanks has become widespread and the method of choice for producing a large part of red wines. It is the easiest and most economical way of mimicking barrel aging, albeit the relatively short extraction period of most of these oak alternatives in wine leads more to aromatization than to aging from a sensory and chemical perspective. Hence, wood quality and its toasting level is more important than the geometric shape and size of the oak fragments. This chapter focuses upon their practical application, sensory results, and a simple method for checking oak chips' quality by bench trials.

11.1. Forms, sizes, and quality differences

The use of oak alternatives aims to replicate the characteristic sensory imprint of oak barrels. They comprise all geometric shapes and sizes of oak wood, which are added to the wine and extracted:

- Chips: Usually they are packed in net bags that are hung in the tanks. For barrels, stainless steel tubes constructed as sieves are available, which are filled with the chips and inserted into the barrel through the bunghole. Loose chips placed in the tank make it difficult to remove the wine without damaging the pump.
- Cubes: Cuboid pieces of wood of different sizes.
- Bullets: Spherical pieces of wood that are easier to remove from used barrels.
- Staves: These are planks of various sizes and thicknesses that are attached to the wall of the tank, placed on frames in the tank, lined up on a string, or added loosely to the wine. Due to their smaller surface area in relation to the volume of wood, they are extracted much more slowly and incompletely than chips; the extraction kinetics are very close to those in barrels. Therefore, they can often be used a second time.
- Sticks: Smaller wooden sticks or ministaves that are inserted into used barrels through the bunghole and attached to the bung with a string.

- Powders: They have dimensions of less than 2 mm and are mainly used to provide ellagitannins during fermentative maceration with the purpose of intensifying color. Since they are untoasted or barely toasted, the sensory characteristics of un-toasted wood can easily be found in the future wine.

The importance of adequate toasting

Oak alternatives such as chips, cubes and staves have developed from the use of oak barrels. Analogously with barrels, they introduce oak aroma compounds and ellagitannins into the wine. Thus, the evaluation of their sensory quality follows the same criteria. Likewise their toasting degree follows that for barrels, whereby it must also be taken into account that the toasting classification does not represent an absolute reference as the technologies used by different producers differ considerably. They comprise:

- direct contact of wood with a heated surface,
- exposure to a heated air jet,
- irradiation with IR rays, which does not allow deep toasting of the wood pieces,
- direct contact with a flame, almost exclusively used for the production of staves.

Generally, the influence of toasting intensity on volatile aroma and ellagitannin composition of oak alternatives is higher than the wood origin, piece size, and drying method. In reality, however, the geographical origin of the oak is often unilaterally overestimated.

Comparison between treatments with oak alternatives and barrel aging

If the quality of the oak used for the production of oak substitutes was comparable with that used for barrels, it would theoretically be irrelevant whether one puts the wine in oak or the oak into the wine. However, whilst in the case of barrels, the wood is obtained from the most valuable parts of the trunk, most oak alternatives are recovered from less valuable parts or remnants of the barrel production. This is one of the reasons why it is generally assumed that it is not possible to obtain wines aged with oak substitutes with sensory characteristics comparable to those of barrel-aged wines. Nevertheless, young wines with oak notes similar to wines aged in new barrels for a short period can be obtained (Ortega-Heras et al. 2010).

After 10 months of aging and using electronic sensors, it was possible to discriminate between wine aged in a barrel and the same wine supplied with an equivalent amount of oak chips of the same origin and toasting level, even when treated with MOX (Apetrei et al. 2012). Wines supplied with various oak substitutes tend to display higher levels of vanillin, acetovanillin, and syringaldehyde than wines aged in barrels. On the basis of such volatiles and their concentration ratios, various sets of analytical criteria have been proposed to differentiate between barrel-aged wines and those treated with oak fragments (Butticaz and Rawyler 2007, Hernández-Orte et al. 2014, Rubio-Bretón et al. 2018). The motivation was to prove counterfeiting when wine treated with oak fragments is commercialized as barrel-aged. This is justified by the enormous cost differences between oak alternatives and real barrels.

Summarized, there are several reasons which explain why it is difficult but not impossible to replicate barrel-aged wines by using oak alternatives:

- The best quality oak can be assumed to be preferably used for barrel manufacturing.
- There are technical differences in the toasting process. Whilst barrel staves are toasted over an open fire, convection toasting by hot air or infrared toasting are frequently used for oak alternatives.
- Wine treatment with oak substitutes soaked in stainless steel tanks does not usually include the oxygen uptake typical of barrels and the wine maturation it triggers. However, this shortcoming can be partially overcome when oak substitutes are used in conjunction with micro-oxygenation or other means of oxygen supply (Chapter 11.3).
- Lastly, any reliable comparison between barrel-aged wines and those treated with oak fragments would require the same wine aged under comparable framework conditions concerning wood quality, its amount, its toasting level, the contact time, the oxygen supply, etc. Comparisons of this type are difficult to perform, are rarely conducted, and place high demands on the interpretation of the results. In particular the aging period in connection with the use of oak alternatives tends to be much shorter than true barrel aging.

In summary, aging in new barrels involves both the supply of oak constituents and maturation promoted by oxygen uptake. In barrels used more than three or four times, only maturation occurs due to the oxygen uptake that continues after exhaustion of extractable oak constituents. Conversely, the addition of oak alternatives to wine stored in a stainless steel tank only causes an aromatization without maturation. The extent to which maturation is additionally brought about depends on the amount of oxygen that is actively supplied. By using oak alternatives in used barrels, the combination of both effects can be almost restored, provided that appropriate oak quality is supplied.

Size effects

The time for exhaustive extraction of oak alternatives depends on their thickness and surface. For example, staves are extracted more slowly than large chips, and these in turn are extracted more slowly than small chips (Rubio-Bretón et al. 2018). Under commercial winemaking conditions, average oak chips can be considered exhaustively extracted after about one month. After that, they may continue to remain in the wine without additional sensory effects.

The size of the oak pieces affects the complexity of the flavor they impart. During their toasting, a temperature gradient is established from their surface to the interior. Along this temperature gradient, different aromas are formed in the outer layers of the wood than in the areas further inside. From a sensory perspective, this effect is relatively pronounced in the case of staves and to some degree in the case of cubes, but is hardly noticeable in the case of chips.

In the following section, sensorially perceptible differences of oak alternatives and application rates are presented on the basis of chips. To a great extent, they are also applicable to other oak substitutes such as staves, cubes, and sticks, provided that their surface and thickness, which controls their extraction rate, is taken into account.

Quality differences in chips

As previously discussed, the sensory quality of chips depends primarily on the intensity of their toasting and to a second order on the oak genus they are obtained from, somewhat analogously to the behavior known from barrels (Cabrita et al. 2011). Good toasting of a medium quality wood results in better quality than poor toasting of a very good wood. The qualitative differences of the chips offered on the market are impressive, even within the range promoted by particular producers. The quality is poorly related to the price or geographical origin.

In order to evaluate the quality of individual chip preparations, chips in amounts of 4 g/L were added to various wines without any previous wood treatment. After one month of extraction under identical storage conditions, a sensory profile analysis of the wines treated in this way was performed. Figure 74 shows the results of five different chips (A-E) of identical size from one and the same producer in a red wine blend.

Figure 74: Influence of five different oak chips preparations (A-E) of identical size on sensory attributes of a red wine blend.
Intensity scoring range 0-5, extraction time = 40 days, dosage = 4 g/L

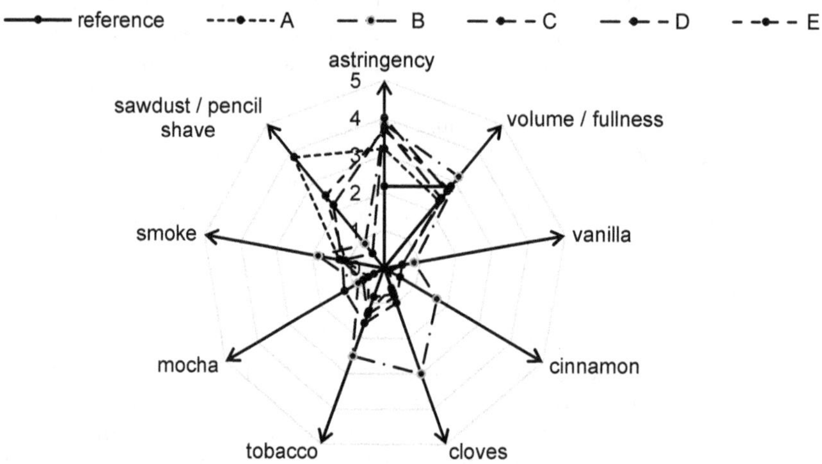

Depending on the chips, the intensity of the individual sensory attributes varies by many integer factors. In this trial, the strongest differences between chips occurred in the intensities of cloves, tobacco, cinnamon, sawdust/pencil shavings, and astringency. The chips preparation coded 'A' would have destroyed any wine under commercial conditions due to the strong 'sawdust/pencil shaving" attribute it imparts.

Olfactory deficiencies due to poor toasting are usually associated with excessive astringency. Some chips are simply unsuitable for use in wine because they do more harm than good. However, all these chip preparations are marketed throughout the world. The enormous differences in quality caused by differences in wood quality and toasting relativize the importance of shape and size in which they are commercialized.

On the other hand, the quality differences between chips also have their advantages, because blending different chip products allows for achieving any degree of complexity. This is an advantage over the use of barrels. How this can be ensured by checking the chips quality in advance is described under the heading 'Practical hints' at the end of the following chapter 11.2. Although using different barrels can also increase the complexity of the wine, this is achieved with less reproducibility than with the use of chips.

Time to summarize: The toasting level of oak alternatives is more decisive for their general quality than their geographical origin. Furthermore, the quality of the wood is more important than its geometric shape and size. The quantity applied has a stronger influence on the wine than the contact time. Deliberate oxygenation brings wines treated with oak alternatives closer to barrel-aged ones. On the other hand, oak substitutes rarely replicate the characteristic imprint of barrels. Results are mixed and appear to depend primarily on oak quality.

Influence of the wine

Additionally, the sensory effect of chips of a given quality is differentiated by wine type and wine matrix. To quantify this effect, four different chip preparations (A-D) in amounts of 3 g/L each were added to two different young red wines varying in anthocyanin content, and to one white wine. None of the wines had undergone previous oak treatment. After an extraction phase of one month, the intensity of general 'oak aroma by smell' was quantified as a single parameter.

Figure 75 shows that the intensity of oak aroma increases from the strongly colored Merlot (750 mg/L anthocyanins) to the weakly colored Pinot Noir (170 mg/L anthocyanins) to the Pinot blanc (0 mg/L anthocyanins). Strongly colored red wines with a high anthocyanin content are able to significantly limit the development of typical oak aromas or - conversely - to better integrate them. A similar behavior can also be observed during barrel aging. There are various tentative explanations. The capability of anthocyanins to quickly bind and deplete DO required for the 'opening' of oak aroma is one of them.

Figure 75: Influence of the wine on the olfactory intensity (0-5) of oak aroma provided by four different oak chip preparations (A-D). Intensity scoring range = 0-5, extraction time = 40 days, dosage = 3 g/L

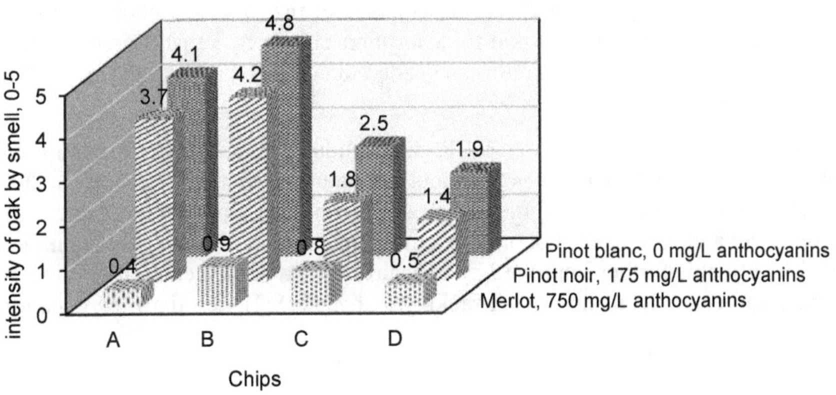

The described experiments with relatively high chip additions to largely clarified wines are very suitable for comparative sensory evaluation of chips quality. For practical purposes, however, these dosages are usually too high since they are not adapted to the type and current stage of the wine.

11.2. Practical application

It goes without saying that the amount of oak to add to a wine depends primarily on the sensory goal sought. At least from this perspective, the use of oak alternatives is not different from that of barrels.

Common wines of the basic segment gain in odor and taste by adding 0.5 to a maximum of 1.0 g/L of suitable chips. Their primary advantage is masking green-vegetative aroma components and improving the complexity of the wine without imparting a noticeable oak aroma that is perceived as such by consumers. Dosages of 2 g/L are the most common, whilst higher dosages make a lasting imprint on the wine and require several months or a year of aging for better sensory integration, preferably under semi-aerobic storage conditions as in wood. The addition of 8 to 12 g/L gives an extraction of oak compounds comparable to that obtained after one year of storage in a new barrel.

In relation to an identical amount of chips, the sensory effect after their addition to cloudy young wines still containing yeast lees is significantly lower than after filtration. This is described as better integration. Only after an early addition will there be enough time to achieve harmonization of the chips' imprint before bottling. Thereby, yeast less and oxygen help in different ways: The yeast reduces excess astringency by directly absorbing ellagitannins and by masking them through the polysaccharides it releases (Chapter 5.3.3). Oxygen opens up the aroma, intensifies the color, and promotes flavor maturation through polymerization of the tannin-anthocyanin complex.

Due to the time required for sensory integration and the beneficial contribution of yeast to that purpose, the early addition of chips during fermentative maceration is preferred. The aim is to use the ellagitannins they release to promote the formation of polymeric pigments as soon as possible (Chapter 4.1.15). However, the chips are inevitably removed upon pressing. As a consequence of the relatively short time they are in contact with the must and due to adsorption effects by yeast, this procedure requires about 4-5 times as much chips as needed to achieve the same sensory effect of an addition post-fermentation.

When chips are added to wine after MLF, their amount is more decisive than their contact time, although the latter is not without importance. After one week of contact, a distinct, albeit somewhat one-dimensional oak aroma can already be noticed. Full extraction over a month results not only in a more intense, but also in a more complex aroma. The effect is further improved by measures of active oxygenation previously described, for example using splash-racking (Chapter 8.2) or micro-oxygenation (Chapter 8.3).

It is always useful to have some unoaked wine available for blending back wines that have become over-oaked. Conversely, there is also the possibility of preparing a kind of 'oak extract' in a timely manner by adding, for example, 50 g/L of oak chips to a small volume of wine. This concentrate can be used for blending with untreated wines in amounts chosen on the basis of previous bench trials and, thus, to precisely adjust them to the desired oak level.

Practical hint: Bench trials for checking chips quality

The previous section has focused upon the enormous quality differences between oak substitutes. Their quality characteristics should not be taken credulously from the product descriptions provided by many suppliers. Hence, they should not be used without previous testing. Whilst staves and sticks cannot be trialed in single-bottle trials in the laboratory, this can be easily done with chips. Extracting 3 grams of chips in a bottle of wine over three to four weeks allows for a satisfactory sensory assessment of their quality. More chips in a correspondingly larger wine sample volume on the pilot scale make the result even more meaningful, because unavoidable differences between individual wood pieces are better compensated and a more reliable average is obtained.

11.3. The combination of oak alternatives with oxygenation

Although red wines do not necessarily have to be characterized by oak, most of them are treated with oak in one or another way to meet widespread consumer expectations. Moreover, with the aim of simulating the effect of barrel aging with lower costs, oak fragments are widely added to wine in combination with micro-oxygenation or other techniques of oxygen supply. Indeed, from a global perspective, the majority of contemporary red wines are produced in this way.

As alluded earlier, the combined use of oak alternatives and deliberate oxygenation would be a perfect solution to replace expensive barrel aging if oak quality was identical in both the barrel and the alternative format. However, this if not always the case. The challenge to replace barrel aging by the use of oak alternatives is further complicated by a multiplicity of additional variables involved. Hence, results of this approach are conflicting and difficult to generalize.

From a sensory perspective, most consumers are not able to clearly distinguish wines treated with oak alternatives in steel containers from those aged in new barrels for short periods of about three to four months. Both are characterized by oak olfactory notes. However, considering the effects of oak wood quality and wood contact time, some evident sensory differences can be observed: Wines aged in used barrels with oak alternatives are much more comparable to those aged in new barrels than wines supplied with oak fragments in tanks. Moreover, the longer they are aged, the more difficult they are to distinguish from those aged in new barrels.

This chapter shows how the combination of oak alternatives and oxygenation can be used to mimic traditional barrel aging as closely as possible.

MOX prior to addition of oak chips does not replace barrel aging

In trials with pre-MLF MOX and subsequent addition of chips on Merlot and Petit Verdot wines, the sole use of MOX promoted polymerization of red pigments and color stability as would be anticipated. It also resulted in higher sensory scores for fruity attributes and caused the presence of new, nutty and sweet fruit attributes. However, MOX prior to the addition of chips also reduced the intensity of the typical oak chips' aroma (Cejudo-Bastante et al. 2011 a, 2011 b). In other studies, MOX showed no effect on the extraction of volatile compounds extracted from subsequently added chips. In contrast, it showed the outstanding importance of chip selection (Rodríguez-Bencomo et al. 2008), or an effect in conjunction with other, wood-related variables (Pizarro et al. 2014).

All in all, MOX prior to chips addition hardly offers the possibility to improve the sensory outcome of the contact with oak fragments in stainless steel tanks. The lack of a continuous supply of oxygen over a longer period of time as it occurs during barrel aging seems to be one of the main causes. Therefore, oak alternatives are frequently used in conjunction with techniques of active oxygen supply. The outcome will depend not only on the individual wine and the amount of oxygen dosed, but also on the size and wood quality of the oak fragments (del Alamo et al. 2010).

Post-MLF MOX plus oak chips in tanks can roughly simulate barrel aging

In one trial conducted on Sangiovese stored in stainless steel, the effects of post-MLF MOX, oak chips addition, and yeast lees addition, each separately or combined, were evaluated. The combined use of all measures resulted in lower amounts of several volatile oak compounds than in the control only treated with oak. On the other hand, it most reduced astringency, increased balance, and enhanced vanilla perception when compared to the untreated wine. Yeast lees played a decisive part in that improvement, and differences were more pronounced on the palate than by smell (Arfelli et al. 2011). These gustative changes are more important than the color improvement, which can be observed after all treatments combining MOX and oak alternatives.

In another trial, the impact of MOX in conjunction with a variety of oak alternatives in stainless steel was compared with traditional barrel aging. Descriptive sensory analysis showed that MOX in combination with oak chips or staves could mimic short term barrel aging of six months in new American and French oak barrels (Oberholster et al. 2015). As so often, the effect of a longer maturation period was also missing in this case. Post-MLF MOX might improve the wine in the short term, but it cannot replace a long term barrel aging.

Oak alternative additions to used barrels can best reproduce barrel aging

When the intention is to imitate barrel-aged wines as closely as possible, the use of oak chips or mini-staves during aging in used barrels is more promising than the combination of oak fragments and MOX in stainless steel tanks. A sensory study on a Monastrell wine showed that when oak chips were added to used barrels, the differences between these wines and those aged in new barrels were still perceptible, but small. Best results were found for wines matured in the presence of oak cubes and after three to six months of contact time. In more general terms, the most pronounced improvement of this approach is expected in mid-range wines, in which it leads to a decrease in vegetal characters, improved complexity, increased volume and mid-palate tannin intensity (Bautista-Ortín et al 2008, Cano-López et al. 2008 b). Obviously, the period of time the wine spends in the barrels plays a major role.

Oak alternatives provide low amounts of oxygen

Wood itself contains air with 20.8 % oxygen. The outgassing of oak from new barrels is known to contribute to the total oxygen uptake of wine during barrel aging. Hence, one can rightly assume that oak fragments also supply the wine with oxygen (Nevares and del Alamo-Sanza 2016). However, contradictions between theoretical considerations and DO measurements in real wine after oak fragment additions gave rise to many speculations.

Figure 76 shows the actual amounts of oxygen released by various types of oak chips flooded with model solution that was previously deoxygenated and free of any oxygen scavengers. The oxygen released was measured as DO increase in the solution using non-invasive fiber optic luminescent technology. Complete outgassing of the chips took place within a few hours. The average oxygen release per 1 g/L of oak chips was 0.19 mg/L DO, but with a large fluctuation around the mean value. It appears that the larger the oak fragments are, the less oxygen they release per unit weight. From a practical perspective, it can be stated that a common chips dosage of 2 to 4 g/L hardly imparts more than 1 mg/L DO to the wine, which should not be overestimated.

Figure 76: Oxygen released by various types of oak chips added to model solution at 1 g/L.
m = medium toasting, dimensions in mm represent the average particle size.

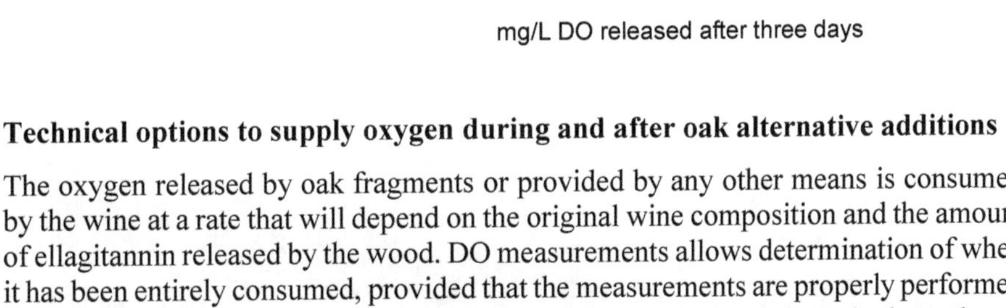

Technical options to supply oxygen during and after oak alternative additions

The oxygen released by oak fragments or provided by any other means is consumed by the wine at a rate that will depend on the original wine composition and the amount of ellagitannin released by the wood. DO measurements allows determination of when it has been entirely consumed, provided that the measurements are properly performed by choosing a representative sampling point of the entire wine volume in the tank and by using a suitable measuring system that can report DO levels lower than 0.1 mg/L. Under commercial winemaking conditions, sensory evaluation is usually performed to determine whether additional oxygen dosing is advisable. If this is the case, four courses of action are possible:

- Macro-oxidation using pumping-over and splashing as outlined in section 8.2.
- Micro-oxygenation according to section 8.3
- Passive oxygen uptake by aging in used barrels after removal of the oak fragments. Moreover, the addition of oak fragments to wine aged in old barrels is an efficient means to approach the sensory profile imparted by new barrels, with the precise outcome depending essentially on oak quality and toasting level. (Bautista-Ortín et al. 2008). It is also a good choice for reusing used barrels under good sanitary conditions. Mini-staves are perfectly suited for that purpose. They are tied together with a string, inserted through the bunghole, and easily pulled out again whenever necessary.
- Passive oxygen uptake by aging in oxygen-permeable HDPE tanks as referred to in sections 8.1.2 and 8.3.7. Reports on results are still rare. In one trial, aging red wine supplied with oak alternatives in one of these tanks resulted in more color due to polymeric pigments and in lower intensities of sweet and cinnamon aromas, as compared to the wine aged in barrels (del Alamo-Sanza et al. 2015). The data suggest that the typical reactions associated with oxidation were more pronounced in the HDPE tank. However, this was a unique result, which does not question the usefulness of such tanks for the stated purpose. The solution to the problem is simple: More trials and experience are needed to optimize the oxygen supply through a suitable volume to wall surface ratio. It makes sense to carry out such trials with the tanks available on one's own winery and with one's own wines. Once the appropriate volume-to-wall surface ratio has been found, the rest is just a matter of identifying suitable oak fragments.

12. Preparing red wine for bottling

In red winemaking, the focus of enological interest, both on the part of the scientific community and of producers, is on technical variables associated with fermentative maceration, micro-oxygenation, and barrel aging. In contrast, less attention is paid to the final steps before bottling, despite their providing a wide range of possibilities to intervene, to shape wine style, and to ensure sensory stability and shelf life in particular. In this section, examples are given of how simple procedures such as fining with proteinaceous agents or minor acidity corrections open the possibility to strongly modulate perceived astringency, body, and other mouthfeel-related attributes. They allow one to better adjust the wine style to marketing requirements or specified benchmarks. Adjustment of free SO_2, its correct measurement and its long-term stability are of particular importance for flavor stability and ageability during bottle storage. SO_2 stability is strongly associated with instantaneous DO as well as with oxygen uptake during bottling, from the bottle headspace, and through the bottle closure. Calculation examples show how post-bottling SO_2 losses can be better controlled and limited. Finally, options of non-interventionist winemaking such as the waiver of SO_2 additions or filtration are also discussed.

12.1. Taste corrections

12.1.1. Modulating astringency by acidity adjustments

Before bottling wines, the question of their sensory quality and consumer acceptance inevitably arises. This concerns both taste and smell characteristics. Whilst aroma is frequently accepted as long as a wine is flawless, taste is often neglected. On the palate, a wine might be considered as too astringent or too soft, as too sour or too flat, as too thin or too concentrated. It depends on the type of wine sought and the cultural context whether a wine is perceived as satisfactory or deficient. Therefore, it is pointless to discuss quality in absolute terms. In contrast, it is important to engage with it and compare it with other wines.

A disturbingly high astringency can be caused by too much tannin for the kind of wine looked for, tannin of poor quality, or too much acidity. Due to the sensory interaction between tannin and acidity (Chapter 5.3.4), a definitive sensory evaluation of the quantity and quality of tannin is only possible after acidity has been brought under control. Acidity corrections downward or upward have an enormous impact on perceived astringency. Albeit easy to perform, they are frequently neglected as a tool of fine tuning when wines are prepared for bottling.

The mere fact that a wine has undergone acidity corrections on the crush pad or that it has completed MLF does not guarantee that sour taste is perceived as optimal. Adding some residual sugar is a poor way to mask excess acidity in dry wines. Whilst 2-3 g/L residual sugar can in some cases still integrate harmoniously into the wine without being perceived as undesirable sugar sweetness, higher contents usually lead to unbalanced wines.

Running trials on acidity adjustments

Quite a few producers are afraid of acidity adjustments at this late stage because they associate it with post-bottling crystal stability issues, though these issues can be easily overcome if they really occur. On the other hand, others expect a loss of quality from such a measure because it has a bad image or because they are actually inexperienced in conducting it properly. Such emotionally driven aversions do not change the fact that many, if not most problems of taste disharmony in red wines can be solved by means of slight acidity adjustments in the range of a small fraction of a g/L of titratable acidity (TA). In this context, it is useful to keep in mind that TA determines perceived sourness much more than pH. However, a one-sided orientation towards analytical numbers is not useful anyway. Simple preliminary bench tests are indispensable.

Table 18 shows how such a preliminary test can be performed for the purpose of deacidification with potassium bicarbonate ($KHCO_3$). It can be easily replaced by potassium carbonate (K_2CO_3), taking into account that 1.0 gram $KHCO_3$ equals 0.69 gram K_2CO_3. Conversely, an analogous test can be performed for acidification, using lactic acid[57] for example. In doing so, 100 mL of lactic acid (80 %) must be made up to 1,000 mL with water. The rest is identical.

Table 18: Preparation of test solution for deacidification of wine and implementation of trials.

Preparation of test solution	Dissolve 100 g of $KHCO_3$ and adjust to 1,000 mL with distilled water.				
Use of the test solution	0.1 mL / 100 mL wine equals in the tank: + 0.1 g/L $KHCO_3$				
	+ 0.25 mL	+ 0.50 mL	+ 0.75 mL	+ 1,00 mL	+ 1.25 mL
Equals in the tank:	+ 0.25 g/L	+ 0.50 g/L	+ 0.75 g/L	+ 1.00 g/L	+ 1.25 g/L

Annotation 1: Before tasting, the carbon dioxide produced upon potassium carbonate addition should be thoroughly removed by shaking. Otherwise its bubbles would considerably detract on the palate.

Annotation 2: For reasons explained in the further text, samples treated with potassium carbonates for deacidification should be frozen over night to remove potassium tartrate produced, whose crystallization causes an additional acidity loss.

[57] In contrast to other acids used in winemaking, lactic acid has the advantage of not affecting crystal stability nor being subject to microbial degradation.

When such tests are performed, the focus should not be on the calculated or pursued final acidity, but on the amount of product – acid or potassium carbonate – required to achieve the desired sensations on the palate. Final TA or pH figures are useful but do not sell the wine.

Every winemaker would be well advised to permanently keep such test solutions at hand. Their frequent use will help understand why minor acidity adjustments are a powerful tool of sensory optimization, indeed much more powerful than a large range of commercial additives usually recommended for that purpose.

Taming astringency via acidity corrections

Whilst the implementation of acidification in practice is possible without major problems, deacidifications require some more considerations with regard to subsequent precipitation of potassium bitartrate. Before going into these details, a look at the sensory outcome is useful. Figure 77 displays the sensory changes on the palate caused by deacidification of a Merlot wine from 5.7 to 4.7 g/L TA using $KHCO_3$. These changes affect much more than only sourness. Since sourness and volume are sensory antipodes, one affecting the other, deacidification can strongly enhance what is perceived as weight and volume on the palate. Another important change is the significant reduction of perceived astringency. It clearly shows that tannin management is inextricably linked to acidity management (Chapter 5.3.4).

Figure 77: Sensory impact of deacidification of a Merlot red wine from 5.7 g/L to 4.7 g/L TA using 1.1 g/L $KHCO_3$ and cold stabilization.
Intensity scoring range = 0–10.

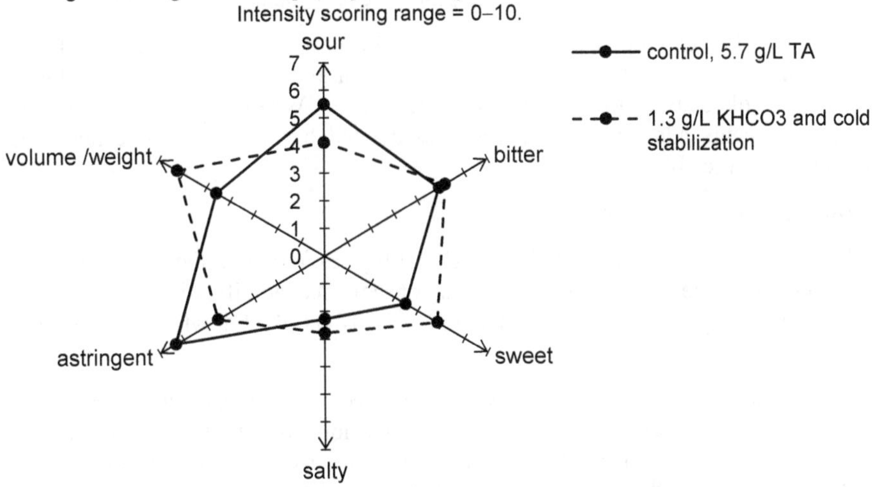

Trials of that kind can show a way to disclose additional potential of flavor improvement even in red wines previously judged as perfect. They help one to better understand the complex relationship between sensory properties and analytical make-up. In numerous red wines, meticulous deacidification provides much more benefit for quality improvement than the popular range of traditional treatments. One of the reasons is that, as explained previously, mitigating sourness reduces astringency and enhances volume.

The amount of 1.1 g/L $KHCO_3$ to reduce TA by 1.0 g/L in this red wine requires an explanation. According to stoichiometric rules and to what one might have experienced with white wines, this amount would be expected to reduce TA almost twice as much. Indeed, 0.67 g/L $KHCO_3$ is commonly supposed to remove 1.0 g/L TA via precipitation of tartaric acid. This calculation may work out in most white wines, but the well-known laws governing potassium bitartrate stability in white wines do not apply to red wines. The reason for this is the tannin-anthocyanin complex, which white wines do not have. It includes molecules that are more or less polymerized and that can reach a colloidal state, in which they act as protective colloids. Anthocyanins are also strong crystallization inhibitors. Thus, these substances partially or entirely inhibit the precipitation of potassium bitartrate produced upon deacidification with $KHCO_3$. Since potassium bitartrate is a sour salt, the acidity loss associated with its precipitation does not occur. As a result, the amount of $KHCO_3$ required to remove a given amount of TA increases. It will be only 0.67 g/L to remove 1.0 g/L TA when it drops out completely, but it will be 1.34 g/L when no precipitation takes at all. Chemical backgrounds and calculations have been largely described (Schneider and Troxell 2018).

The extent of potassium bitartrate precipitation in red wines is quite unpredictable, and so also is the efficiency rate of a given amount of $KHCO_3$ with regard to TA decrease. For that reason, it is advisable to freeze the samples obtained by deacidification trials as outlined in table 18 overnight and to remove bitartrate crystals before sensory evaluation takes places the next day. In this way, a possible subsequent reduction in acidity due to bitartrate precipitation can be taken into account.

However, whether with or without cold stabilization, potassium added as $KHCO_3$ quite often remains partially or completely in solution for a long period. This also applies to cold stabilization on the commercial scale. Whenever some bitartrate precipitation occurs, additional acidity losses associated therewith are most often modest, though unpredictable.

When pH becomes critical

All winemakers are concerned about high pH figures, although there is no consensus on what constitutes a truly high pH. As a consequence, acidification as a kind of astringency management is always welcome as it lowers pH, but what about deacidification measures?

Any reduction in acidity is inevitably accompanied by an increase in pH. From a microbiological perspective, this is not a problem as long as the wines are bottled under sterile conditions as it occurs in most wineries. However, experience has shown that pH figures above 3.8 are detrimental to wine shelf life in a chemical sense for two reasons:

- Acids drive the pH down, mineral cations and especially potassium drive it up. Therefore, pH is the result of an interaction of TA and potassium content. Potassium has a soapy aftertaste regardless of the actual wine acidity level. This inherent taste becomes more evident during wine aging, when the masking effects of the astringency of young red wines are reduced.

- The higher the pH, the more phenols are present in their dissociated form, which accelerates their oxidation and polymerization (Chapters 6.4 and 7.2). The extent to which this happens depends on the oxygen ingress rate through the bottle closure (Chapter 12.5). It is negligible when hermetically sealing screw caps are used as bottle closures, but it gains in significance when corks are used. Therefore, in premium wines intended for a long bottle storage, pH should kept under control and by no way exceed 3.8. Slightly higher figures of up to 4.0 are only acceptable in wines expected to be rapidly marketed and drunk.

In many red wines, problems with apparently too much tannin or too much astringency can be solved by a slight deacidification, without the need for smoothening fining agents aiming at reducing tannins. But there are exceptions. Especially when ellagitannin are involved, for example after aging in new barrels or after a late treatment with oak alternatives.

Caution with surplus carbon dioxide

Whenever carbonates are used for sensory fine tuning, care should be taken to remove the carbon dioxide (CO_2) they inevitably produce. But even without such a treatment, wines may still contain residual CO_2 from fermentation at the moment of bottling, especially when they have been stored in stainless steel tanks. Therefore, attention should be paid to the CO_2 level in any case before bottling.

All still wines contain CO_2 at sub-saturation concentrations, which cause a sensation of fizziness when they exceed some 0.6 g/L. Whilst this fizzy or 'spritz' sensation may be desired in some styles of white wine, it is considered absolutely undesirable in red wines of any kind. It is not clear how a perceptible CO_2 level impacts astringency and bitterness, but there is general agreement that it accentuates the harsh character of high-phenol red wines and reduces perceived viscosity and fullness. Hence, it is advisable to make sure that red wines do not contain more than 0.5 g/L CO_2 at bottling. Lower levels are welcome.

Measuring CO_2 in the winery is easy and fast using the 'Carbodoseur' method: 100 mL of wine are sampled in a graduated cylinder and shaken to degas the CO_2. The shrunken wine volume and temperature are then measured, and a handy chart translates the result into mg/L CO_2. Accuracy is ± 0.1 g/L and satisfactory for practical purposes. For those who like it even easier, fill a bottle half full with the wine, closes the rim with the thumb, and shake it vigorously. If a distinct pressure is felt when removing the thumb, the wine contains too much CO_2.

There are elaborate methods to remove excess CO_2 such as sparging with nitrogen, etc., but a simple splash-racking by pumping the whole wine volume from one into another vessel is the most useful for red wines. Since it entails some oxygen uptake, it should be performed in a timely manner to allow enough time for stabilizing free SO_2 (Chapter 12.4.2).

12.1.2. Lowering astringency with mannoprotein additions

In practice, the question of reducing astringency in the short term often arises shortly before bottling. The techniques of micro- or macro-oxygenation (Chapters 8.2 and 8.3) are fundamentally unsuitable for that purpose because they would require a period of several weeks or months, apart from the fact that they are of little use at this relatively advanced stage of wine development. They can by no means be shortened by increasing oxygen supply rates because intensive oxygenations are always counter-productive (Chapter 7.4). They are not an option to make up for what was previously missed or to accelerate maturation in a time-lapse mode during the last weeks before bottling. Thus, other corrective treatments come into play.

In chapter 5.3.3, the sensory effect of mannoproteins extracted from yeasts has been extensively discussed. They are able to better integrate tannins, to enhance weight and volume on the palate, and to mask excessive astringency to a certain extent in some wines. Numerous commercial preparations obtained from yeast are available, but their effect cannot be predicted since they greatly differ in their chemical composition and sensory properties. Therefore, preliminary tests should be performed with increasing additions of such preparations according to manufacturer's instructions. Sensory diligence is required.

Care should also be taken with regard to filtration. Some mannoprotein preparations cause a considerable pressure increase, in particular when membranes are used.

It is indispensable to keep in mind that the euphemistic product descriptions used in marketing sometimes have little in common with reality, and that what works out on one wine might not work out on another wine. This applies in particular to the various preparations of gum arabic, which are advertised as having comparable properties. Their softening effect is below the sensory threshold. Therefore, they are unsuitable to reduce surplus astringency (Constantin et al. 2011).

If trials with mannoproteins fail, fining agents will probably come into play as the next and last step. Unlike acidity management or mannoprotein additions, which improve the sensory expression of a given amount of tannin, fining causes its reduction, precipitating a part of it.

12.1.3. Lowering astringency with gelatin finings

Fining with gelatin or similar fining agents has been traditionally associated with the promotion of self-clarification, and many older textbooks continue to overstate that effect although it occurs to a certain extent. In the meantime, the role of clarification has been largely taken over by modern and abundant filter equipments, which also provide the required microbiological stability post-bottling. For this reason, finings focus on flavor improvements and sometimes on stability against post-bottling hazes. In red wines, they are usually aimed at reducing astringency.

Traditionally hen's egg white was used to remove the most astringent tannin fractions and to simultaneously promote clarification before bottling. More recently, it has largely been replaced by a wide range of commercially available ready-to-use fining agents. Gelatin, caseinates, and polyvinylpolypyrrolidone (PVPP) are the most popular of them. Since people have become increasingly critical of animal proteins, proteins of plant origin are also becoming more widespread. The efficiency and application rates of these fining agents vary considerably.

Frequently, such finings are performed on the basis of empiricism or preliminary tests. However, the emotional closeness to one's own wine and the preferred fining agent are often at the expense of sensory objectivity. Therefore, key analytical data can become helpful to back up sensory evaluation. For this purpose, the first step is the simple measurement of total phenols (Chapter 3.1), perhaps complemented by a more specific measurement of the tannin fraction (Chapter 3.4). The knowledge and proper interpretation of the data obtained facilitates the rational decision on measures to take for reducing phenols and astringency. In some way, this approach is similar to the determination of pH and total acidity, which provides guidance on how to perform acidity corrections. However, it cannot replace sensory evaluation.

Gelatin is a proteinaceous fining agent of porcine origin. Among all proteinaceous fining agents, it is the most widely used and the most effective one in red wines. As it preferentially binds with large phenol molecules, it can have a dramatic effect on tannin reduction in older wines as they contain a greater percentage of large tannins. Hence, its use can result in stripped wines and some color losses when added in excess. Accordingly, preliminary trials are required.

Figure 78 shows the effects of two fining agents, gelatin and PVPP, on the total phenol content of two red wines with quite different phenolic constitutions. Sensory evaluation suggested a need for fining in the low-phenol Merlot due to the poor quality of its unripe tannins, whilst the high-phenol Pinot noir was considered too astringent as a consequence of its high tannin level due to over-extraction. Data for gelatin represent means of three different types of gelatin. The following features can be identified:

Figure 78: Decrease of total phenols induced by increasing amounts of gelatin and PVPP in two red wines.

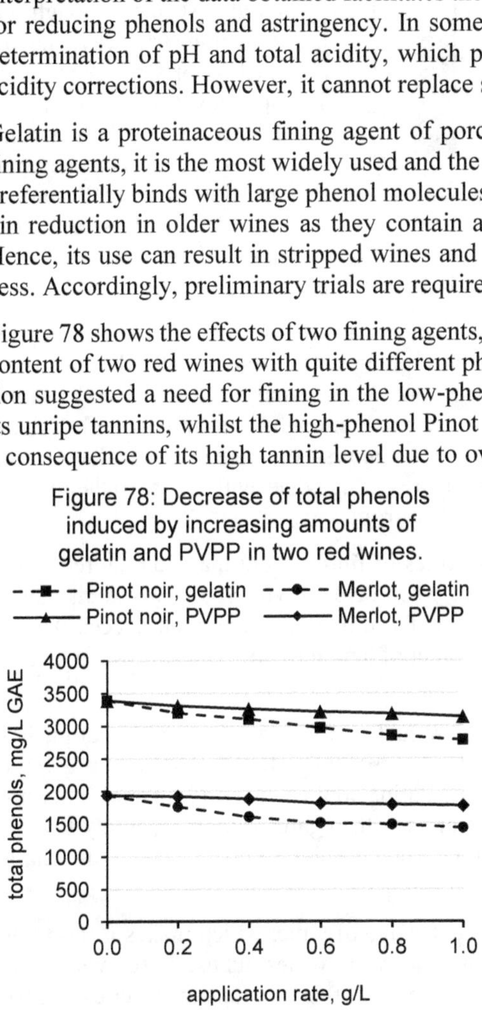

- At comparable application rates, the gelatins in both wines led to a greater reduction in total phenolic content than the PVPP. In contrast to white wines, for which PVPP has considerable interest to remove low-molecular flavonoid phenols on a low concentration range, the efficiency of PVPP in red wines is of less practical importance.

- Additions of fining agent that might be considered high under commercial winemaking conditions result only in a relatively small percent reduction in total phenol content.

However, much more important than any analytical data is the sensory outcome, in this case the effect of fining on perceived astringency. An example is given in figure 79 for a gelatin applied in increasing amounts to a high-phenol red wine blend. In this assay, astringency gives the mean scoring of 12 panelists on an intensity scale ranging from 0 to 10, and tannins were measured using methyl cellulose (Chapter 3.4.4). Whilst total phenols and tannins are reduced with increasing gelatin additions in an almost linear though not stoichiometric way, the reduction of astringency does not show this linear pattern. Instead, it displays a relatively strong decrease upon addition of a small amount of gelatin, whilst this sensory effect, related to a given amount of gelatin, decreases with increasing gelatin additions.

Figure 79: Effect of increasing amounts of gelatin on astringency and total phenols in a red wine blend.

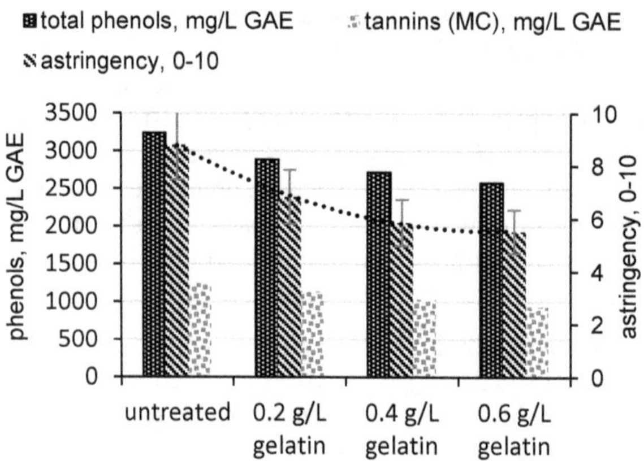

The reason why low additions reduce astringency disproportionately more than higher amounts has a simple explanation: Gelatin precipitates preferentially the most polymerized and most aggressive tannin fractions. Macro-analytical assays such as total phenol or tannin measurements are not sensitive enough to accurately reflect these sensory differences. However, this does not diminish the informative value of total phenol or tannin measurements for roughly assessing wines with regard to their tannin level and the astringency to be expected therefrom.

Notwithstanding this, considerable quantities of fining agent may be required to noticeably reduce astringency, in particular when dealing with young wines displaying high total phenol levels. In extreme cases, these amounts may well exceed 0.5 g/L gelatin. The precipitation of a certain amount of tannin requires an equivalent amount of fining agent.

There are measurable differences in the reduction of astringency and total phenols between individual gelatins when added at comparable rates. An example is given in figure 80, which displays results of two gelatins applied to a high-phenol Pinot noir red wine. Such differences between gelatins with regard to their reactivity towards tannins have been known for a long time. They are due to differences in the charge density and average molecular weight.

Gelatin carries a positive charge in the pH range of wine, which makes it react with the negatively charged tannins. In contrast to white wines, its use in red wines does not require any co-fining with a negatively charged flocculation aid. After charge balancing, the complex produced is insoluble and precipitates. Gelatins with high charge density and low molecular weight preferentially precipitate higher polymerized procyanidins and those esterified with gallic acid. These are the most astringent tannin fractions. High molecular weight gelatin also extends its effect to less polymerized procyanidins (Yokoktsuka and Singleton 1995, Lagune and Glories 1996, Versari et al 1999, Sarni-Manchado et al 1999 b, Maury et al 2001, Cosme et al 2009).

Figure 80: Impact of two gelatins on astringency and total phenols of a red wine blend.

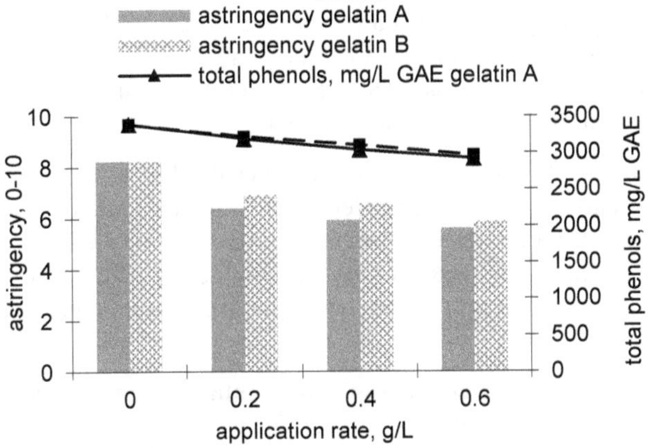

Unfortunately, these differences are neither clear-cut nor directly transferable from one wine to another. Instead, the sensory effects resulting therefrom depend considerably on the phenolic composition of the respective wine. Thus, they are not exactly predictable in practice. Accordingly, preliminary bench fining trials are highly recommended.

Bench fining trials

These trials are typically performed on a magnetic stirrer enabling continuous mixing of the wine sample, for example 100 mL. The gelatin is previously dissolved in hot water of 40-50° C (104-122° F) to make up a 10 %-solution if commercially available liquid gelatin is not used. The dissolved gelatin is slowly added to the wine while the magnetic stirrer is running at medium speed. Mixing is continued for another two to three minutes after the gelatin addition is completed.

It is important to note that the sensory effect can only be assessed in the clarified wine. As long as the gelatin-tannin complexes are still in suspension, reactive sites of the tannin molecules still interact with the proteins of the saliva and continue causing astringency. This applies to both bench trials and commercial-scale application. Sample centrifuging is recommended to accelerate decision-making when preliminary lab tests are run. Otherwise sedimentation overnight will be required.

After fining, color losses are measurable, but are rarely relevant to the human eye.

Practical application of gelatin fining

0.2 g/L gelatin is a common application rate for the purpose of reducing excessive astringency. This amount represents a rough orientation and does not dispense from performing preliminary bench trials. In order to achieve a sensorially perceptible effect, 0.1 g/L is required in most wines. Liquid gelatin preparations are also useful for the above purpose, taking into account their effective concentration, usually 20 %.

Low molecular weight (low Bloom strength) gelatins are preferred to high molecular weight (high Bloom strength) gelatins, which need to be dissolved in hot water before incorporation into the wine. Although there are differences between gelatins, they should not be overstated. More important than these differences between gelatins is the way they are used.

It is crucial to ensure that when the gelatin solution is added, it is added slowly at the point of very intense agitation such that a small amount of gelatin is immediately mixed into a large bulk of wine that is intensively stirred. Any local overconcentration of gelatin in a partial wine volume should be avoided. In the absence of other reaction partners, the effect of gelatin would thus also extend to low molecular weight tannins of low astringency which are worth preserving. Furthermore the part of the gelatin thus inappropriately reacted is no longer available for precipitation of the more aggressive tannin fraction in the remaining volume (Schneider 2009 b). This principle applies in the broadest sense to all proteinaceous fining agents.

On the other hand, the risk of overfining is minimal. However the fining is carried out, red wines always contain enough tannin to completely and almost spontaneously precipitate the amounts of gelatin required for sensory optimization - up to 1 g/L - without the risk of residual precipitation post-bottling. In contrast to white wines, which contain almost no tannins, for reds no co-fining is required to ensure complete precipitation.

The mixing intensity is essential for the effectiveness of the fining. Pumping-over is the worst solution, although it is still sometimes practiced. The use of compressed air or inert gas for mixing is not much better. Tank mounted propeller mixers screwed onto a tank fitting, e.g. a shut-off valve near the bottom of the tank, provide the most efficient mixing. Variable speed mixers provide additional benefits in any winery. In the case of very large tanks, in-line dosing by means of a metering pump or Venturi tube may be useful.

The height of fall and the temperature have considerable influence on the settling rate of the flocculated gelatin-tannin complex. In containers of the height that is commonly used in wineries, it settles within one to five days. However, when the wine is left for more time on the fining slurry, the fining lees become increasingly compact and eventually assume an almost granular structure comparable to pebbles. Once the fining lees are settled, they should not be resuspended by a subsequent mixing. This would cause a breaking of the flocculated particles into finer ones, which are hardly able to settle again. Expenses with filtration would be accordingly higher. In particular cross-flow filtration reacts to residual gelatin-tannin particles with a strong decline in performance.

12.1.4. Lowering astringency with alternative finings

Casein is another proteinaceous fining agent of animal origin and the principal protein of milk, from which it is obtained. Fining with pure casein requires previous and laborious dissolution in hot (~ 45° C / 113° F) or alkalized (pH 8 with K_2CO_3) water. Furthermore, it flocculates instantaneously as soon as it comes into contact with the acidic pH environment of wine, thus complicating its homogeneous incorporation into the vessel. Therefore, it is commonly commercialized as potassium caseinate, which is easy to use but less effective than pure casein. In red wines, all casein preparations are less effective than gelatin. Similarly to PVPP, the focus of their application is on white wines.

The ethically justified rejection of animal proteins and the obligation to declare potentially allergenic proteins from eggs and milk led to the development of alternative fining agents. These are obtained from wheat, peas, rice, potatoes, lupines, soy, and other vegetal sources. They act similarly to gelatin, but most display a lower degree of action (Maury et al. 2003). The effect of proteins from potatoes, peas, and soy is closest to that of gelatin; proteins from potatoes and rice are most suitable substitutes for gelatin and PVPP (Kang et al. 2018). The development of such preparations is ongoing and further innovations can be expected.

12.2. Filtration and bottling unfiltered wines

Keeping clarity and stability in bottled wines has been essential for winemakers since filtration became available. Filtration prior to bottling is generally considered essential to provide wine stability under both aesthetic and microbial aspects for the vast majority of wines. It eliminates turbidity, impurities and microorganisms, and it would be ridiculous to suggest that these substances make a positive contribution to flavor.

More recently, however, voices have been raised in certain customer segments questioning the value of perfect clarity and stability, claiming that filtration strips the wine of its integrity and compromises its quality. Immediately, a number of producers adopted this view and began bottling unfiltered and cloudy wines, referring to their naturalness. In narrowly defined markets, such wines can achieve considerable recognition and prices because their turbidity is interpreted as genuineness and a return to nature, as a response to society's increasing drift towards things that are considered natural and authentic, regardless of possible wine flaws and associated health aspects.

This chapter is not concerned with laypersons' esoteric ideas, hostility to technology, or modern filtration techniques. Nor does it question that taste properties of a cloudy young wine inevitably changes upon filtration simply due to the plain fact that the yeast, bacteria, and tissue fragments wines might still contain at bottling contribute to body and texture. The question is rather whether filtration leads to a reduction of tannin, polysaccharides, the gustative impressions associated therewith, and ultimately also of aroma. This question can only be answered by filtration of wines that have already undergone a relatively extensive process of self-clarification. The answer is conflicting. It seems to depend more on the wine than on the filtration system used, although the majority of studies in this field have been performed using cross-flow filters. On the other hand, the topic is hotly debated in both conventional and non-interventionist winemaking: Hence, it warrants a closer look.

Conflicting sensory results

Let's review the conclusions of some peer-reviewed research and technical trials.

Pilot-scale filtration through sterilizing filter sheets reduced total phenols by 10 % (Cattaruzza et al. 1987). It is not specified how much of this reduction should be attributed to color losses commonly observed at the beginning of any sheet filtration.

Static membrane filtration using a 1.2-µm prefilter followed by a 0.65-µm final filter on pilot scale reduced color and polyphenolic profile, whilst aroma was only slightly affected (Arriagada-Carrazana et al. 2005). However, results from pilot-scale filtrations must be interpreted with caution, since the filtration media adsorb some wine compounds, in particular color, at the beginning of the filtration until they become saturated in the further course of filtration.

Cross-flow (CF) filtration of red wines from Burgundy was reported to have caused losses of color, tannins, polysaccharides, and sensory quality in all wines without exceptions. The extent of these losses depended on the age of the wine, the type of membrane, and the temperature (Feuillat et al. 1987).

Sequential filtration of one Portuguese red wine through 5 µm, 1 µm, and 0.6 µm fiberglass membranes followed by a 0.45-µm polyethersulfone membrane before bottling yielded a filtrate that could not be sensorially distinguished from the unfiltered control two months after bottling. The decrease in total phenols was barely significant (Canas et al. 2011).

Sheet filtration of one Croatian red wine showed losses of general quality, which were the greater the tighter the sheet grade ratings were (Sito et al. 2013).

Commercial-scale CF filtration of one South African Pinotage wine decreased the mean degree of polymerization of tannins by 25 % similarly to gelatin and egg albumin fining, but it removed more color. As it removed the same highly polymerized tannins as finings, it was considered able to replace such finings (Oberholster et al. 2013).

Pre-bottling CF filtration of one Californian red wine using a 0.22-µm polyethersulfone membrane in comparison with a control bottled without any filtration changed scoring of 6 out of 16 sensory attributes. The unfiltered wine was scored higher in earthy, grassy, oak, and smoke aromas, and lower in fruit attributes two months post-bottling. Although there were significant changes in color and the phenolic profile, mouthfeel perceptions were considered negligible and possibly not great enough to be sensorially detected (Buffon et al. 2014).

The effect of CF filtration using a ceramic membrane was compared with egg albumin fining and self-clarification on four red monovarietal wines from Spain. CF filtration produced the highest reduction of tannins and polysaccharides in all wines, mainly highly polymerized procyanidins and yeast-derived mannoproteins (Martínez-Lapuente et al. 2017).

In a commercial-scale study comprising four Australian red wines processed in two bottling facilities, wines were filtered sequentially through four filtration grades: cross-flow (0.2 µm), 0.65-µm nylon membrane, dual grade 60/30 lenticular filters (0.5 µm), and 0.45-µm nylon membrane cartridge filters. Polysaccharide concentration was decreased by 0.45-µm membrane filtration, but no significant filtration effects could be observed on color, tannin concentration, and tannin composition. The average particle size in all wines decreased significantly with CF filtration. However, in two of the four wines, the average particle size of the filtered and unfiltered aliquot was similar after 18 months of bottle aging, suggesting that the age and composition of red wines contribute to the recovery of wine colloids after filtration. Sensory analysis showed no consistent filtration-related trends in textural attributes across all wines, although there were some significantly different aromas or flavor attributes for samples of different filtration grades within each wine. These results suggested that commercial filtration practices have only minimal effects on the color and sensory profile of red wines (McRae et al. 2017).

An attempt to clarify contradictions

How to explain these inconsistent results? Sensory bias might occur in some cases, but there is also some evidence that the filter tightness and colloid retention rate plays a role since it is obvious that a coarse-grade filter adsorbs less high-molecular material with sensory importance than a tight-grade one. Adsorption effects modulated by variable filter medium hydrophobicity have an additional impact (Ulbricht et al. 2009). More important, however, seems the composition of the wine.

In wines containing low concentrations of high-molecular tannins and polysaccharides, filtration-induced losses of these compounds are moderate and barely detectable by sensory means. In contrast, these losses have been shown to become more important in concentrated red wines, in particular in aged ones with tannin-polysaccharide complexes that have already turned to the colloidal state (Serrano et al. 1988). This suggests that a closer look at colloids would be informative.

Colloids are nanosized particles with at least one dimension ranging from 1 nm to 1 µm. In wine, they result from a physical aggregation of macromolecules such as polymerized tannins aggregated between themselves and/or with polysaccharides. To be able to aggregate into colloids, tannins need not necessarily reach a high degree of polymerization (Saucier et al. 1997 a); their concentration plays a greater role as the driving force in this process. Low-tannin wines form few colloids.

However the resulting colloids are composed, they show glue-like characteristics, have a very large surface area, refract light, and can be denatured and precipitated by addition of a large surplus of alcohol. Nanoparticle tracking analysis applied to precipitates obtained by ethanol addition and to the ethanol-soluble material showed that the average particle size of the tannin fraction was 75 to 89 nm and increased to ~200 nm when combined with the precipitate (Bindon et al. 2016). Thus, they can reach dimensions larger than the nominal pore size of CF filters (200 nm) and other filter media with a comparable filtration grade. This allows them to be broken down or even retained by filter sieving effects.

When colloids are only broken down into particles of a smaller particle size, in some wines they can recover their original size during aging (McRae et al. 2017). Ultrafiltration (< 0.1 µm) of aged red wines induced significant losses of tannins and polysaccharides, but the associated sensory losses were no longer detectable after one year (Serrano et al. 1988). These observations confirm the occasional occurrence of what is commonly called bottle sickness or bottle shock and its temporary nature. However, when colloids are retained in the filter by sieving or, more likely, by adsorption phenomena, their loss will be irreversible. In any case, sensory consequences can only be observed when such losses are substantial. Indeed, losses of mouthfeel and texture after filtration are primarily experienced in top-end wines, whilst ordinary wines can hardly lose what they do not contain. In the majority of cases, however, no significant decreases in macro-analytical parameters such as total phenols or tannin can be observed, suggesting that polysaccharides play a major role in the colloidal material affected by filtration of some wines.

Since filtration-induced quality losses do actually occur in a few wines, some winemakers will wonder when they can do without filtration or with less filtration before bottling, at least without sterile filtration.

When can we do without filtration?

Bottling unfiltered is essentially the art and knowledge of controlling microbial risks post-bottling. It is easy to understand that when fermentable sugar and malic acid are completely depleted, no more fermentation can occur post-bottling. Depletion is given when fermentable sugar (fructose + glucose) is less than 0.5 g/L and malic acid less than 0.1 g/L, both measured by enzymatic analysis or HPLC. Furthermore, Acetobacter bacteria cannot produce volatile acidity or ethyl acetate because they lack the necessary amounts of oxygen in the sealed bottles to do so. This might be counter-intuitive, but basic stoichiometric rules teach that as much as 52 mg/L O_2 are required to produce only 0.1 g/L acetic acid by oxidation of alcohol.

Sugars and malic acid can be easily measured, but fermentable sugar and malic acid are rarely depleted in the first year post-fermentation. There is general agreement that a wine must spend at least two years or two warm seasons in the barrel to deplete all organic matter, which can be microbially broken down. This is the safest way to allow microbial activity to come close to balance before bottling. However, it does not guarantee absolute microbial stability. In a high-pH medium like red wine, and in particular at low or zero levels of free and molecular SO_2, there are always some bacteria and yeast cells with subliminal activity. The key point is post-bottling storage temperature up to the point of consumption by the customer.

Storage temperature explains why wines of extraordinary quality can be produced both with and without filtration. Many years of experience have shown that 12° C (54° F) is the critical upper limit of storage temperature; whilst even lower temperatures are welcome. Sensory assessments performed on high-end wines from Burgundy revealed that best classified wines were unfiltered and stored at 12° C, but the same wines tended to be the worst ones when stored at 18° C (65° F), obviously for microbial reasons. On the other hand, among the wines stored at 12° C, the lowest rated wines were the most filtered ones (Tupinier and Gerbeaux 2002). Without any doubt, whilst filtration can negatively affect some wines, high storage temperature also does.

Whoever does not feel compelled to meet the demands of the natural wine movement and still wants to avoid over-processed or overly 'clean' wines might be able to find a balance. The bottom line is that wine quality does not necessarily equate with non-filtration but rather with the grade and quality of filtration. It is well known that high filter differential pressures in conjunction with uncontrolled aeration can churn a wine. A one-step polishing filtration with a pore size rating of only 5 microns instead of sterile filtration can meet the requirements for microbial stability of many wines when malic acid is less than 0.3 g/L and fermentable sugar less than 2 g/L.

Centrifugation is not able to achieve microbial stability, though it might provide a satisfactory level of clarity in some wines. Colloids and color are not retained.

Pasteurization as an alternative?

Pasteurization would be another possibility. For a pasteurization temperature of 70° C (158° F), a heat holding time of 20 seconds is required. In connection with bottling, a filling temperature of 60° C (140° F) is feasible when it is followed by a slow cooling over the subsequent hours. However, for quality reasons, heat exposure over several days should be avoided by cooling back in a timely manner.

The alcohol content facilitates working with this relatively low pasteurization temperature. In contrast to white wines, red wines bear that temperature quite well and might even improve. Nevertheless, many producers experience psychological inhibitions in doing so.

Preservatives as an alternative?

In most countries, two preservatives other than SO_2 are legalized for wine. There use is controversial, because they clearly meet the demand of consumers and producers of natural wines even less than filtration. Nevertheless, they are worth mentioning in order to correctly assess their value. They comprise potassium sorbate and dimethyl dicarbonate.

Potassium sorbate is the soluble potassium salt of sorbic acid, a short-chained fatty acid widely used in the food industry. It is effective against most yeasts and other fungi, but not against bacteria. In a medium containing alcohol and free SO_2 such as wine, roughly 200 mg/L are required to prevent sweet wines from refermentation in the bottle provided that the wine contains no more than 200 viable yeast cells per bottle. This is hardly feasible without filtration. For this reason, sorbate is preferably used in sweet wines by wineries that do not adequately master sterile filtration. The need for filtration is further reinforced by its ineffectiveness against bacteria.

Sorbate has subtle sensory characteristics reminiscent of flowers, but a portion of the population finds it particularly offensive. This odor intensifies when packages that have been opened for a longer period of time are utilized. The chemical background of this behavior is not known. The intrinsic odor must not be confused with the more potent, geranium-like off-odor of 2-ethoxyhexa-3,5-diene, to which sorbate can be rearranged by lactic acid bacteria. For all these reasons, sorbate is not very suitable for stabilizing red wines with their known sensitivity to bacterial growth, and even less suitable for replacing filtration.

Dimethyl dicarbonate (DMDC) is another preservative approved for soft drinks and also for wine in accordance with national or international OIV regulations. It is marketed under the trade name Velcorin™. It is an ester in liquid form that breaks down into its parent components carbon dioxide and methanol within a few hours of being added to wine. At an approved maximum permitted dose of 200 mg/L, the wines natural levels of carbon dioxide and methanol are increased only insignificantly without affecting taste and odor properties. Since DMDC is subsequently undetectable and non-persistent, it is not subject to labeling requirements and is often classified as a technical adjuvant rather than an additive. An additional advantage is its effectiveness against all wine-spoilage microorganisms including Brettanomyces and acetic bacteria.

However, its use is costly. It is difficult to dissolve as it tends to form layers or even to sediment during simple stirring of the wine. For this reason, it is only commercialized in conjunction with a dosing system supplied by its manufacturer, which must be integrated into the existing bottling line. This practically excludes its use in small and medium-sized operations.

12.3. Causes and prevention of colloidal haze formation

As outlined in the previous section, polyphenol polymers, polysaccharides and some grape-derived proteins contained in red wines are macromolecules that can aggregate progressively during aging. Thereby they can form colloidal particles with dimensions between 1 nm and 1 µm. Such particles play a determined role in quality owing to their impact on body. However, in contrast to macromolecules that are dissolved components forming colloidal solutions, colloidal particles form colloidal dispersions, which are biphasic systems. This results in two consequences:

- Colloids are largely responsible for premature filter fouling when they are present in sufficiently high concentrations. Under these conditions, a portion of them is removed during filtration, causing a sensorially perceived decrease of body and volume in some wines that present really high levels of colloidal material.
- The colloids arising from the aggregation of macromolecules in red wines can lead to post-bottling haze formation. It will depend on the average size and concentration of these particles whether or not visually perceptible hazes or deposits will appear. If they appear during aging of bottled red wine, they present a red or brown color

It is important to note that the majority of contemporary red wines are not subject to post-bottling haze formation for the following reasons:

- Filtration before bottling removes a substantial amount of the unstable colloidal material.
- The presence of free sulfur dioxide in bottled red wines slows down regenerative polymerization of phenols and the subsequent formation of phenolic polymers large enough to precipitate alone or in association with polysaccharides.

Testing for proneness to haze formation

However, there are concentrated and rich wines in which it makes sense to check colloidal stability and to assess the risk of haze formation. For that purpose, two tests are available, both of them based on the use of extreme temperatures:

- The most widely used way to assess the risk of colloidal instability are cold tests. Low temperatures accelerate the aggregation and precipitation of poorly soluble macromolecules and colloids. For the practical implementation, a preclarified wine sample is held over several days at a temperature around 4° C (39° F); the turbidity measured after the test correlates to colloidal coloring matter instability.
- Alternatively, heat tests are also used. In these tests, the preclarified wine sample is held at 80° C (176° F) for at least two hours and subsequently cooled back under running tap water. After another waiting period of at least one hour, the sample is checked for turbidity. When a turbidity meter is used, turbidity should not exceed 2 NTU[58] in stable wines. This test is particularly sensitive to proteins. Although red wines are considered devoid of proteins due to their flocculation with tannins during the winemaking process, there is some evidence that residual proteins may be found in some red wines, even if only in very low amounts. This probability increases in wines treated with proteinaceous fining agents.

Preventing colloidal haze formation in unstable wines

When a wine transpires to be unstable with regard to colloidal haze formation, various treatments are feasible:

- Filtration should first be checked for its effectiveness, because it is used for the vast majority of wines anyway. After filtration on a laboratory scale through a filter medium close to that of the filter used on the technical scale, the filtrate should be checked again for haze formation. If the wine continues to be unstable, one of the following additional methods can be used before filtration:
- Fining with proteins such as gelatin may be sufficient to remove unstable pigments and to stabilize wines. This is a fortuitous side effect of such finings, which are more often used to reduce excessive astringency (Chapter 12.1.3).
- For wines assumed to have a high content of colloidal coloring matter, the use of bentonite can be more appropriate. Bentonites are negatively charged clay particles that react with positively charged compounds. They are mainly used to remove unstable proteins from white musts and wines. However, they also react with anthocyanins and polymeric pigments with an intensity that is usually higher than the one observed with protein finings. Their disadvantage is that they lead to a strong reduction in color intensity. This makes them less suitable for slightly colored red wines, which on the other hand also show little proneness to form colloidal hazes.
- Besides fining, cold treatments as performed for bitartrate stabilization are also effective to remove unstable colloidal matter before bottling. However, their effectiveness strongly depends on the wine, the applied temperature, and the treatment length. It can be assumed that colloidal stability is achieved when bitartrate stability is also achieved.
- Lastly, the use of an additive, gum arabic, added at bottling at amounts of 0.1 to 0.2 g/L, is another possibility. Gum arabic acts as a protective colloid able to coat

[58] NTU = nephelometric turbidity units.

wine colloidal particles, preventing their growth and, thus, their flocculation as visible hazes or deposits.

Last but not least, it is crucial to keep in mind that all measures to ensure colloidal stability do not prevent haze formation caused by microbial growth and activity. Hence, they cannot replace filtration (Chapter 12.2).

12.4. Stabilizing free SO_2 prior to bottling

12.4.1. Methods for measuring free SO_2

Free SO_2 adds with bound SO_2 to yield the total SO_2 according to the generally known formula

$$SO_2 \text{ free} + SO_2 \text{ bound} = \text{total } SO_2$$

Legal limits must be observed for total SO_2, which are relatively low in dry red wines. From a sensory perspective, there is an interest in bottling with the lowest possible levels of free SO_2, while still ensuring a minimum of it throughout the life of the wine. Without free SO_2, there is a serious risk of the appearance of the typical smell of free acetaldehyde reminiscent of bruised apples and sherry. On the other hand, red wines with high levels of free SO_2 taste thin and astringent, because correspondingly more anthocyanins are bleached. Only color-active anthocyanins contribute to volume and weight (Chapter 5.3.2.). For these reasons, precise SO_2 measurements are of vital importance.

There are basically three traditional approaches to free SO_2 measurement, which are widely used and comprise

- iodine titration by the Ripper method,
- the aeration-oxidation method, and
- flow-injection analysis using para-rosaniline to produce a colored complex with SO_2 by means of automated analyzers.

All these methods show serious drawbacks. Because of their far-reaching significance for redox management, microbial stability and wine ageability, they are discussed in detail below. Finally, new and considerably more reliable methods are presented.

Shortcomings of iodine titration and possible corrections

The determination of free as well as total SO_2 in red wines by the traditional iodine titration, also known as Ripper method, encounters a fundamental difficulty that is less known from white wines.

In addition to SO_2, any wine contains other constituents, which react with iodine and which are sometimes referred to as reductones. They simulate more SO_2 than the wine actually contains. This error amounts to some 7 to 8 mg/L of apparent SO_2 in average white wines that do not contain ascorbic acid. It increases by a few more mg/L in the presence of colorless tannins at concentrations encountered in red wines.

However, a much larger increase in apparent free SO_2 is caused by the partial dissociation of the anthocyanin-bisulfite complexes during the analysis. As outlined in chapter 2.2.4, these complexes represent loose and reversible bindings being in a dynamic equilibrium with free SO_2. Thus, following the progressive loss of free SO_2 during iodometric titration, simultaneous release of anthocyanin-bound SO_2 partially replenishes free SO_2 and re-establishes the initial equilibrium. This explains the sluggish endpoint in the iodine titration of SO_2 in anthocyanin-rich red wines.

The extent of this reaction depends on the speed at which the iodometric titration is performed, as does the resulting error. In common red wine, the release of anthocyanin-bound SO_2 usually simulates 10-20 mg/L more free SO_2 than the wine actually contains, but in young and anthocyanin-rich wines this error can exceed 50 mg/L. The error is greater in young red wines than in old ones because young wines contain more monomeric anthocyanins (Pontallier 1981).

Summarizing, substances other than SO_2 in addition to SO_2 released from the anthocyanin-bisulfite complex are responsible for artificially high levels of free SO_2 measured by iodine titration. In light red wines, this methodological error can be corrected for with some expense. For this purpose, an additional sample of the wine is mixed with acetaldehyde or another aldehyde in excess, which binds the free SO_2 within five minutes. However, this does not detract from the fact that the Ripper method is largely unsuitable for SO_2 measurements in red wines.

Indicating the titration endpoint by amperometric titration using a platinum electrode instead of using the more traditional starch color change might facilitate endpoint indication in strongly colored red wines, but it does not substantially improve results as it does not remove the methodological sources of error.

Advantages and disadvantages of the aeration-oxidation method

In an attempt to circumvent the interferences described above, methods are available that first liberate the SO_2 from the wine matrix before it is measured. In such methods, free SO_2 is separated from the acidified sample before being quantified by titrimetric, colorimetric, or electrochemical techniques. Such separations can be accomplished by the gas-permeable membranes used in flow-injection analysis or by sparging the free SO_2 into a reaction vessel containing hydrogen peroxide (H_2O_2). The latter approach constitutes the basis of the popular aeration-oxidation method.

In the aeration-oxidation method, in order to carry the free SO_2 to the downstream reaction vessel, a gas stream (air or nitrogen) is conducted through the acidified wine sample at 10 to 15° C (50 to 59° F) for 15 minutes. The SO_2 is trapped in the H_2O_2 solution and oxidized to sulfuric acid, which is then titrated with a standardized sodium hydroxide solution to determine the amount of acid formed and, hence, the original amount of free SO_2 in the wine.

This method is widely used and provides more realistic and consistent results in red wines than iodine titration. Details of its implementation are sufficiently described in the standard literature. As it liberates the free SO_2 from the sample matrix, it is fairly robust in terms of interferences from other reducing wine compounds such as tannins, ascorbic acid, etc. However, it does not impede a portion of SO_2 bound to anthocyanins from being released and also measured as free SO_2 during the 15-minute sample sparging period. The reason is easy to understand: As free SO_2 is transferred from the sample to the H_2O_2 solution, simultaneously hydrolysis of anthocyanin-bound SO_2 partially replenishes free SO_2 in the sample to re-establish equilibrium (Burroughs 1975, Bertrand 1976, Buechenstein and Ough 1978). However, in comparison to the Ripper method, the error thereby caused is somewhat smaller.

A closer look at the anthocyanin-bisulfite equilibrium

Regardless of whether free SO_2 is measured by iodometric titration, the aeration-oxidation method or by flow-injection analysis using para-rosaniline, previous acidification and dilution of the sample is required. This is the very first step resulting in a partial release of SO_2 from its bound forms. Correspondingly higher free SO_2 readings are obtained. This effect is almost negligible if bound SO_2 occurs predominantly in its strong binding with acetaldehyde. However, the error it causes becomes much larger if SO_2 is also present in less stable binding with anthocyanins in red wines.

Without any doubt, acetaldehyde is the strongest binder of SO_2. It binds with free SO_2 at a ratio of 1 mg acetaldehyde to 1.45 mg SO_2 as long as free SO_2 is available. The resulting hydroxysulfonate addition product is odorless. The small dissociation constant of $K_D = 1 \times 10^{-6}$ of this bond explains why the equilibrium favors the formation of product. Furthermore, it explains why 99.9999 % of all acetaldehyde is bound to SO_2 as long as free SO_2 is available to undergo this reaction. Other binding partners such as pyruvate and ketoglutarate are somewhat more weakly bound to SO_2 with $K_D = 10^{-5}$. Now to the decisive point: Anthocyanins also bind with SO_2 with $K_D = 10^{-5}$. Therefore, in red wines containing free SO_2, 70-85 % of the monomeric anthocyanins are present in their colorless form bound to SO_2 (Somers and Evans 1974).

Compared to free SO_2 measurements, the determination of total SO_2 is a relatively simple task. Total SO_2 occurs at a concentration that is clearly defined regardless of how much of it is bound to wine constituents. The task of the measurement method is to capture it completely and without analytical interferences. The various distillation methods are particularly suited for that purpose (Fazio and Warner 1990).

In contrast, the determination of free SO_2 in red wines by conventional methods is much more demanding. As we have seen, the reason is that the equilibrium between anthocyanin-bisulfite complexes and actual free SO_2 is perturbed when these methods are used. Thus, artificially high free SO_2 readings are obtained, with the size of the error depending on the method used. Therefore, results require careful interpretation and must be reported along with the method used to obtain them. They cannot be directly compared when obtained from different labs if the method used is not specified. There is no other parameter in wine analysis with such a strong variation between methods. This makes the determination of free SO_2 in red wine one of the most complex topics in wine analysis. Since none of the results reflect the actual content of free SO_2, it has always been a matter of convention, personal definition, and empiricism to determine what is a high or low content. There is no way to get around this relatively subjective interpretation until the new generation alternative methods are widely adopted.

Headspace measurements

The search for a more specific method led to the measurement of gaseous SO_2 in the headspace of a closed vessel containing the wine sample. In a first attempt, SO_2 was carried by a CO_2 gas flow from the acidified wine samples through commercial SO_2 detection tubes. The length of the tube that changes color is proportional to the free SO_2. Results obtained by that approach are similar to those obtained by the aeration-oxidation method, because both methods rely on previous sample acidification causing anthocyanin-bound SO_2 to be released (Pegram et al. 2013).

Further progress was made by using a syringe to sample untreated wine and to create a closed headspace in the syringe. After allowing time for setting the equilibrium between the SO_2 dissolved in the wine and the molecular, gaseous SO_2 present in the headspace of fixed volume, the headspace is then expelled through calibrated gas detection tubes. These tubes contain a colorimetric SO_2-selective reagent such that the length of discoloration on the tubes is proportional to the SO_2 headspace concentration. The latter can be related to the molecular SO_2 concentration in the wine by observing Henry's law coefficient (H)[59], which can be further converted in "truly" free SO_2 by pK_{a1} of HSO_3^- and the Henderson-Hasselbach equation[60] using simple formulae. For this purpose, the ethanol concentration, pH, and sample temperature must be known to establish the correct values of pK_{a1} and H (Coelho et al. 2015).

Ultimately, this latter approach is a simplified version of the headspace gas chromatography (GC) procedure. It has the crucial advantage of not involving pH shifts, sample dilution, or temperature changes, and thus not perturbing the equilibrium between free and bound SO_2 during the measurement. Only the free SO_2 present in the intact sample is recorded. Furthermore, there is no interference from other reducing wine constituents or sluggish endpoints. Inevitably, the values obtained in this way are significantly lower than those obtained by methods that alter the wine matrix by dilution and acidification. Conversely, free SO_2 levels as measured by the aeration-oxidation method are overestimated by 24-76 % in red wines due to the interference of the anthocyanin-bound SO_2 (Coelho et al. 2015).

These findings also have far-reaching consequences for validity and meaningfulness of molecular SO_2 data. These data impact both the reductive power (Chapter 6.4) and microbial protection exerted by free SO_2. Depending on authors and textbooks, 0.5 or 0.8 mg/L molecular SO_2 are required for what is considered microbial safety. The whole wine industry worships these values more than any bible, and many wineries are keen to obtain them by rash pH adjustments regardless of the sensory outcome.

Traditionally, molecular SO_2 is calculated from pH and free SO_2 as measured by conventional methods. However, when free SO_2 data are wrong, molecular SO_2 data cal-

---

[59] According to Henry's law, the amount of gas that dissolves in a liquid is proportional to the partial pressure of that gas in equilibrium with that liquid. H is the ratio between the concentration of the gas in the liquid phase and in the gaseous phase.

[60] The Henderson-Hasselbach equation shows the relationship between the pH of a solution, the pK_a (acid dissociation constant), and the concentration of the chemical species involved. In this context, pK_{a1} is the dissociation constant of the acid group of bisulfite (HSO_3^-).

culated therefrom must also be wrong. Furthermore, the validity of microbial protection provided by molecular SO_2 data calculated in the traditional way we have discussed has been reevaluated. Thus, it has been shown that it must actually exceed 2.0 mg/L to markedly inhibit the growth of Saccharomyces cerevisiae strains such as EC 1118 (Howe et al. 2018). This is three to four times more than the 0.5 or 0.8 mg/L molecular SO_2 required by traditional teaching.

Saccharomyces cerevisiae control by SO_2 is not an issue with dry reds. However, these results are very indicative of the elevated free SO_2 levels as measured by techniques such as aeration-oxidation that are required to control spoilage yeasts such as Brettanomyces (Agnolucci et al. 2010, 2014).

This finding has brought down an enological doctrine that is more than half a century old. It proves that conventional approaches to measuring molecular SO_2 are ill-suited for predicting the microbial stability of red wines. Additionally, it demonstrates that the anthocyanin-bisulfite complex has negligible antimicrobial activity. From a more practical perspective, it suggests that most attempts to lower pH to levels considered 'safe' have in reality failed in doing so. Nevertheless, hardly any of these wines has become microbially flawed. After sterile bottling wines have no possibility to do so at all. Hence, the widespread pH fetish, predicated by a perception of necessary conditions for molecular SO_2 to facilitate microbial security, should be seriously questioned.

12.4.2. Impact of oxygen uptake on post-bottling SO_2 stability

Every wine loses variable amounts of SO_2 after bottling. This can lead to highly detrimental sensory consequences if free SO_2 disappears completely and oxidation damage occurs. The precise determination of the "true" content of free SO_2 alone cannot solve the problem of uncontrolled SO_2 losses. More important than the instantaneous content of free SO_2 is its stability over time. It depends essentially on the oxygen entrapped in the bottle.

Chapter 6.4 shows that 1 mg/L DO oxidizes on average 2.1 mg/L SO_2, and that this ratio is not a stoichiometric one. The SO_2 losses due to oxidation are somewhat greater at high initial free SO_2 contents and less at low initial contents. Thereby, free and total SO_2 decrease simultaneously, although not to the same extent. This process is not related to any binding of free SO_2. It can be regarded as a chemical desulfitation. The SO_2 oxidized to sulfate has simply disappeared and is not measure any more. Thus, it can be easily replaced by further SO_2 additions without taking the risk of coming closer to the legal limit for total SO_2.

Oxygen uptake during preparation for bottling

In any red wine, losses of free SO_2 increase proportionally with the oxygen dissolved in the wine during bottling (Jakobs 1976). In commercial practice, wines are often pumped, stirred, filtered, stored in tanks that are not completely filled, or even transported to external bottling facilities just before bottling. The associated oxygen uptake rates are given in table 11; the depletion of DO by binding to wine constituents is discussed in chapter 6.1.

The oxygen uptake associated with each of the measures identified above can accumulate to considerable amounts of DO. Under extreme conditions, these can come close to the solubility limit of 8 mg/L DO if the measures are performed in rapid succession without allowing time for the dissolved oxygen to bind and disappear between measures.

DO meters are widely available and work without major problems, but much care should be taken when sampling wine for DO measurements. When taking a sample, the aim is to exclude any oxygen from the sample bottle. The easiest way to achieve this is by replacing the oxygen in the bottle with an inert gas. Ideally the sample bottle should then be lowered rapidly into the tank from the top, so that the bottle sinks and the sample is drawn from the middle of the tank. Glass bottles are more efficient for this as they sink faster. If taking a sample through the tank lid is not possible, then the sample will need to be taken from the sample valve extended by a thin sample hose. In this case, some sample should run out first into a separate container and then filled into the gassed sample bottle with the sampling hose submerged as deep into the bottle as possible.

When a DO meter is not available, an approximate evaluation of the amount of oxygen that enters wine during processing operations can be achieved by measuring the decrease in total SO_2 after each operation. Knowing total SO_2 values before individual cellar operations, and then measuring the total SO_2 again one week after the operation, can give a rough guide of the extent of oxygen pick-up. A written record can be extremely useful should a problem arise later.

Oxygen pickup during filling

In addition to the DO already present in the wine before bottling, there is further oxygen pickup during the actual bottle filling process. Many wines from small and medium-sized wineries are bottled using vacuum or normal pressure fillers without the protection of inert gas. Under these conditions, the wine experiences oxygen pickup in the filler bowl and when it enters the bottle of 1.0 to 3.0 mg/L DO, with a tendency toward values as low as 1.0 mg/L DO with modern filling equipment and a smoothly running filling process according to industry best practice conditions (Table 11). These values can be further reduced by pre-evacuation of the bottles, inert gas injection into the bottles before filling and before sealing, and inertization of the filler bowl headspace. Thus, oxygen pickup during the actual filling process must not be overestimated.

The first two or three dozens of bottles, subject to filler head number, should be isolated and returned to the wine in tank. DO measurements, whenever they are run, should only start in subsequent bottles with the bottling line working smoothly.

Oxygen pickup from the bottle headspace

After the bottles have been sealed, a variable headspace is present which either contains air that leads to further oxygen pickup, or has been depleted of oxygen more or less efficiently by headspace treatment using vacuum or inert gas. The oxygen contained in the headspace gradually dissolves in the wine.

Headspace oxygen is of great importance: In standard bottles, 1 cm of headspace height corresponds to approximately 2.9 mL of gas volume. If this is pure air, it contains 0.86 mg of oxygen. This calculation is based on the fact that air contains 20.8% oxygen that has a density of 1.43 mg/mL. If this amount of oxygen is trapped in a bottle of 750 mL, it provides the bottled wine with 0.86 / 0.75 = 1.15 mg/L O_2. Expressed in mg/L, it is higher in smaller bottles and lower in larger bottles.

Total pack oxygen

A headspace of 5 cm (1.97") height, typical for screw capped bottles, contains approx. 15 mL of air, unless it has been treated with vacuum or inert gas. This air in turn contains 3.12 mL or 4.48 mg of O_2. Depending on the bottle size, this amount of oxygen impacts the wine to a varying degree, for example with 4.48 mg/L O_2 in the case of bottles with 1,000 mL nominal volume, and with 5.97 mg/L for 750 mL nominal volume. To these amounts of headspace oxygen, one must add the oxygen already dissolved in the wine. The sum is expressed as "total pack oxygen" (TPO) and is an important technical parameter in the beverage industry. It corresponds to the total amount of oxygen trapped in a packaging unit at the moment of bottling. These relationships and their calculation are illustrated in table 19.

Table 19: Calculation of total pack oxygen (TPO) in bottles of different nominal volumes and identical headspace volume (15 mL) containing air and filled with wine containing 3.0 mg/L DO at the moment of filling.

bottle nominal volume (mL)	375	750	1,000	1,500
oxygen (mg O_2) in the headspace containing 15 mL air	4.48	4.48	4.48	4.48
+ amount of oxygen (mg O_2) already dissolved in the wine volume containing 3.0 mg/L DO	1.13	2.25	3.00	4.50
= total of oxygen (mg O_2) in the bottle (TPO)	5.61	6.73	7.48	8.98
total of oxygen converted to mg/L	14.96	8.97	7.48	5.99

With comparable headspace, small bottles are clearly at a disadvantage. Or at an advantage, if the TPO is expected to promote favorable aging of red wines. With increasing bottle size, the relative influence of the headspace on the total pack oxygen (TPO) is reduced in any case.

Decrease of oxygen and SO_2 after bottling

The DO in the bottled wine gradually decreases due to its binding with wine constituents, provided that it takes no more oxygen up (Chapter 6.1). However, if an air-containing headspace is left after bottle sealing, oxygen diffuses from the headspace into the wine, because headspace oxygen has always a higher partial pressure than dissolved oxygen (Vidal and Moutounet 2006). Thus, it adds to the DO, acting as an oxygen reservoir.

Under the condition that the bottle closure represents an absolute oxygen barrier, the headspace oxygen plus dissolved oxygen are completely depleted after some time. Subsequently the inside of the bottle will be free of oxygen. From this moment on, SO_2 losses come to a standstill. Figure 81 shows an example of the decrease of DO and headspace oxygen in a red wine after bottling and hermetically sealing with screw caps fitted with a tin-Saran liner.

Figure 81: Decrease of headspace oxygen and dissolved oxygen after bottling of a standard red wine sealed with hermetically sealing tin/Saran-lined screw caps.

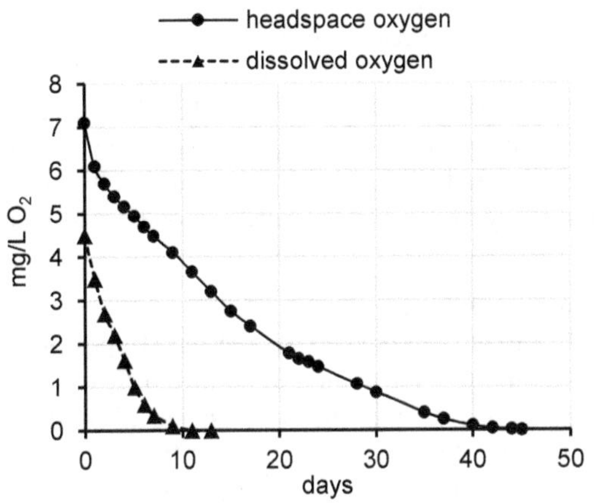

In this wine, the initial DO was relatively quickly bound down to a residual level of < 0.1 mg/L after approximately 9 days. The oxygen in the headspace, however, had only completely disappeared after 44 days. At the end of this period, a TPO of zero had been reached and SO_2 decrease had come to a standstill.

This behavior completely changes if additional oxygen is constantly supplied via less gas-tight bottle closures as it is the case with most cork closures. In this case, neither DO nor headspace oxygen reach an absolute zero, because it is bound more slowly than it diffuses through the bottle closure (Waterhouse et al. 2016). The reason is as follows:

The diffusion rate of headspace oxygen into the liquid phase is limited by the small wine surface area. It is slower than the binding of DO, but faster than the diffusion of oxygen from and through the bottle closure. It controls the rate of TPO decrease and associated SO_2 losses more than bottle closure OTR during the first weeks and months after bottling (Vidal and Moutounet 2006).

With a certain margin of fluctuation, all the oxygen (TPO) trapped in the bottle at the moment of bottling will have reacted and disappeared after storage of about one and a half to three months. Accordingly, the strongest SO_2 losses of about 20 to 25 mg/L SO_2 are also observed within the first three months post-bottling and largely regardless of the closure type used, provided that no inertization of the headspace takes place (Scimgeour and Wilkes 2014, Jung and Schüssler 2016). Only after the oxygen trapped in the bottle has been chemically bound, a significantly slower SO_2 decrease follows, which is now exclusively dependent on the oxygen transmission rate (OTR) through the bottle closure.

When the closure OTR is constant as in the case of technical corks or synthetic closures, the rate of the OTR-dependent SO_2 loss is linear even after free SO_2 becomes undetectable and only strongly bound SO_2 remains, essentially SO_2 bound to acetaldehyde. It is independent of free or total SO_2 concentration and not significantly affected by wine composition. Under these conditions, as the oxygen consumption rate is faster than the oxygen ingress rate, a very low (≪ 0.1 mg/L) steady-state level of DO can be observed (Sacks et al. 2020).

Oxygen transmission rate (OTR) of bottle closures

The various closure types exhibit different permeability to atmospheric oxygen. This behavior is characterized as "oxygen transmission rate" (OTR, in mg O_2/year or µg O_2/day).

Most screw caps and technical corks are at the lower end of the variation range with an OTR of 0.0 to 1.5 mg O_2/year, while synthetic corks have long tended to display high OTR values of 3 to over 10 mg O_2/year (Karbowiak et al. 2010, Diéval et al. 2011). More recently, there has been a steady development towards synthetic and technical stoppers and screw caps with different OTR values adapted to specific wine types. At the same time, the original tendency of synthetic stoppers to one-sided high OTR values have begun to decrease.

The natural cork stoppers occupy an intermediate position with OTR values ranging from 1 to 4.8 mg O_2/year during the first twelve months after bottling (Lopes et al. 2006, 2007, Silva et al. 2011, Oliveira et al. 2013). Oxygen uptake from the cork is understood as diffusion of atmospheric oxygen through the cork tissue and/or along the cork-glass interface (Casey 1989, Oliveira et al. 2013). During the first year post-bottling, natural corks can provide increased amounts of oxygen that were originally trapped in the cork cells and are slowly pushed into the bottle after compression of the cork when it is inserted into the bottle neck (Waters et al. 1996).

Figure 82: Oxygen transmission rate (mg O_2/year) of natural cork stoppers during the first year of upright storage at 10-15° C. Fluctuation ranges of three pieces per batch.

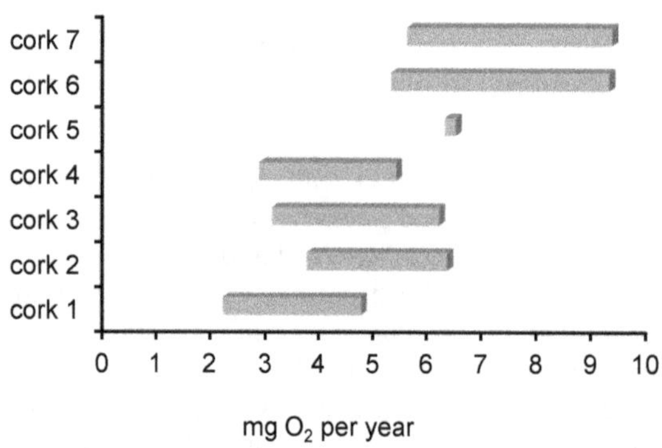

Cork is a natural product. Therefore its properties are subject to biological and technological variability. This also applies to its OTR. Large differences in OTR can be observed between batches of cork as well as between individual corks within a batch (Caloghiris et al. 1997, Oliveira et al. 2013). Examples of variations between individual pieces within a batch are shown in figure 82.

It can be stated that natural cork stoppers can provide the wine with highly variable amounts of oxygen. These amounts are significantly higher in the first year post-bottling than in subsequent years. Therefore, after bottling with natural cork, oxidation-related SO_2 losses are particularly high in the first year. They vary from batch to batch and even between individual corks within a batch. Figure 83 gives an example of different SO_2 decay rates after bottling and sealing with natural cork.

Figure 83: Decrease of total SO_2 by oxidation during bottle storage of a wine sealed with two different natural corks.

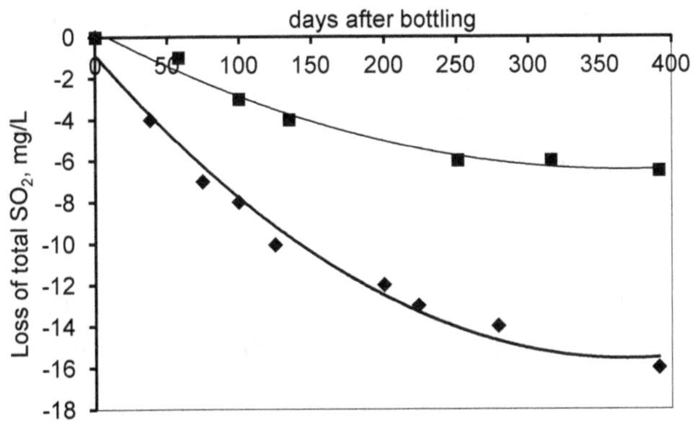

Oxygen during and after bottling determines SO_2 losses during bottle storage

Summarizing, oxygen exposure of bottled wine can be attributed to five oxygen sources:

- the oxygen already dissolved in the wine before bottling,

- the oxygen absorbed during the actual bottling process,

- the oxygen trapped in the bottle head space,

- the oxygen diffusing through the bottle closure,

- in the case of natural corks, the oxygen contained in the cork tissue.

The total pack oxygen (TPO) calculated according to the example given in table 19 can thus be continuously increased by the oxygen diffusing through the bottle closure. This leads to the concept of extended TPO. Figure 84 illustrates these relationships.

Figure 84: Oxygen in bottled wine: The extended concept of "total pack oxygen" (TPO).

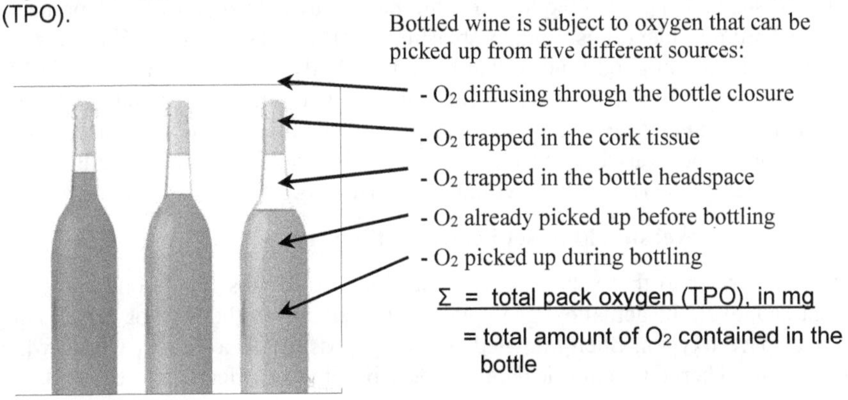

When free SO_2 is adjusted before bottling, these five subsets of TPO must be taken into account. Their sum controls the post-bottling SO_2 decrease. Therefore, knowing the exact level of free SO_2 before bottling is only as valuable as one is able to evaluate the amount of oxygen available in the bottle. Of particular importance is the oxygen picked up by the wine immediately before bottling via the associated processes like fining, blending, filtration, etc. For this reason, an increasing number of winemakers tend to combine both SO_2 and DO measurements. This approach facilitates precise SO_2 dosing when SO_2 is accurately measured. In this context, portable DO meters find an important field of application in order to better predict future SO_2 losses and compensate for them in advance.

Chapter 6.4 explains that DO in wine is not completely scavenged by SO_2 in the stoichiometric ratio of 4 mg SO_2 per 1 mg O_2, but only to approx. 50 %. This is due to additional oxidizable wine constituents, which compete for DO. Therefore, each mg/L of DO must be compensated by approx. 2 mg/L more free SO_2 than one wishes to find after some time of bottle storage. If, for example, 5 mg/L DO is still measured directly before bottling, this alone results in a loss of 5 x 2 = 10 mg/L SO_2, which can be compensated by a correspondingly higher SO_2 dosage. In this regard, a better knowledge and consideration of the bottle closure OTR would be highly desirable, in particular when wines intended for long-term bottle maturation are bottled. It will be difficult to commercialize top-end aged wines smelling of free acetaldehyde after their free SO_2 has disappeared prematurely.

Practical hints

In the absence of technical equipment for on-site DO measurements, the time factor must be taken into account in order to initiate SO_2 adjustments in good time before bottling. It takes about 10 days for a red wine to consume about 90 % of the DO still present at room temperature (Figure 35). If the wine is stored unagitated under oxygen exclusion for such a period of time, the DO-based reactions leading to SO_2 losses that normally take place in the bottle will take place in the tank instead. This provides the opportunity to monitor free SO_2 stability and to compensate for SO_2 losses in good time before bottling.

However, such an approach requires that all measurements associated with preparing wine for bottling - blendings, finings, splashings, tartrate stabilization, filtrations, etc. - are also initiated in good time so that there is actually a 10-days' time spell left for SO_2 stabilization. Storage vessels that are not completely topped are not well suited for that final stage before bottling as headspace blanketing is rarely as perfect as desirable. Conversely, variable capacity tanks with floating lids are very useful for that purpose, at least when small wine volumes are handled.

What free SO_2 level should be set before bottling?

A precise answer to this very essential question is only possible if one knows the current DO level, the actual oxygen uptake expected during the bottling process, and the headspace oxygen. Such knowledge is only possible in a perfect world, which barely exists. Therefore, empirical data of post-bottling SO_2 decrease are useful.

After about two to three months post-bottling, the total oxygen trapped in the bottles is largely consumed. Then the associated SO_2 decrease comes to a halt when hermetically sealing bottle closures are used. Under these conditions, it is possible to approximately calculate the SO_2 decrease caused by TPO.

When natural or synthetic cork closures are used, there will be further oxygen uptake from the cork tissue or by diffusion of atmospheric oxygen through the closure as such. This uptake rate in the order of 0.5 to 5 mg O_2 per year is a power of ten lower than in micro-oxygenation and several powers of ten lower than in macro-oxygenation. This is why we refer to it as nano-oxygenation.

Nano-oxygenation is known to mitigate the formation of post-bottling reduction flavor in wines that are prone to it (Limmer 2005). However, only a few sensory studies are available on its sensory effect on phenol-related mouthfeel attributes such as astringency, body, texture, etc. It is to be expected that the results cannot be generalized and are considerably dependent on wine matrix, its reductive strength (total phenols and free SO_2) as well as the stage of development of the wine at bottling.

The oxygen uptake through and from natural corks is mostly unknown in practice. Simplified, it can be assumed that the oxygen uptake from the cork tissue and the associated SO_2 losses come to a halt after about one year (Figure 83). However, oxygen uptake through such closures does not have a fixed end point. This makes it difficult to decide upon the ideal free SO_2 content at bottling.

For sensory reasons, the free SO_2 should be as low as possible to maintain the level of ionized, color-active anthocyanins (Chapter 5.3.2). On the other hand, it should be high enough to ensure a minimum of it throughout the life span of the wine. When the last mg/L of free SO_2 has disappeared, the first mg/L of acetaldehyde will be released from its bound form and adversely affect aroma and wine's shelf life. This phenomenon is sufficiently known from white wines that have completely lost their free SO_2, but chemistry is slightly more favorable to red wines.

Release of SO_2 from binding to anthocyanins during oxidation

High-quality red wines have a decisive advantage over simple red wines and white wines. They generally contain large amounts of tannins and anthocyanins, which counteract the formation of free acetaldehyde. This is due to two reactions that complement each other:

1. A large portion of the SO_2 is reversibly bound as its bisulfite ion (HSO_3^-) to the positively charged C ring of the anthocyanins (A^+). This is a weak form of bonding, which is in a dynamic equilibrium with the free SO_2 according to the scheme

   $$A\text{-}HSO_3 \leftrightarrow A^+ + HSO_3^-$$

   (Chapters 2.2.4). It breaks down when SO_2 is removed from the system, for example by oxidation when oxygen is consumed. In this process, SO_2 is released from the binding. This release is faster than the rate of SO_2 oxidation during oxygen consumption (Sacks et al. 2020). In analytical terms, this can be observed by the fact that the loss (Δ) of free SO_2 is less than that of total SO_2. The release of SO_2 from its bound form can thus be easily calculated according to the formula

   $$SO_2 \text{ released} = \Delta \text{ total } SO_2 - \Delta \text{ free } SO_2$$

Figure 85 shows, using a set of 10 red wines as an example, that under these conditions the release of SO_2 indeed increases with the anthocyanin content. Therefore, after the consumption of 10 mg/L O_2, a loss of only 6 mg/L of free SO_2 occurred, while the loss of total SO_2 amounted to 21 mg/L (Chapter 6.4). No such relationship exists with tannin and total phenolics. In other words: Anthocyanins hold a reservoir of potentially free SO_2. This reservoir is missing in white wines.

Figure 85: Release of SO_2 (mg/L) from its bound form after consumption of 10 mg/L O_2 in red wines.
SO_2 measured by flow-injection analysis.

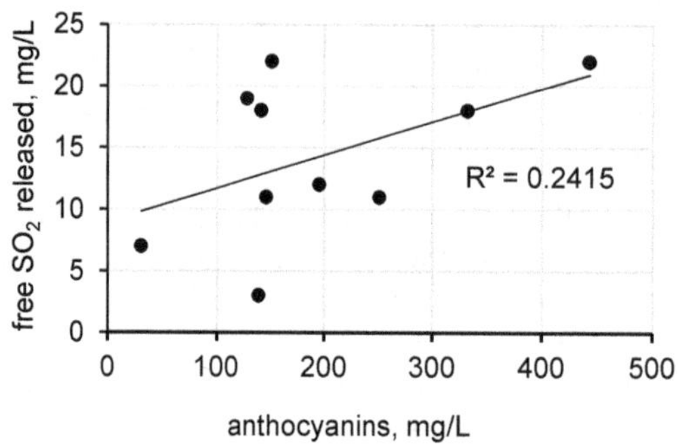

2. Tannin itself acts as a reducing agent and is able to bind acetaldehyde via the ethyl bridge (Chapters 2.2.3 and 7.2). This reaction starts when, after oxidation of the last mg/L of free SO_2, the release of acetaldehyde from its SO_2 bond begins. Ideally, formation and binding of free acetaldehyde balance each other. Its accumulation to odor-active levels, as known from white wines after the complete loss of free SO_2, is thus reduced.

The additive effect of the two reactions described explains why high-phenol red wines can be relatively stable against oxidative damage without containing analytically detectable amounts of free SO_2. It is the basis for the production of red wines without added sulfites. Such wines do not necessarily have to deviate from the usual odor and taste patterns, as long as their microbial stability is guaranteed. However, low-phenol red wines do not comply with this rationale. During post-bottling maturation, they require free SO_2 and/or oxygen exclusion by the use of well-sealing bottle closures to protect them from oxidation and formation of free acetaldehyde.

All these observations and differences known from practice find their explanation in the phenolic composition of the red wines. They show that the easily measured contents of total phenols and anthocyanins make a very significant, if not the most significant, contribution to the differentiation of red wines among themselves. Nevertheless, it is an unknown variable under many commercial winemaking conditions. Therefore, many highly scored wines intended for long bottle storage develop oxidative flaws faster than expected.

On the other hand, it is deceptive to believe that wine quality and phenolic strength alone determine wines' post-bottling shelf life. In the long term, both the free SO_2 level at bottling and the bottle closure OTR are crucial variables. Hence, most wineries adjust free SO_2 to 20-40 mg/L before bottling as measured by the aeration-oxidation method. Thereby, the level of 0.5 or 0.8 mg/L molecular SO_2 generally recommended for microbial safety is rarely achieved in the high pH range of most red wines without losses of quality or exceeding legal limits of total SO_2. This is not a problem since most wines are sterile bottled. In the absence of germs, no measures are required to inhibit them. However, it can become a problem when wines are bottled unfiltered and stored in the warmth.

As a general rule, the style of wine being produced also needs to be considered when SO_2 concentrations at bottling or other forms of packaging are determined. All other things being equal, if a wine is intended for medium to long term storage it will require a greater concentration of free SO_2 than a wine intended for consumption over a shorter time frame. However, it should be remembered that wine packed in alternative packaging systems as bag-in-box packs, for instance, may be intended for early consumption, but may also rapidly lose SO_2 after packing due to the oxygen permeability of the bag-in-box material. Critical points of oxygen ingress are the tap, the neck, and the weld connecting the EVOH film to the neck.

12.4.3. Red wines without added sulfites

Red wines without added sulfites meet consumer demands for wines obtained by non-interventionist winemaking, similar to unfiltered wines. Consumer health concerns are one of the reasons why they are produced here and there. However, conventional wines only contain a small fraction of the amounts of SO_2 contained in many other foods, while less than 1 % of the population is reported to be affected by a sulfite allergy. It is debatable whether such an allergy can exist at all since the human body produces and metabolizes more than a gram of sulfites per day, which corresponds to approximately ten times the amount of sulfites contained in a bottle of wine. Thus, ethical considerations might play a greater role as dispensing with SO_2 additions is frequently associated with organic or biodynamic winegrowing.

As a matter of fact, wines without any SO_2 do not exist. During primary fermentation, yeasts produce commonly some 20 mg/L SO_2 with deviations upward and downward. A level of 10 mg/L total SO_2 is the upper limit below which the warning "contains sulfites" need not be stated on the label. There are only a very few selected yeast strains producing amounts of SO_2 below this limit under optimal nutritional conditions. Indigenous yeasts carrying out spontaneous fermentation usually exceed it.

In principle, winemaking without added SO_2 is feasible. In this context, feasibility means that such wines should not display oxidation flaws such as browning or the smell of free acetaldehyde or of other oxidation-related compounds. However, to achieve this objective and prevent oxidation, much more knowledge and experience is required than for conventional winemaking with SO_2. Winemaking without added SO_2, while maintaining quality and hygiene standards, requires control of virtually every facet of production from the crush pad to the bottled wine.

It is important to emphasize that the basic issue in producing wines without added SO_2 is not so much the microbial stability SO_2 additions would provide. Microbial safety can easily be obtained by physical means such as sterile filtration, cooling, and pasteurization. After all, the technical facilities for that purpose are almost universally available in the world of modern enology. Conversely, the issue is first and foremost about redox management and excessive oxidation rather than microbial stability.

Elaborating red wines without free SO_2 requires a complete rethink in the weighting of oxygen scavengers (Chapter 6). It does not exclude micro-oxygenation and other approaches of oxygen supply, but the orders of magnitude are inevitably different. Since SO_2 is lacking as an oxygen scavenger, working with post-fermentation yeast lees (Chapter 6.3) takes on much greater importance.

When such wines have run smoothly through barrel aging without microbial flaws or browning, they are most easily distinguished from conventional wines by their smell of free acetaldehyde. Therefore, care must be taken to ensure that its level is reduced as early as possible before bottling, and that it also remains at a level as low as possible. For that purpose, headspace inertization and the use of bottle closures with a low oxygen ingress rate are additional tools. Working with a DO meter is highly recommended to check the amounts of DO picked up and to understand where it comes from.

Alternatively, red wines without SO_2 additions can also be obtained by handling them after fermentations as if they were fruity white wines, i.e. under oxygen exclusion, without oak additions and, thus, dispensing with any aging process and the redox implications it entails. Wines obtained by this approach are less exposed to microbial risks than barrel-aged ones, but stylistically fairly different from standard red wines. They display a distinctive varietal aroma and are intended to be drunk young.

12.5. Bottle closures and post-bottling redox management

Redox management extends beyond bottling. Besides free sulfur dioxide, oxygen ingress from and through the bottle closure as well as from the bottle headspace plays a role. Whilst the oxygen uptake from the headspace is predominant in the short term up to several months post-bottling, the oxygen uptake via the bottle closure modulates the evolution and maturation of wines in the long term. However, compared to the amounts of oxygen taken up inadvertently during cellar operations and barrel aging or deliberately supplied during micro-oxygenation, the amounts provided by the bottle closure are incomparably lower and in the range of what is called nano-oxygenation (Chapter 8.3.1).

Chapter 12.4.2 provides an overview of the oxygen uptake through bottle closures and from other sources such as the bottle headspace. It shows that the oxygen transfer rate (OTR) of bottle closures, expressed in mg O_2/year, is highly variable. In a nutshell, the OTR of most screw caps and technical corks is at the lower end with less than 2 mg/year O_2 and can reach zero in extreme cases, whilst some synthetic stoppers have become known for their very high OTR of up to 20 mg/year O_2. The OTR of natural corks shows strong fluctuations between these two extremes.

White wines are known to respond to nano-oxygenation as modulated by bottle closure OTR in a very sensitive way. The reactions induced thereby can be responsible for either oxidative or reductive aging (Schneider 2019). In contrast, red wines are expected to respond with less sensitivity to post-bottling oxidative spoilage due to their reductive strength caused by their high levels of oxygen-consuming phenolics.

Nevertheless, this can become an issue when high-tier red wines are bottle-aged over longer periods. On the other hand, the extremely low or nil oxygen ingress provided by some bottle closures can be expected to trigger what is called post-bottling reduction flavor. Hence, the impact of bottle closures on red wines deserves a closer look. Of particular interest is the question of whether red wines necessarily require oxygen for their maturation in the bottle.

How bottle closure OTR affects wine quality in the medium term

In a study (Kwiatkowski et al. 2007) performed on Cabernet Sauvignon from Australia, the wine was bottled with screw caps, synthetic corks, and natural corks as well as with different air-containing headspace volumes. For the screw cap, headspace volumes of 4, 16, and 64 mL air were allowed, whilst the bottles sealed with corks had a headspace of 6 mL. Within the first year after bottling, the lots sealed with screw cap with either 4 or 64 mL headspace volume were clearly different from each other and the other treatments. The bottles with the screw cap and the largest headspace showed the lowest concentrations of SO_2, anthocyanins, but highest levels of polymeric pigments and color intensity. Results obtained on the bottles with the screw cap and the smallest headspace were just the opposite. The wines bottled under the more commercially relevant conditions such as screw cap with 16 mL headspace, synthetic stopper or natural cork, were more similar in both composition and sensory characters. It seemed to be the variable headspace volume under the screw cap that caused the most difference. Reduction flavor was observed in some cases after 11 months, with highest scores for this attribute in the wine sealed under screw cap with the smallest headspace. Since the headspaces were not the same, the effect of the closures could not be compared as would have been desirable.

In a further study (Caillé et al. 2010) performing bottling under controlled atmosphere, the impact of post-bottling oxygen exposure was investigated on four Grenache red wines from France. Different bottle closures with controlled OTR's were used, providing the bottled wines with oxygen equivalents of 0.22, 0.52, 2.19, and 3.26 mg/L·year O_2. As was to be expected, after 5 and 10 months of bottle storage, the loss of free SO_2, flavan-3-ol monomers and the conversion rate of anthocyanins to polymeric pigments was shown to increase with increasing OTR levels. The simultaneous increase of color was partially explained by the losses of SO_2. In sensory analysis, the modalities with higher oxygen ingress were rated fruitier and perceived as less animal than modalities with lower oxygen, whilst oxygen only had a minor impact on 'in mouth' attributes.

In another investigation (McRae et al. 2013), a Cabernet Sauvignon wine from Australia was adjusted to three pH levels (3.2, 3.5, and 3.8) and bottled under screw caps with either low-OTR liner (tin-Saran) or medium-OTR liner (Saranex). Bottle storage over 24 months under both closures led to a significant decrease of tannin concentration, tannin structure, and tannin polymerization (mDP), all of which can potentially contribute to a softening of astringency with aging. As could be expected, the Saranex liner with higher OTR accelerated the transformation of anthocyanins into polymeric pigments, confirming results of the previously discussed trial. However, there was an interdependency between closure and pH. Increasing pH caused less losses of anthocyanins and a lower formation of polymeric pigments during aging. Tannin structure and wine color of the lower pH wines bottled under Saranex screw caps changed more rapidly with aging than those of the higher pH wines bottled under tin-Saran. Conversely, this means that wines with very high pH values are better protected from negative aging as perceived on the palate if they are bottled with well-sealing closures such as screw caps fitted with tin-Saran liner.

In a study investigating the effects of bottle closures in connection with gas management at bottling (Vidal et al. 2017), a Merlot/Tannat wine from France with 27 mg/L free SO_2 was bottled using eight closures (four screw caps, two synthetic stoppers and two technical stoppers) with an OTR ranging from 1.8 to 24 mg/year O_2. Bottling using extensive inertization procedures allowed for an average total O_2 content (TPO) of 2 mg/L immediately after bottling. After 17 month of bottle storage, average SO_2 losses were negligible. In sensory terms, a graduation of OTR was observable based upon the 'animal' attribute by smell and color intensity. The 'animal' note points to what is commonly referred to as reduction flavor and confirms other results obtained on low-OTR closures. On the other hand, even with the most oxygen-permeable closures, the wines were not systematically characterized by oxidation or aging attributes, a result that would certainly be different for white wines.

Ultimately, however, wine aroma is too multifaceted to be reduced to generalizing concepts such as oxidative or reductive. In simplified chemical terms, oxidative spoilage is caused by free acetaldehyde and increased levels of higher aldehydes (Escudero et al. 2000 a, b, Bueno et al. 2018), and reduction taints by those of volatile sulfur compounds (Limmer 2005). Nevertheless, there are more compositional variables explaining subtle differences in aroma, which are influenced by oxygen but do not sensorially manifest themselves in terms of oxidation or reduction: In Cabernet Sauvignon wines from China bottled with six different closures and scrutinized up to 18 months post-bottling, 103 volatile aroma compounds were quantified. Sensorially perceived aroma intensity and aroma persistence correlated positively with the contents of ethyl octanoate, 1-octen-3-ol, and ethyl dodecanoate. In contrast, volatiles that had the most negative correlation with aroma quality were decanol, (Z)-3-hexen-1-ol, 2-methoxyphenol, isobutyric acid, octanoic acid, and isoamyl acetate (Liu et al. 2016).

Summary and practical teaching

From the entirety of these studies it is apparent that oxygen is not a vital component for the ongoing evolution and maturation of red wines after bottling as far as in-mouth perceptions are concerned. Red wines continue to evolve post-bottling without measurable oxygen ingress via the closure by anaerobic reactions, but the access to small amounts of oxygen accelerates maturation through oxidative reactions. Similar observations were also made in young red wines before bottling (Somers and Evans 1986).

Bottled red wine responds to nano-oxygenation in basically the same way as to prebottling micro or macro-oxygenation, albeit to a much lesser extent. Moreover, differences in the OTR of commercially relevant bottle closures are much lower than the quantities of oxygen introduced during typical MOX treatment. Hence, they do not affect phenol and color composition to a sensorially significant extent during the first two years post-bottling, but may well affect the long-term evolution of wines intended for longer periods of bottle maturation. In the short term of up to two years post-bottling, headspace oxygen can have more impact than closure OTR.

As for the olfactory perceptions, they are more impacted by bottle closure OTR than in-mouth perceptions. The most common observation is that low-OTR closures enhance the perception of post-bottling reductive taints in red wines if they are prone to develop them, similarly to what can be observed in red wines. The use of closures with an OTR ranging from 1 to 2 mg/year O_2 best meets the oxygen requirements of most bottled red wines. They comprise high-quality natural corks, but they can also be found among technical corks and screwcaps that are available with a graduated scale of defined OTRs. Hermetically sealing screwcaps like those fitted with a tin-Saran liner should only be used when their liner contains immobilized copper able to scavenge the volatile sulfur compounds causing post-bottling reduction flavor (Schneider et al. 2017).

Regardless of the aromatic evolution perceived by smell, not all wines age in a positive way. Some of them become meager, dry out, or produce a sour aftertaste that cannot be explained by changes in total acidity or pH. The reasons for such a detrimental aging are poorly understood.

In general, the higher the total phenol content, the more likely the wine is to age well, but the opposite is not always true. The structure of the tannins extracted during maceration, the tannin-anthocyanin ratio, the polysaccharide content, and the oxygen ingress through the bottle closure play a role. Wines with a high seed-tannin content age more rapidly than those with a high skin-tannin content. A wine that tastes hard and astringent at the moment of bottling will generally retain that character over years, albeit wishful thinking often leads to the opposite assumption.

Aging proceeds very slowly at temperatures around 10° C (50° F), but much faster at 20° C (68° F). This property can be used to prepare medium-quality wines for sale in the short term. Color is generally considered to be a marker for aging. A brick-red color is characteristic of a wine that has aged for some time, but a visible browning indicates that the wine is ‘over the hill’. Rapid oxidation as caused by a porous natural cork degrades the wine in a very short spell of time.

13. Literature

Agnolucci M., Rea F., Sbrana C., Cristani C., Fracassetti D., Tirelli A., Nuti M., 2010. Sulphur dioxide affects culturability and volatile phenol production by Brettanomyces/Dekkera bruxellensis. Int. J. Food Microbiol. 143 (1-2): 76-80.

Agnolucci M., Cristani C., Maggini S., Rea F., Cossu A., Tirelli A., Nuti M., 2014. Impact of sulphur dioxide on the viability, culturability, and volatile phenol production of Dekkera bruxellensis in wine. Ann. Microbiol. 64 (2): 653-659.

Aleixandre-Tudo J.L., Lizama V., Álvarez I., Nieuwoudt H., Garcia M.J., Aleixandre J.L., du Toit W.J., 2016. Effect of acetaldehyde addition on the phenolic substances and volatile compounds of red Tempranillo wines. Aust. J. Grape Wine Research 22 (2): 205-214.

Aleixandre-Tudo J.L., Buica A., Nieuwoudt H., Aleixandre J.L., du Toit W., 2017. Spectrophotometric analysis of phenolic compounds in grapes and wines. J. Agric. Food. Chem. 65 (20): 4009-4026.

Alpeza I., Kovačević Ganić K., Vanzo A., Herjavec S., 2017. Improved chromatic and sensory characteristics of Plavac Mali wines – Efficiency of maceration enzymes. Czech. J. Food Sci, 35 (3): 236-245.

Amerine M.A., 1955. Further studies on controlled fermentations. Am. J. Enol. Vitic. 6 (1): 1-16.

Amerine M.A. and Ough C.S., 1957. Studies on controlled fermentations, III. Am. J. Enol. Vitic. 8 (1): 18-30.

Anli R.E. and Cavuldak Ö.A., 2012. A review of microoxygenation application in wine. J. Inst. Brew. 118: 368-385.

Apetrei I.M., Rodríguez-Méndez M.L., Apetrei C., Nevares I., del Alamo-Sanza M., de Saja J.A., 2012. Monitoring of evolution during red wine aging in oak barrels and alternative method by means of an electronic panel test. Food Res. Int. 45 (1): 244-249.

Arapitsas P., Scholz M., Vrhovsek U., di Blasi S., Biondi Bartolini A., Masuero D., et al. (2012). A metabolomic approach to the study of wine micro-oxygenation. PLoS ONE 7 (5): e37783. https://doi.org/10.1371/journal.pone.0037783.

Arfelli G., Sartini E., Corzani C., Fabiani A., 2011. Chips, lees, and micro-oxygenation: Influence on some flavors and sensory profile of a bottled red Sangiovese wine. Eur. Food. Res. Technol. 233: 1-10.

Arnold R.A., Noble A.C., Singleton V. L., 1980. Bitterness and astringency of phenolic fractions in wine. J. Agric. Food Chem. 28 (3): 675-678.

Arriagada-Carrazana J.P., Sáez-Navarrete C., Bordeu E., 2005. Membrane filtration effects on chromatic and phenolic quality of Cabernet Sauvignon wines. J. Food Eng. 68 (3): 363-368.

Atanasova V., Fulcrand H., Cheynier V., Moutounet M., 2002. Effect of oxygenation on polyphenol changes occurring in the course of wine-making. Anal. Chim. Acta. 458 (1): 15-27.

Auw J.M., Blanco V., O'Keefe S.F., Sims C.A., 1996. Effect of processing on the phenolics and color of Cabernet Sauvignon, Chambourcin, and Noble wines and juices. Am. J. Enol. Vitic. 47 (3): 279-286.

Bai B., He F., Yang L., Chen F., Reeves M.J., Li J., 2013. Comparative study of phenolic compounds in Cabernet Sauvignon wines made in traditional and Ganimede fermentors. Food Chem. 141 (4): 3984-3992.

Balboa-Lagunero T., Arroyo T., Cabellos J.M., Aznar M., 2011. Sensory and olfactometric profiles of red wines after natural and forced oxidation processes. Am. J. Enol. Vitic. 62 (4): 527-535.

Balík J., 2006. Dynamics of changes in total anthocyanins during the fermentative maceration of grapes. Hort. Sci. 33 (3): 103-107.

Bakker J., Preston N.W., Timberlake C.F., 1986. The determination of anthocyanins in aging red wines. Comparison of HPLC and spectral methods. Am. J. Enol. Vitic. 37 (2): 121-126.

Bakker J., Bridle P., Bellworthy S.J., Garcia-Viguera C., Reader H.P., Watkins S.J., 1998. Effect of sulphur dioxide and must extraction on colour, phenolic composition and sensory quality of red wine. J. Sci. Food Agric. 78 (3): 297-307.

Bakker J., Bellworthy S.J., Reader H.P., Watkins S.J., 1999. Effect of enzymes during vinification on color and sensory properties of Port wines. Am. J. Enol. Vitic. 50 (3): 271-276.

Bartolini A.B., Cavini F., Basquiat M., 2008. Oxygène et vin. Du rôle de l'oxygène à la technique de la micro-oxygénation. Parsec Edition, Firenze 2008.

Bautista-Ortín A.B., Martínez-Cutillas A., Ros-García J.M., López-Roca J.M., Gómez-Plaza E., 2005. Improving colour extraction and stability in red wines: the use of maceration enzyme and enological tannins. Int. J. Food Sci. Technol. 40 (8): 867-878.

Bautista-Ortín A.B., Romero-Cascales I., Fernández-Fernández J.I., López-Roca J.M., Gómez-Plaza E., 2007. Influence of the yeast strain on Monastrell wine colour. Innov Food Sci. Emerg. Technol. 8 (3): 322-328.

Bautista-Ortín A.B., Lencina A.G., Cano-López M., Pardo-Minguez F., López-Roca J.M., Gomez-Plaza E., 2008. The use of oak chips during the ageing of a red wine in stainless steel tanks or used barrels: Effect of the contact time and size of the oak chips on aroma compounds. Aust. J. Grape Wine Res. 14 (2): 63-70.

Beelman R.B. and McArdle F.J., 1974. Influence of carbonic maceration on acid reduction and quality of Pennsylvania dry red table wine. Am. J. Enol. Vitic. 25 (4): 219-221.

Belda I., Conchillo L.B., Ruiz J., Navascués E., Marquina D., Sanatos A., 2016. Selection and use of pectinolytic yeasts for improving clarification and phenolic extraction in winemaking. Int. J. Food. Microbiol. 223: 1-8. https://doi.org/10.1016/j.ijfoodmicro.2016.02.003

Bernath K., Flueler T., Hühn T., 2003. Mikrooxigenation, II. Schweiz. Z. Obst-Weinbau 01/03: 8-10.

Bertrand G.L., 1976. Free sulfur dioxide in red wine: A comparison of analytical methods in relation to the thermodynamic activity. Am. J. Enol. Vitic. 27 (3): 106-110.

Bindon K.A., Carew A.L., Mierczynska-Vasilev A., Kassara S. Kerslake F., Smith P.A., 2016. Characterization of macromolecular complexes in red wine. Composition, molecular mass distribution, and particle size. Food Chem. 199: 838-846.

Bindon K.A., Kassara S., Solomon M., Bartel M., Bartel C., Smith P.A., Barker A., Curtin C., 2019. Commercial *Saccharomyces cerevisiae* yeast strains significantly impact Shiraz tannin and polysaccharide composition with implications for wine colour and astringency. Biomolecules 9 (466). https://doi:10.3390/biom9090466

Blateyron L. and Sablayrolles J.M., 2001. Stuck and slow fermentations in enology: Statistical study of causes and effectiveness of combined additions of oxygen and diammonium phosphate. J. Biosci. Bioeng. 91 (2): 184-189.

Blazquez-Rojas I., Smith P.A., Bartowsky E.J., 2012. Influence of choice of yeasts on volatile fermentation-derived compounds, colour and phenolics composition in Cabernet Sauvignon wine. World J. Microbiol. Biotechnol. 28 (12): 3311-3321. https://doi:10.1007/s11274-012-1142-y

Blouin J., Llorca L., Montreau F.R., Dufour J.H., 1972. Études des conditions optimales pour la détermination des composés phénoliques totaux par le réactif de Folin-Ciocalteu. Conn. Vigne Vin 6 (4): 405-413.

Blouin J., Papet N., Stonestreet E., 2000. Étude de la structure polyphénolique des vins rouges par analyses physico-chimiques et sensorielles. J. Int. Sci. Vigne Vin 34 (1): 33-40.

Bobinaité R., Pataro G., Lamanauskas N., Šatkauskas S., Viškelis P., Ferrari G., 2014. Application of pulsed electric field in the production of juice and extraction of bioactive compounds from blueberry fruits and their by-products. J. Food Sci. Technol. 52 (9): 5898-5905.

Bosso A., Guaita M., Panero L., Borsa D., Follis R., 2009. Influence of two winemaking techniques on polyphenolic composition and color of wines. Am. J. Enol. Vitic. 60 (3): 379-385.

Bosso A., Panero L, Petrozziello M., Follis R., Motta S., Guaita M., 2011. Influence of submerged-cap vinification on polyphenolic composition and volatile compounds of Barbera wines. Am. J. Enol. Vitic. 62 (4): 503-511.

Boulet J.C., Ducasse M.A., Cheynier V., 2017. Ultraviolet spectroscopy study of phenolic substances and other major compounds in red wines : Relationship between astringency and the concentration of phenolic compounds. Aust. J. Grape Wine Res. 23 (2) : 193-199.

Boulton R., 2001. The copigmentation of anthocyanins and its role in the color or red wines. A critical review. Am. J. Enol. Vitic. 52 (1): 67-87.

Brannan G.D., Setser C.S., Kemp K.E., 2001. Effectiveness of rinses in alleviating bitterness and astringency residuals in model solutions. J. Sens. Stud. 16 (3): 261-275.

Brooks L., McCloskey L., McKesson D., Sylvan M., 2008. Adams-Harbertson protein precipitation-based wine tannin method found invalid. J. AOAC International 91 (5): 1090-1094.

Buechsenstein J.W. and Ough C.S., 1978. SO_2 determination by aeration-oxidation: A comparison with ripper. Am. J. Enol. Vitic. 29 (3): 161-164.

Bueno M., Marrufo-Curtido A., Carrascón V., Fernández-Zurbano P., Escudero A., Ferreira V., 2018). Formation and accumulation of acetaldehyde and Strecker aldehydes during red wine oxidation. Front. Chem. 6: 20. doi: 10.3389/fchem.2018.00020

Buffon P., Heymann H., Block D.E., 2014. Sensory and chemical effects of cross-flow filtration on white and red wines. Am. J. Enol. Vitic. 65 (3): 305-314.

Burdach K.J., 1988. Geschmack und Geruch. Verlag Hans Huber Editions, Bern (Switzerland) 1988.

Burkhardt R., 1976. Analytische Bestimmung phenolischer Inhaltsstoffe von Mosten und Wein. Mitt. II: Trennung in niedermolekulare, nicht gerbende phenolische Bestandteile und gerbende oligomere Derivate. Mitteilungsblatt G. D. Chem., Fachgruppe Lebensmittelchemie und gerichtl. Chemie, 30: 206-213.

Burroughs L.F., 1975. Determining free sulfur dioxide in red wine. Am. J. Enol. Vitic. 26 (1): 25-29.

Butticaz S.and Rawyler A., 2007. Différenciation analytique des vins élevés en fût de chêne et macérés avec des copeaux de chêne. Rev. Suisse Vitic. Arboric. Hortic. 39 (6): 367-373.

Cabrita M.J., Barrocas Dias C., Costa Freitas A.M., 2011. Phenolic acids, phenolic aldehydes and furanic derivatives in oak chips: American vs. French oak. S. Afr. J. Enol. Vitic. 32 (2): 204-210.

Cáceres-Mella A., Peña-Neira A., Narváez-Bastias J., Jara-Campos C., López-Solís R., Canals J.M., 2013. Comparison of analytical methods for measuring proanthocyanidins in wines and their relationship with perceived astringency. Int. J. Food Sci. Tech. 48 (12): 2588-2594.

Cagnasso E., Rolle L., Caudana A., Gerbi V., 2008. Relationship between grape phenolic maturity and red wine phenolic composition. Ital. J. Food Sci. 20 (3): 365-381.

Caillé S., Samson A., Wirth J., Diéval J.-B., Vidal S., Cheynier V., 2010. Sensory characteristics changes of red Grenache wines submitted to different oxygen exposures pre und post bottling. Anal. Chim. Acta 660 (1-2): 35-42.

Calderón J. F., Del Alamo-Sanza M., Nevares I., Laurie V. F., 2014. The influence of selected winemaking equipment and operations on the concentration of dissolved oxygen in wines. Cien. Inv. Agr. 41 (2): 273-280.

Caloghiris M., Waters E. J., Williams P. J., 1997. An industry trial provides further evidence for the role of corks in oxidative spoilage of bottled wines. Aust. J. Grape Wine Res. 3: 9-17.

Canals R., Llaudy C.M., Canals J.M., Zamora F., 2008. Influence of the elimination and addition of seeds on the colour, phenolic composition and astringency of red wine. Eur. Food Res. Technol. 226 (5): 1183-1190.

Canas S., Sun B., Coimbra L., Barroca J., 2011. Efeitos da filtração esterilizante na estabilidade microbiológica, composição química e características sensoriais do vinho tinto. Ciência Téc. Vitic. 26 (2): 77-83.

Cano-López M., Pardo-Mínguez F., Schmauch G., Saucier C., Teissedre P.-L., López-Roca J.M., Gómez-Plaza E., 2008 a. Effects of micro-oxygenation on color and anthocyanin-related compounds of wines with different phenolic contents. J. Agric. Food Chem. 56 (14): 5932-5941.

Cano-López M., Bautista-Ortín A.B., Pardo-Mínguez F., López-Roca J.M., Gómez-Plaza E., 2008 b. Sensory descriptive analysis of a red wine aged with oak chips in stainless steel tanks or used barrels: Effect of the contact time and size of the oak chips. J. Food Qual. 31 (5): 645-660.

Canuti V., Puccioni S., Giovani G., Salmi M., Roso I., Bertuccioli M., 2012. Effect of oenotannin addition on the composition of Sangiovese wines from grapes with different characteristics. Am. J. Enol. Vitic. 63 (2): 220-231.

Carew A.L., Smith P., Close D.C., Curtin C., Dambergs R.G., 2013. Yeast effects on Pinot noir wine phenolics, color, and tannin composition. J. Agric. Food Chem. 61 (41): 9892-9898.

Casassa L.F., Beaver C.W., Mireles M., Larsen R.C., Hopfer H., Heymann H., Harbertson J.F., 2013 a. Influence of fruit maturity, maceration length, and ethanol amount on chemical and sensory properties of Merlot wines. Am. J. Enol. Vitic. 64 (4): 437-449.

Casassa L.F., Beaver C.W., Mireles M.S., Harbertson J.F., 2013 b. Effect of extended maceration and ethanol concentration on the extraction and evolution of phenolics, colour components and sensory attributes of Merlot wines. Aust. J. Grape Wine Res. 19 (1): 25–39.

Casassa L.F., 2017. Flavonoid phenols in red winemaking. In: Intech Open Science, Chapter 6. http://dx.doi.org/10.5772/67452

Casey J. A., 1989. Closures for wine bottles – a user's viewpoint. Austral. Grapegrower and Winemaker 4: 99-107.

Castellari M., Simonato B., Tornielli G.-B., Spinelli P., Ferrarini R., 2004. Effect of different enological treatments on dissolved oxygen in wine. Ital. J. Food Sci. 16 (3): 387-396.

Cattaruzza A., Peri C., Rossi M., 1987. Ultrafiltration and deep-bed filtration of a red wine: Comparative experiments. Am. J. Enol. Vitic. 38 (2): 139-142.

Cayla L., Cottereau P., Renard R., 2002. Estimation de la maturité phénolique des raisins rouges par la méthode I.T.V. standard. Rev. Fr. d'Œnologie 193: 10-16.

Cejudo Bastante M.J., Hermosín-Gutiérrez I., Pérez-Coello M.S., 2011 a. Micro-oxygenation and oak chip treatments of red wines: Effects on colour-related phenolics, volatile composition and sensory characteristics. Part I: Petit Verdot wines. Food Chem. 124 (3): 727-737.

Cejudo Bastante M.J., Hermosín-Gutiérrez I., Pérez-Coello M.S., 2011 b. Micro-oxygenation and oak chip treatments of red wines: Effects on colour-related phenolics, volatile composition and sensory characteristics. Part II: Merlot wines. Food Chem. 124 (3): 738-748.

Celotti E. and Carcereri de Prati G., 2005. The phenolic quality of red grapes at delivery: Objective evaluation with colour measurements. S. Afr. J. Enol. Vitic. 26 (2): 75-82.

Chatonnet P. and Dubourdieu D., 1998. Comparative study of the characteristics of American white oak (Quercus alba) and European oak (Quercus petraea and Q. robur) for production of barrels used in barrel aging of wines. Am. J. Enol. Vitic. 49 (1): 79-85.

Chatonnet P., 1999. Discrimination and control of toasting intensity and quality of oak wood barrels. Am. J. Enol. Vitic. 50 (4): 479-494.

Chauvet S., Sudraud P., Vivas N., Glories Y., 1992. Les tanins œnologiques. Caractérisation de la nature des produits commerciaux. Revue des Œnologues 64: 8-10.

Chen L.Y., Cheng C.W., Liang J.Y., 2015. Effect of esterification condensation on the Folin-Ciocalteu method for the quantitative measurement of total phenols. Food Chem. 170: 10-15.

Cheynier V., Fulcrand H., Brossaud F., Asselin C., Moutounet M., 1998. Phenolic constitution as related to red wine flavor. *In:* Chemistry of Wine Flavor, ACS Symposium Series 714: 124-141.

Cheynier V., Dueñas-Paton M., Salas E., Maury C., Souquet J.-M., Sarni-Manchado P., Fulcrand H., 2006. Structure and properties of wine pigments and tannins. Am. J. Enol. Vitic. 57 (3): 298-305.

Chira K., Jourdes, M., Teissedre P.L., 2012. Cabernet Sauvignon red wine astringency quality control by tannin characterization and polymerization during storage. Eur. Food Res. Technol. 234 (2): 253-261.

Cliff M.A., Brau N., King M.C., Mazza G., 2002. Development of predictive models for astringency from anthocyanin, phenolic and color analyses of British Columbia wines. J. Int. Sci. Vigne Vin 36 (1): 21-30.

Coelho J.M., Howe P.A., Sacks G.L., 2015. A headspace gas detection tube method to measure SO_2 in wine without disrupting SO_2 equilibria. Am. J. Enol. Vitic. 66 (3): 257-265.

Colonna A.E., Adams D.O., Noble A.C., 2004. Comparison of procedures for reducing astringency carry-over effects in evaluation of red wines. Aust. J. Grape Wine Research 10 (1): 26-31.

Constantin C., Patz C.D., Dietrich H., 2011. Die Wirkung von Gummi arabicum im Wein. Deut. Weinbau 06: 30-35.

Cosme F., Ricardo-da-Silva J.M., Laureano O., 2009. Effect of various proteins on different molecular weight proanthocyanidin fractions of red wine during wine fining. Am. J. Enol. Vitic. 60 (1): 74-81

Cuenat P., Lorenzini F., Brégy C., Zufferey E., 1996. La macération préfermentaire à froid du Pinot noir. Aspects technologiques et microbiologiques. Rev. Suisse Vitic. Arboric. Hortic. 28 (4) : 259-265.

Cuinier C., 1997. Ceppi di lievito e composizione fenolica dei vini rossi. Vignevini 7/8 : 39-42.

Dallas C., Ricardo-da-Silva J.M., Laureano O., 1995. Degradation of oligomeric procyanidins and anthocyanins in a Tinta Roriz red wine during maturation. Vitis 34 (1): 51-56.

Danilewicz J.C., 2007. Interaction of sulfur dioxide, polyphenols, and oxygen in a wine-model system: Central role of iron and copper. Am. J. Enol. Vitic. 58 (1): 53-60.

Danilewicz J.C., 2012. Review of oxidative processes in wine and value of reduction potentials in enology. Am. J. Enol. Vitic. 63 (1): 1-10.

Danilewicz J.C., 2015. Folin-Ciocalteu, FRAP, and DPPH* Assays for measuring polyphenol concentration in white wine. Am. J. Enol. Vitic. 66 (4): 463-471.

Danilewicz J. C., 2016. Reaction of oxygen and sulfite in wine. Am. J. Enol. Vitic. 67 (1): 13-17.

Danilewicz J.C. and Standing M.J., 2018. Reaction mechanisms of oxygen and sulfite in red wine. Am. J. Enol. Vitic. 69 (3): 189-195.

de Beer D., Harbertson J.F., Kilmartin P.A., Roginsky V., Barsukova T., Adams D.O., Waterhouse A.L., 2004. Phenolics: A comparison of diverse analytical methods. Am. J. Enol. Vitic. 55 (4): 389-400.

de Miglio P. and Pickering G.J., 2008. The influence of ethanol and pH on the taste and mouthfeel sensations elicited by red wine. J. Food Agric. Environ. 6 (3-4): 143-150.

de Santis D. and Frangipane M.T., 2010. Effect of prefermentative cold maceration on the aroma and phenolic profiles of a Merlot wine. Ital. J. Food Sci. 22 (1): 47-53.

del Alamo M., Nevares I., Gallego L, Fernández de Simón B., Cadahía E., 2010. Micro-oxygenation strategy depends on origin and size of chips or staves during accelerated aging. Anal. Chim. Acta 660 (1-2): 92-101.

del Alamo-Sanza M. and Nevares I., 2014. Recent advances in the evaluation of the oxygen transfer rate in oak barrels. J. Agric. Food Chem. 62 (35): 8892-8899.

del Alamo-Sanza M., Laurie V.F., Nevares I., 2015. Wine evolution and spatial distribution of oxygen during storage in high-density polyethylene tanks. J. Sci. Food Agric. 95 (6): 1313-1320.

del Alamo-Sanza M., Cárcel L.M., Nevares I., 2017. Characterization of the oxygen transmission rate of oak wood species used in cooperage. J. Agric. Food Chem. 65 (3): 648-655.

del Alamo-Sanza M. and Nevares I., 2018. Oak wine barrel as an active vessel: A critical review of past and current knowledge. Crit. Rev. Food Sci. Nutr., 58 (16): 2711-2726.

del Barrio-Galán R., Pérez-Magariño S., Ortega-Heras M., 2011. Techniques for improving or replacing ageing on lees of oak aged red wines: The effects on polysaccharides and the phenolic composition. Food Chem. 127 (2): 528-540.

del Barrio-Galán R., Pérez-Magariño S., Ortega-Heras M., Guadalupe Z., Ayestarán B., 2012. Polysaccharide characterization of commercial dry yeast preparations and their effect on white and red wine composition. Food Sci. Technol. 48 (2): 215-223.

del Barrio-Galán R., Cáceres-Mella A., Medel-Maraboli M., Peña-Neira Á., 2015. Effect of selected Saccharomyces cerevisiae yeast strains and different aging techniques on the polysaccharide and polyphenolic composition and sensorial characteristics of Cabernet Sauvignon wines. J. Sci. Food Agric. 95 (10): 2132-2144.

Delcour J. A., Vandenberghe M. M., Corten P. F., Dondeyne P., 1984. Flavor thresholds of polyphenolics in water. Am. J. Enol. Vitic. 35 (3): 134-136.

Delteil D., 2000. Utilisation des tanins œnologiques sur les raisins et les vins méditerranéens. Rev. Fr. d'Œnologie, 181, 20-22.

Deshaies S., Cazals G., Enjalbal C., Constantin T, François Garcia F., Mouls L., Saucier C., 2020. Red wine oxidation: Accelerated ageing tests, possible reaction mechanisms and application to Syrah red wines. Antioxidants, MDPI, 2020, 10.3390/antiox9080663.hal-02941501.

Devatine A., Chiciuc I., Mietton-Peuchot M., 2011. The protective role of dissolved carbon dioxide against wine oxidation: A simple and rational approach. J. Int. Vigne Vin 45 (3): 189-197.

Diéval J.-B., Vidal S., Aagaard O., 2011. Measurement of the oxygen transmission rate of co-extruded wine bottle closures using a luminescence-based technique. Packag. Technol. Sci. 24 (7): 375-385.

Ducruet V., Flanzy C., Bourzeix M., Chambroy Y, 1983. Les constituants volatils des vins jeunes de macération carbonique. Sci. Aliments 3: 413-426.

Ducruet J., Dong A., Canal-Llaubères R.M., Glories Y., 1997. Influence des enzymes pectolytiques sélectionnées pour l'œnologie sur la qualité et la composition des vins rouges. Rev. Fr. d'Œenologie 155: 16-19.

Durner D. and Fischer U., 2009 a: Mikrooxygenierung von Rotwein, Teil II. Makro- oder Mikrooxygenierung – Welches Verfahren birgt Potenzial? Deut. Weinmagazin 10: 42-45.

Durner D. and Fischer U., 2009 b. Mikrooxygenierung von Rotweinen, Teil III. Welche Rebsorten eignen sich? Deut. Weinmagazin 11: 28-31.

Durner D. and Fischer U., 2009 c. Mikrooxygenierung von Rotweinen, Teil V. Holzfass, Reifetank und Mikrooxygenierung – ein Vergleich. Deut. Weinmagazin 18: 23-27.

Durner D., Weber F., Fischer U., 2009. Mikrooxygenierung von Rotweinen, Teil IV. Mikrooxygenierung auch bei der Maischeerhitzung? Deut. Weinmagazin 13: 10-14.

Durner D., Weber F., Neddermeyer J., Koopmann K., Winterhalter P., Fischer U., 2010 a. Sensory and color changes induced by microoxygenation treatments of Pinot noir before and after malolactic fermentation. Am. J. Enol. Vitic. 61 (4): 474-485.

Durner D., Ganss S., Fischer U., 2010 b. Monitoring oxygen uptake and consumption during microoxygenation treatments before and after malolactic fermentation. Am. J. Enol. Vitic. 61 (4): 465-473.

Durner D., Nickolaus P., Weber F., Trieu H.-L., Fischer U., 2015. Evolution of anthocyanin-derived compounds during micro-oxygenation of red wines with different anthocyanin-flavanol ratios. ACS Symposium Series, Vol. 1203: 253-274.

Dykes S. and Kilmartin P. (2007). Micro-oxygenation: Optimising the maturation process. Aust. Wine Ind. J. 22: 31–41.

Elias R. J. and Waterhouse A. L., 2010. Controlling the Fenton reaction in wine. J. Agric. Food Chem. 58 (3): 1699-1707.

Escot S., Feuillat M., Dulau L., Charpentier C., 2001. Release of polysaccharides by yeasts and the influence of released polysaccharides on colour stability and wine astringency. Aust. J. Wine Grape Research 7 (3): 153-159.

Escudero A., Ferreira V., Hernández P., Cacho J. F., 2000 a. Wine flavor oxidation: Changes in the aroma profiles during oxidation and their potential sensory significance. In: Œnologie 99 (A. Lonvaud-Funel, coord.), Éditions Tec & Doc, Paris 2000, pp. 422-424.

Escudero A., Hernández-Orte P., Cacho J., Ferreira V., 2000 b. Clues about the role of methional as character impact compound on some oxidized wines. J. Agric. Food. Chem. 48: 4268-4272.

Escudero A., Asensio E., Cacho J., Ferreira V., 2002. Sensory and chemical changes of young white wines stored under oxygen. An assessment of the role played by aldehydes and some other important odorants. Food Chem. 77: 325-331.

Escudier J.L. and Moutounet M., 2001. Prétraitement des raisins par Flash-Détente sous vide. Incidence sur la qualité des vins. Rev. Fr. d'Œnologie 9: 19.22.

Fabre S., 1993. Destination de l'oxygène consommé par le vin. Revue des Œnologues 71: 23-26.

Fanzone M., Peña-Neira Á., Jofre V., Assof M., Zamora F., 2013. Efecto de la técnica de sangrado sobre la composición fenólica de vinos cv. Malbec. Rev. FCA UNCUYO 45 (1): 199-209.

Favre G., Peña-Neira Á., Baldi C., Hernández N., Traverso S., Gil G., Gonzáles-Neves G., 2014. Low-molecular weight phenols in Tannat wines made by alternative winemaking procedures. Food Chem. 158: 504-512.

Fazio T. and Warner C.R., 1990. A review of sulphites in foods: Analytical methodology and reported findings. Food Addit. Contam. 7 (4): 354-354.

Ferreia V., Carrascon V., Bueno M., Ugliano M., Fernandez-Zurbano P., 2015. Oxygen consumption by red wines. Part I: Consumption rates, relationship with chemical composition, and role of SO_2. J. Agric. Food Chem. 63 (51): 10928-10937.

Feuillat M., Peyron D., Berger J.L., 1987. Influence de la microfiltration tangentielle des vins sur leur composition physico-chimique et leurs caractères sensoriels. Application aux vins de Bourgogne. Bull. OIV 673-674: 227-244.

Feuillat M., 1996. Vinification du Pinot noir en Bourgogne par macération préfermentaire à froid. Revue des Œnologues 82: 29-31.

Feuillat M., Escot S., Charpentier C., Dulau L., 2001. Élevage des vins rouges sur lies fines – Intérêt des interactions entre polysaccharides de levures et polyphénols du vin. Revue des Œnologues 98: 17-18.

Fischer U., Boulton R.B., Noble A.C., 1994 a. Physiological factors contributing to the variability of sensory assessments: Relationship between salivary flow rate and temporal perception of gustatory stimuli. Food Qual. Pref. 5 (1-2): 55-64.

Fischer U. and Noble A.C., 1994 b. The effect of ethanol, catechin concentration, and pH on sourness and bitterness of wine. Am. J. Enol. Vitic. 45 (1): 6-10.

Fischer U., 1997 a. Neue Trends bei der Maischegärung von Rotwein, Teil 1: Farbe bekennen. Deut. Weinmagazin 13: 24-29.

Fischer U., 1997 b. Maischegärung ist nicht gleich Maischegärung. Deut. Weinbau 18: 44-52.

Fischer U., Strasser M., Gutzler K., 2000. Impact of fermentation technology on the phenolic and volatile composition of German red wines. Int. J. Food Sci. Tech. 35 (1): 81-94.

Fontoin H., Saucier C., Teissedre P.L., Glories Y., 2008. Effect of pH, ethanol and acidity on astringency and bitterness of grape seed tannin oligomers in model wine solution. Food Qual. Prefer. 19: 286-291.

Fornairon C., Mazauric J.P., Salmon J.M, Moutounet P., 1999. Observations sur la consommation de l'oxygène pendant l'élevage des vins sur lies. J. Int. Sci. Vigne Vin 33 (2): 79-86.

Fornairon-Bonnefond C., Camarasa C., Moutounet M., Salmon J.M., 2002. New trends on yeast autolysis: A bibliographic review. J. Int. Sci. Vigne Vin 36 (2): 49-69.

Fornairon-Bonnefond C. and Salmon J.-M., 2003. The impact of oxygen consumption by yeast lees on the autolysis phenomenon during simulation of wine aging on lees. J. Agric. Food Chem. 51 (9), 2584-2590.

Frost S.C., Blackman J.W., Hjelmeland A.K., Ebeler S.E., Heymann H., 2018. Extended maceration and cap management impacts on the phenolic, volatile, and sensory profiles of Merlot wine. Am. J. Enol. Vitic. 69 (4): 360-370.

Fulcrand H., Atanasova V. Salas E., Cheynier V., 2004. The fate of anthocyanins in wine: Are there determining factors? ACS Symposium Series, Vol. 886: 68-88.

Gambuti A., Han G., Peterson A.L., Waterhouse A.L., 2015. Sulfur dioxide and glutathione alter the outcome of microoxygenation. Am. J. Enol. Vitic. 66 (4): 411-423.

Gambuti A., Siani T., Picariello L., Rinaldi A., Lisanti M.T., Ugliano M., Diéval J.B., Moio L., 2016. Oxygen exposure of tannins-rich red wines during bottle aging. Influence on phenolics and color, astringency markers and sensory attributes. Eur. Food Res. Technol. 243 (4): 669-680.

Gambuti A., Pacariello L., Rolle L., Moio L., 2017. Evaluation of the use of sulfur dioxide and glutathione to prevent oxidative degradation of malvidin-3-monoglucoside by hydrogen peroxide in the model solution and real wine. Food Res. Int. 99 (1): 454-460.

Gao L., Girard B., Mazza G., Reynolds A., 1997. Changes in anthocyanins and color characteristics of Pinot noir wines during different vinification processes. J. Agric. Food Sci. 45 (6): 2003-2008.

García-Estevez I., Alcalde-Eon C., Rivas-Gonzalo J. C., Escribano-Bailón M. T., Fernández A., Nevares I., del Alamo-Sanza M., 2016. An approach to study the interactions between ellagitannins and oxygen during barrel aging. *In*: Groupe Polyphénols (Ed.), 28th International Conference on Polyphenols, Vienna. University of Technology, Vienna.

Garrido-Bañuelos G.G., Buica A., Sharp E., de Villiers A., du Toit W.J., 2019. The impact of different tannin to anthocyanin ratios and of the oxygen on the phenolic polymerisation over time in wine-like solutions. S. Afr. J. Enol. Vitic. 40 (2): 1-1.

Gawel R., 1998. Red wine astringency: a review. Aust. J. Grape Wine Res. 4 (2): 74-95.

Gawel R., Oberholster A., Francis I.L., 2000. A 'mouth-feel wheel': Terminology for communicating the mouth-feel characteristics of red wine. Aust. J. Grape Wine Res. 6 (3): 203-207.

Gawel R., Iland P.G., Leske P.A., Dunn C.G., 2001. Compositional and sensory differences in Syrah wines following juice runoff prior to fermentation. J. Wine Res. 12 (1): 5-18.

Gawel R. Francis L., Waters E.J., 2007. Statistical correlations between the in-mouth textural characteristics and the chemical composition of Shiraz wines. J. Agric. Food Chem. 55 (7): 2683-2687.

Geffroy O., Lopez R., Serrano E., Dufourcq T., Gracia-Moreno E., Cacho E., Ferreira V., 2013. Incidence de cinq techniques de macération sur les charactéristiques analytiques, aromatiques et sensorielles des vins rouges. Revue des Œnologues 149: 55-59.

Geffroy O., Lopez R., Serrano E., Dufourcq T., Gracia-Moreno E., Cacho J., Ferreira V., 2015. Changes in analytical and volatile compositions of red wines induced by pre-fermentation heat treatments of grapes. Food Chem. 187: 243-253.

Gerbaux V., 1993. Étude de quelques conditions de cuvaison susceptibles d'augmenter la composition polyphénolique des vins de Pinot noir. Rev. Fr. d'Œnologie 69: 15-18.

Gerbaux V., Vincent B., Briffos C., 2003. Optimisation de la maceration finale à chaud, intérêt d'une maceration sous chapeau immerge pour la vinification du Pinot noir. Rev. Fr. d'Œnologie 201: 16-21.

Ghanem C., Taillandier P., Rizk Z. Nehme N., Souchard J.P., El Rayess Y., 2019. Evolution of polyphenols during Syrah grapes maceration: Time versus temperature effect. Molecules 24, 2845. https://doi: 10.3390/molecules24152845.

Giovani G., Canuti V., Rosi I., 2010. Effect of yeast strain and fermentation conditions on the release of cell wall polysaccharides. Int. J. Food Microbiol. 137 (2-3): 303-307.

Girard B., Kopp T.G., Reynolds A.G., Cliff M., 1997. Influence of vinification treatments on aroma constituents and sensory descriptors of Pinot noir wines. Am. J. Enol. Vitic. 48 (2): 5- 14.

Girard B., Yuksel D., Cliff M.A., Delaquis P., Reynolds A.G., 2001. Vinification effects on the sensory, color, and GC profiles of Pinot noir wines from British Columbia. Food Res. Int. 34 (3): 483-499.

Giusti M. M. and Wrolstad R.E., 2001.Characterization and measurement of anthocyanins by UV-visible spectroscopy. Current Prot. Food Anal. Chem., 00(1): F1.2.1.–F1.2.13.

Glories Y., 1974. Recherches sur la structure et les propriétés des composés phénoliques polymérisés des vins rouges. I. Précipitation par l'aldéhyde formique et par l'acide chlorhydrique concentré. Conn. Vigne Vin 1: 57-79.

Glories Y. 1984. La couleur des vins rouges. Conn. Vigne Vin 18 (3): 195-217 and 18 (4) : 253-271.

Glories Y. and Saucier C., 2000. Tannin evolution from grape to wine. Effects on wine taste. *In*: Proceedings ASEV 50th Anniversary Annual Meeting, Seattle 2000.

Gonzáles-Arenzana L., Santamaria R., Escribano-Viana R., Portu J., Garijo P., López-Alfaro I., López R., Santamaría P., Gutiérrez A.R., 2020. Influence of the carbonic maceration winemaking method on the physicochemical, colour, aromatic and microbiological features of Tempranillo red wines. Food Chem. 319: 126569. doi: 10.1016/j.foodchem.2020.126569

Gonzáles-Manzano S., Rivas-Gonzalo J.C., Santos-Buelga C., 2004. Extraction of flavan-3-ols from grape seed and skin into wine during simulated fermentation. Anal. Chim. Acta 513 (1): 283-289.

Gonzáles-Manzano S., Santos-Buelga C., Pérez-Alonso J.J., Rivas-Gonzalo J.C., Escribano-Bailón M.T., 2006. Characterization of the mean degree of polymerization of proanthocyanidins in red wines using liquid chromatography-mass spectrometry (LC-MS). J. Agric. Food Chem. 54 (12): 4326-4332.

Gonzáles-Royo E., Esteruelas M., Kontoudakis N., Fort F., Canals J.M., Zamora F., 2017. The effect of supplementation with three commercial inactive dry yeasts on the colour, phenolic compounds, polysaccharides and astringency in a model wine solution and red wine. J. Sci. Food Agric. 97 (1): 172-181.

González-Sanjosé M.L., Ortega-Heras M., Pérez-Magariño S., 2008. Microoxygenation treatment and sensory properties of young red wines. Food Sci. Tech. Int. 14 (5): 123-130.

Gordillo B., López-Infante M.I., Ramírez-Pérez P., Gonzáles-Miret M.L., Heredia F.J., 2010. Influence of prefermentative cold maceration on the color and anthocyanic copigmentation of organic Tempranillo wines elaborated in a warm climate. J. Agric. Food Chem. 58 (11): 6797-6803.

Guadalupe Z., Palacios A., Ayestarán B., 2007. Maceration enzymes and mannoproteins: A possible strategy to increase colloidal stability and color extraction in red wines. J. Agric. Food Chem. 55 (12): 4854-4862.

Guadalupe Z. and Ayestarán B., 2008. Effect of commercial mannoproteins addition on polysaccharide, polyphenolic, and color composition of red wines. J. Agric. Food Chem. 56 (19): 9022-9029.

Guadalupe Z., Martínez L., Ayestarán B., 2010. Yeast mannoproteins in red winemaking: Effect on polysaccharide, polyphenolic, and color composition. Am. J. Enol. Vitic. 61 (2): 191-200.

Guérin L., Sutter D.-H., Demois A., Chereau M., Trandafir G., 2009. Determination of activity profiles of the main commercial enzyme preparations used in winemaking. Am. J. Enol. Vitic. 60 (3): 322-331.

Guerrand D. and Gervais J.P., 2002. Extraction of red wine phenolics during fermentation: A new pectinase preparation. Aust. N.Z. Grapegrower & Winemaker 466: 61-64.

Guilloux-Bénatier M., Guerreau J., Feuillat M., 1995. Influence of initial colloid content on yeast macromolecule production and on the metabolism of wine organisms. Am. J. Enol. Vitic. 46 (4): 486-492.

Guinard J.-X., Pangborn R.M., Lewis M.J., 1986 a. The time-course of astringency in wine upon repeated ingestion. Am. J. Enol. Vitic. 37 (3): 184-189.

Guinard J.-X., Pangborn R.M., Lewis M.J., 1986 b. Preliminary studies on acidity-astringency interactions in model solutions and wines. J. Sci. Food Agric. 37 (8): 811-817.

Gunata Z. Pineau J., Cordonnier R., 1987. Détermination de la qualité de la vendange par sa richesse en composés phénoliques. Rev. Fr. d'Œnologie 107: 7-13.

Guymon J.F. and Crowell E.A., 1977. The nature and cause of cap-liquid temperature differences during wine fermentation. Am. J. Enol. Vitic. 28 (2) : 74-79.

Hagerman A.E. and Butler L.G., 1978. Protein precipitation method for the quantitative determination of tannins. J: Agric. Food Chem. 26 (4): 809-812.

Hanlin R.L., Kelma M.A., Wilkinson K.L., Downey M.O., 2011. Detailed characterization of proanthocyanidins in skin, seeds, and wine of Shiraz and Cabernet Sauvignon wine grapes (*Vitis vinifera*). J. Agric. Food Chem. 59 (24): 13265-13276.

Harbertson J.F., Kenneday J.A., Adams D.O., 2002. Tannin in skins and seeds of Cabernet Sauvignon, Sytah, and Pinot noir berries during ripening. Am. J. Enol. Vitic. 53 (1): 54-59.

Harbertson J.F., Picciotto E.A., Adams D.O., 2003. Measurement of polymeric pigments in grape berry extracts and wines using a protein precipitation assay combined with bisulfite bleaching. Am. J. Enol. Vitic. 54 (4): 301-306.

Harbertson J.F. and Spayd S., 2006. Measuring phenolics in the winery. Am. J. Enol. Vitic. 57 (3): 280-288.

Harbertson J.F., Hodgins R.E., Thurston L.N., Schaffer L.J., Reid M.S., Landon J.L., Ross C.F., Adams D.O., 2008. Variability of tannin concentrations in red wines. Am. J. Enol. Vitic. 59 (2): 210-214.

Harbertson J.F., Mireles M.S., Harwood E.D., Weller K.M., Ross C.F., 2009. Chemical and sensory effects of saignée, water addition, and extended maceration on high brix musts. Am. J. Enol. Vitic. 60 (4): 450-460.

Harbertson J.F., Parpinello G.P., Heymann H., Downey M.O., 2012. Impact of exogenous tannin additions on wine chemistry and wine sensory character. Food Chem. 131 (3): 999-1008.

Haslam E., 1980. In vino veritas: Oligomeric procyanidins in the aging of red wines. Phytochemistry 19: 1577-1582.

Herderich M.J. and Smith P.A., 2005. Analysis of grape and wine tannins: Methods, applications and challenges. Aust. J. Grape Wine Research 11 (3): 205-214.

Héritier J., Bach B., Schönenberger P., Gaillard V., Ducruet J., Segura J.-M., 2016. Quantification of the production of hydrogen peroxide H_2O_2 during accelerated wine oxidation. Food Chem. 211: 957-962.

Hernández-Jiménez A., Kennedy J.A., Bautista-Ortín A.B., Gómez-Plaza E., 2012. Effect of ethanol on grape seed proanthocyanidin extraction. Am. J. Enol. Vitic. 63 (1): 57-61.

Hernández-Orte P., Franco E., González Huerta C., Martínez García J., Cabellos M., Suberviola J., Orriols I., Cacho J., 2014. Criteria to discriminate between wines aged in oak barrels and macerated with oak fragments. Food Res. Int. 37 (234-241)

Holt H., Cozzolino D., McCarthy J., Abrahamse C., Holt S., Solomon M., Smith P., Chambers P.J., Curtin C., 2013. Influence of yeast strain on Shiraz wine quality indicators. Int. J. Food Microbiol. (3) 165, 302–311.

Howe P.A., Worobo R., Sacks G.L., 2018. Conventional measurements of sulfur dioxide (SO_2) in red wine overestimate SO_2 antimicrobial activity. Am. J. Enol. Vitic. 69 (3): 210-220.

Ingledew W.M. and Kunkee R.E., 1985. Factors influencing sluggish fermentations. Am. J. Enol. Vitic. 36 (1): 65-76.

Jackson M.G., Timberlake C.F., Bridle P., Vallis L., 1978. Red wine quality: Correlations between colour, aroma and flavour and pigment and other parameters of young Beaujolais. J. Sci. Food Agric. 29 (8): 715-272.

Jakobs D.D., 1976. Effect of dissolved oxygen on free sulfur dioxide in red wines. Am. J. Enol. Vitic. 27 (1): 42-43.

Jensen J.S., Blachez B., Egebo M., Meyer A.S., 2007. Rapid extraction of polyphenols from grapes. Am. J. Enol. Vitic. 58 (4): 451-561.

Joutei K.A. and Glories Y., 1994. Étude en conditions modèles de l'extractibilité des composés phénoliques des pellicules et des pépins de raisins rouges. J. Int. Sci. Vigne Vin 28 (4): 303-317.

Julien A., Roustan J.L., Dulau L., Sablayrolles J.M., 2000. Comparison of nitrogen and oxygen demands of enological yeasts: Technological consequences. Am. J. Enol. Vitic. 51 (3): 215-222.

Jung R. and Schüssler C., 2016. Alternative Verschlüsse für Wein. Deut. Weinmagazin 21: 16-19.

Jurd L., 1966. Anthocyanins and related compounds. 10. Peroxide oxidation products of 3-alkylflavylium salts. Tetrahedron 22, 2913-2921.

Kallithraka S., Bakker J., Cliffored M.N., 1997. Evaluation of bitterness and astringency of (+)-catechin and (-)-epicatechin in red wine and in model solution. J. Sensory Studies 12 (1): 25-37.

Kallithraka S., Bakker J., Clifford M.M., 1998. Evidence that salivary proteins are involved in astringency. J. Sens. Stud. 13 (1): 29-43.

Kang W., Niimi J., Bastian S.E.P., 2018. Reduction of red wine astringency perception using vegetable protein fining agents. Am. J. Enol. Vitic. 69 (1): 22-31.

Karbowiak T., Gougeon R. D., Alinc J.-B., Brachais L., Debeaufort F., Voilley A., Chassagne D., 2010. Wine oxidation and the role of cork. Crit. Rev. Food Sci. Nutr. 50 (1): 20-52.

Kassara S., Kennedy J.A., 2011. Relationship between red wine grade and phenolics. 2. Tannin composition and size. J. Agric. Food Chem. 59 (15): 8409-8412.

Kelly M. and Wollan D., 2003 Micro-oxygenation of wine in barrels. Australian NZ Grapegrower & Winemaker, Annual Technical Issue, 473: 29-32.

Kennedy J.A., Matthews M.A., Waterhouse A.L., 2000. Changes in grape seed phenols during fruit ripening. Phytochemistry 55 (1): 77-85.

Kennedy J.A., Ferrier J., Harbertson J.F., Peyrot des Gachons C., 2006. Analysis of tannins in red wine using multiple methods: Correlation with perceived astringency. Am. J. Enol. Vitic. 57 (4): 481-485.

Kilmartin P.A., 2009. The oxidation of red and white wines and its impact on wine aroma. Chemistry in New Zealand 73 (1): 18-22.

King E.S., Dunn R.L., Heymann H., 2013. The influence of alcohol on the sensory perception of red wines. Food Qual. Prefer. 28: 235-243.

Köhler H.J., Geßner M., Hermann J., 2005: Vergleich verschiedener Mazerationsverfahren: Den Rotweinstil optimieren. Deut. Weinmagazin 19: 14-21.

Kovać V., Alonso E., Revilla E. 1995. The effect of adding supplementary quantities of seeds during fermentation on the phenolic composition of wines. Am. J. Enol. Vitic. 46 (3): 363-367.

Koyama K., Goto-Yamamoto N., Hashizume K., 2007. Influence of maceration temperature in red wine vinification on extraction of phenolics from berry skins and seeds of grape (Vitis vinifera). Biosci. Biotechnol. Biochem. 71 (4): 958-965.

Kramling T.E. and Singleton V.L., 1969. An estimate of the nonflavonoid phenols in wine. Am. J. Enol. Vitic. 20 (2): 86-92.

Kwiatkowski M., Skouroumounis G., Lattey K., Waters E., 2007. The impact of closures, including screw cap with three different headspace volumes, on the composition, colour and sensory properties of a Cabernet Sauvignon wine during two years' storage. Aust. J. Grape Wine Res. 13 (2): 81-94.

Lagune L. and Glories Y., 1996. Les nouvelles données concernant le collage des vins rouges avec les gélatines œnologiques. Rev. Fr. d'Œnologie 158: 18-20.

Laurie V.F., Law R., Joslin W.S., Waterhouse A.L., 2008. In situ measurements of dissolved oxygen during low-level oxygenation in red wines. Am. J. Enol. Vitic. 59 (2): 215-219.

Laurie V.F., Salazar S., Campos M.I., Cáceres-Mella A., Peña-Neira A., 2014. Periodic aeration of red wine compared to microoxygenation at production scale. Am. J. Enol. Vitic. 65 (2): 254-260.

Lawless H.T., Horne J., Giasi P., 1996. Astringency of organic acids is related to pH. Chem. Senses 21 (4): 397-403.

Lee C.B. and Lawless H.T., 1991. Time-course of astringent sensations. Chem. Senses 16 (3): 225-238.

Lee J., Kennedy J.A., Devlin C., Redhead M., Rennaker C., 2008. Effect of early seed removal during fermentation on proanthocyanidin extraction in red wine: A commercial production example. Food Chem. 107: 1270-1273.

Lee C.A. and Vickers Z.M., 2009. Discrimination among astringent is affected by the choice of palate cleanser. Food Qual. Pref. 21 (1): 93-99.

Lempéreur V., Blateyron L., Labarbe B., Saucier C., Kelebek H., Glories Y., 2002. Groupe national de travail sur les tanins œnologiques: premiers résultats. Rev. Fr. d'Œnologie 196: 23-29.

Lerno L.A., Reichwage M., Ponangi R., Hearne L., Block D.E., Oberholster A., 2015. Effects of cap and overall fermentation temperature on phneolic extraction in Cabernet Sauvignon fermentations. Am. J. Enol. Vitic. 66 (4) : 444-453.

Lerno L., Reichwage M., Panprivech S., Ronangi R., Hearne L., Oberholster A., Block D.E., 2017. Chemical gradients in pilot-scale Cabernet Sauvignon fermentations and their effect on phenolic extraction. Am. J. Enol. Vitic. 68 (4) : 401-411.

Lerno L.A., Panprivech S., Ponangi R., Hearne L., Blair T., Oberholster A., Block D.E., 2018. Effect of pump-over conditions on the extraction of phenolic compounds during Cabernet Sauvignon fermentation. Am. J. Enol. Vitic. 69 (3) : 295-301.

Li S., Wilkinson K.L., Bindon K.A., 2018. Compositional variability in commercial tannin and mannoprotein products. Am. J. Enol. Vitic. 69 (2): 176-181.

Limmer A., 2005. Do corks breathe? Or the origin of SLO. Austral. & New Zealand Grapegrower & Winemaker, Annual Technical Issue, 89-98.

Liu D., Xing R.-R., Li Z., Yang D.-M., Pan Q.-H., 2016. Evolution of volatile compounds, aroma attributes, and sensory perception in bottle-aged red wines and their correlation. Eur. Food. Res. Technol. 242: 1937-1948.

Llaudy M.C., Canals R., Canals J.M., Rozea N., Arola S.L., Zamora F., 2004. New method for evaluating astringency in red wines. J. Agric. Food Chem. 52 (4): 742-746.

Llaudy M.C., Canals R., Gonzáles-Manzano S., Zamora F., 2006. Influence of micro-oxygenation treatment before oak aging on phenolic compounds composition, astringency, and color of red wines. J. Agric. Food Chem. 54 (12): 4246-4252.

Llaudy M.C., Canals R., Canals J.M., Zamora F., 2008. Influence of ripening stage and maceration length on the contribution of grape skins, seeds and stems to phenolic composition and astringency in wine-simulated macerations. Eur. Food Res. Technol. 226 (3): 337-344.

Lopes P., Saucier C., Teissedre P.-L., Glories Y., 2006. Impact of storage position on oxygen ingress through different closures into wine bottles. J. Agric. Food Chem. 54 (18): 6741-6746.

Lopes P., Saucier C., Teissedre P.-L., Glories Y., 2007. Main routes of oxygen ingress through different bottle closures into wine bottles. J. Agric. Food Chem. 55 (13): 5167-5170.

Lorenzini F., 2001. Comparaison de six preparations de levures œnologiques sur du Pinot noir. Rev. Suisse Vitic. Arboric. Hortic. 33 (2): 73-76.

Lorincz G. and Vas G., 1998. Examination of volatile components in carbonic macerated wines. Wein-Wissenschaft 53 (1): 18-21.

Lorrain B., Ky I., Pechamat L., Teissedre P.L., 2013. Evolution of analysis of polyphenols from grapes, wines, and extracts. Molecules 18: 1076-1100.

Luengo E., Franco E., Ballesteros F., Àlvarez I., Raso J., 2014. Winery trial on application of pulsed electric fields for improving vinification of Garnacha grapes. Food Bioprocess Technol. 7 (5): 1457-1464.

Lukić I., Budić-Leto I., Bubola M., Damijanić K., Staver M., 2017. Pre-fermentative cold maceration, saignée, and various thermal treatments as options for modulating volatile aroma and phenol profiles of red wine. Food Chem. 224: 251-261.

Lyman B.J. and Green B.G., 1990. Oral astringency: Effects of repeated exposure and interactions with sweeteners. Chem. Sens. 15 (2): 151-164.

Madrigal-Galan B. and Heymann H., 2006. Sensory effects of consuming cheese prior to evaluating red wine flavor. Am. J. Enol. Vitic. 57 (1): 12-22.

Magalhães L.M., Santos F., Segundo M.A., Reis S., Lima J.L.F.C., 2010. Rapid microplate high-throughput methodology for assessment of Folin-Ciocalteu reducing capacity. Talanta 83 (2): 441-447.

Marais J., 2003 a. Effect of different wine-making techniques on the composition and quality of Pinotage wine. I. Low-temperature skin contact prior to fermentation. S. Afr. J. Enol. Vitic. 24 (2): 70-75.

Marais J., 2003 b. Effect of different wine-making techniques on the composition and quality of Pinotage wine. II. Juice/skin mixing practices. S. Afr. J. Enol. Vitic. 24 (2): 76-79.

Martínez-Lapuente L., Guadalupe Z., Ayestarán B., 2017. Effect of egg albumin fining, progressive clarification and cross-flow microfiltration on the polysaccharide and proanthocyanidin composition of red varietal wines. Food Res. Int. 96: 235-243.

Mattivi F., Vrhovsek U., Masuero D., Trainotti D., 2009. Differences in the amount and structure of extractable skin and seed tannins amongst red grape varieties. Austral. J. Grape Wine Res. 15 (1): 27-35.

Maury C., Sarni-Manchado P., Lefebvre S., Cheynier V., Moutounet M., 2001. Influence of fining with different molecular weight gelatins on proanthocyanidin composition and perception of wines. Am. J. Enol. Vitic. 52 (2): 140-145.

Maury C., Sarni-Manchado P., Lefebvre S., Cheynier V., Moutounet M., 2003. Influence of fining with plant proteins on proanthocyanidin composition of red wines. Am. J. Enol. Vitic. 54 (2): 105-111.

McRae J.M. Dambergs R.G., Kassara S., Parker M., Jeffery D.W., Herderich M.J., Smith P.A., 2012. Phenolic composition of 50 and 30 year sequences of Australian red wine: The impact of wine age. J. Agric. Food Chem. 60 (40): 10093-10102.

McRae J., Kassara S., Kennedy J.A., Waters E.J., Smith P.A., 2013. Effect of wine pH and bottle closure on tannins. J. Agric. Food Chem. 61 (47): 11618-11627.

McRae J.M., Kirby N., Mertens H.D.T., Kassara S., Smith P.A., 2014. Measuring the molecular dimensions of wine tannins: Comparison of small-angle X-ray scattering, gel-permeation chromatography and mean degree of polymerization. J. Agric. Food Sci. 62 (29): 7216-7224.

McRae J.M., Day M.P., Bindon K.A., Kassara S. Schmidt S.A., Schulkin A., Kolouchova R., Smith P.A., 2015. Effect of early oxygen exposure on red wine colour and tannins. Tetrahedron 71 (20): 3131-3137.

McRae J.M., Mierczynska-Vasilev A., Soden A., Barker A.M., Day M.P., Smith P.A., 2017. Effect of commercial-scale filtration on sensory and colloidal properties of red wines over 18 months bottle aging. Am. J. Enol. Vitic. 68 (3): 263-274.

Mercurio M.D. and Smith P.A., 2008. Tannin quantification in red grapes and wine. Comparison of polysaccharide- and protein-based tannin precipitation techniques and their ability to model wine astringency. J. Agric. Food Chem. 56 (14): 5528-5537.

Mercurio M., Dambergs R.G., Cozzolino D., Herderich M.J., Smith P.A., 2010. Relationship between red wine grades and phenolics. 1. Tannin and total phenolics concentration. J. Agric. Food Chem. 58 (23): 12313-12319.

Moenne M.I., Saa P., Laurie V.F., Pérez-Correa J.R., Agosin E., 2014. Oxygen incorporation and dissolution during industrial-scale red wine fermentations. Food Bioprocess Tech. 7 (9): 2627-2636.

Montreau F.R., 1972. Sur le dosage des composés phénoliques totaux dans les vins par la méthode Folin-Ciocalteu. Conn. Vigne Vin 6 (4): 397-404.

Morata A., Gómez-Cordovés M.C., Colomo B., Suárez J.A., 2005. Cell wall anthocyanin adsorption by different Saccharomyces strains in the fermentation of Vitis vinifera L. cv Graciano grapes. Eur. Food. Res. Technol. 220 : 341-346.

Morata A., Loira I., Suárez L.J.A., 2016. Influence of yeasts in wine colour. *In:* Grape and Wine Biotechnology, A. Morata and I. Loira (editors), IntechOpen, https://doi: 10.5772/65055.

Morel-Salmi C., Souquet J.M., Bes M., Cheynier V., 2006. Effect of Flash Release treatment on phenolic extraction and wine composition. J. Agric. Food Chem. 54 (12): 4270-4276.

Moutounet M., 1981. Dosage des polyphénols des moûts de raisin. Conn. Vigne Vin 15 (4): 287-301.

Moutounet M., Mazuric J.P., Saint-Pierre B., Hanocq J.F., 1998. Échanges gazeux des vins logés en barriques. J. Sci. Tech. Tonnellerie, 4, 115-129.

Nagel C.W. and Glories Y., 1991. Use of a modified dimethylaminocinnamaldehyde reagent for analysis of flavanols. Am. J. Enol. Vitic. 42 (4) : 364-366.

Navarro M., Kontoudakis N., Gómez-Alonso S., García-Romero E., Canals J.M., Hermosín-Gutíerrez I., Zamora F., 2016. Influence of the botanical origin and toasting level on the ellagitannin content of wines aged in new and used barrels. Food Res. Int. 87 : 197-203.

Nel A.P., Louw L., Lambrechts M.G., van Rensburg P., 2015. The influences of different winemaking techniques on the mouthfeel of Shiraz grapes. S. Afr. J. Enol. Vitic. 36 (1) : 71-93.

Nevares L., del Alamo M., Cárcel L.M., Crespo B., Martin C., Gallego L., 2009. Measure the dissolved oxygen consumed by red wines in aging tanks. Food Bioprocess Technol. 2: 328-336.

Nevares I., del Alamo-Sanza M., Gonzales-Muñoz C., 2010. Dissolved oxygen distribution during micro-oxygenation. Determination of representative measurement points in hydroalcoholic solution and wines. Anal. Chim. Acta 660 (1-2): 232-239.

Nevares I., Crespo R., Gonzáles C., del Alamo-Sanza M., 2014. Imaging of oxygen permeation in the oak wood of wine barrels using optical sensors and a colour camera. Aust. J. Grape Wine Res. 20 (3): 353-360.

Nevares I. and del Alamo-Sanza M., 2015. Oak stave oxygen permeation: A new tool to make barrels with different wine oxygenation potentials. J. Agric. Food Chem. 63 (4):1268–1275.

Nevares I. and del Alamo-Sanza M., 2016. Wine aging technologies. *In*: Recent Advances in Wine Stabilization and Conservation Technologies, 1st ed., pp. 209-245. Jordão A.M. and Cosme F., Eds., Nova Science Publishers, Inc., New York.

Nickolaus P. and Durner D., 2016. Wie viel Sauerstoff braucht der Wein? Deut. Weinbau 25-26: 47-53.

Nicolini G., Mattivi F., Larcher R., Volpini A., 2003. Vinificazioini in rosato ed in rosso con lieviti selezionati : observazioni circa il colore. Enologo 39 (5) : 79-83.

Noble A.C., 1994. Bitterness in wine. Physiol. Behav. 56: 1251-1255.

Noble A.C., 1995. Application of time-intensity procedures for the evaluation of taste and mouthfeel. Am. J. Enol. Vitic. 46 (1): 128-133.

Oberholster A., Carstens L.M., du Toit W.J., 2013. Investigation on the effect of gelatine, egg albumin and cross-flow microfiltration on the phenolic composition of Pinotage wines. Food Chem. 138 (2-3): 1275-1281.

Oberholster A., Elmendorf B.L., Lerno L.A., King E.S., Heymann A., Brennemann C.E., Boulton R.B., 2015. Barrel maturation, oak alternatives and micro-oxygenation: Influence on red wine aging and quality. Food Chem. 173: 1250-1258.

Obreque-Slier E., Peña-Neira A., López-Solís R., Zamora-Marín F., Ricardo da Silva J.M., Laureano O., 2010 a. Comparative study of the phenolic composition of seeds and skins from Carmenere and Cabernet Sauvignon grape varieties (Vitis vinifera L.) during ripening. J. Agric. Food Chem. 58 (6): 3591-3599.

Obreque-Slier E., Peña-Neira A., López-Solís R., 2010 b. Enhancement of both salivary protein−enological tannin interactions and astringency perception by ethanol. J. Agric. Food Chem. 58 (6): 3729–3735.

Oliveira C.M., Ferreira A,C.S., de Freitas V., Silva A.M.S., 2011. Oxidation mechanisms occurring in wines. Food Res. Int. 44 (5): 1115-1126.

Oliveira V., Lopes P., Cabral M., Pereira H., 2013. Kinetics of oxygen ingress into wine bottles closed with natural cork stoppers of different qualities. Am. J. Enol. Vitic. 64 (3): 395-399.

Ortega-Heras M., Rivero-Pérez M.D., Pérez-Magariño S., Gonzáles-Huerta C., Gonzáles-San José M.L., 2008. Changes in the volatile composition of red wines during aging in oak barrels due to microoxygenation treatment applied before malolactic fermentation. Eur. Food Res. Technol. 226 (6): 1485-1493.

Ortega-Heras M., Pérez-Magariño S., Cano-Mozo E., Gonzáles-San José M.L., 2010. Differences in the phenolic composition and sensory profile between red wines aged in oak barrels and wines aged with oak chips. LWT-Food Sci. Technol. 43 (10): 1533-1541.

Ough C.S. and Berg H.W., 1974. The effect of two commercial pectic enzymes on grape musts and wines. Am. J. Enol. Vitic. 25 (3): 208-211.

Ough C.S., Noble A.C., Temple D., 1975. Pectic enzyme effects on red grapes. Am. J. Enol. Vitic. 26 (4): 195-200.

Paissoni M.A., Waffo-Teguo P., Ma, W., Jourdes M., Giacosa S., Rio Segade S., Rolle L., Teissedre P.L., 2020. Sensory assessment of grape polyphenolic fractions: an insight into the effects of anthocyanins on in-mouth perceptions. Oeno One 54 (4): 1059-1075.

Palomero F., Morata A., Benito S., Gonzáles M.C., Suárez-Lepe J.A., 2007. Conventional and enzyme-assisted autolysis during ageing over lees in red wines: Influence on the release of polysaccharides from yeast cell walls on wine monomeric anthocyanin content. Food Chem. 105 (2): 838-846.

Parker M., Smith P.A., Birse M., Francis I.L., Kwiatkowski M.J., Lattey K.A., Liebich B., Herderich M.J., 2007. The effect of pre- and post-ferment additions of grape derived tannin on Shiraz wine sensory properties and phenolic composition. Aust. J. Grape Wine Res., 13 (1): 30-37.

Parley A.L., Vanhanen L., Heatherbell D., 2001. Effects of pre-fermentation enzyme maceration on extraction and colour stability in Pinot noir wine. Aust. J. Grape Wine Res. 7 (3): 146-152.

Parpinello G.P., Plumejeau F., Maury C., Versari A., 2012. Effect of micro-oxygenation on sensory characteristics and consumer preference of Cabernet Sauvignon wine. J. Sci. Food Agric. 92 (6): 1238-1244.

Pegram Z., Kwasniewski M.T., Sacks G.L., 2013. Simplified method for free SO_2 measurement using gas detection tubes. Am. J. Enol. Vitic. 64 (3): 405-410.

Peleg H., Bodine K.K., Noble A.C., 1998. The influence of acid on astringency of alum and phenolic compounds. Chem. Senses 23 (3): 371-378.

Pellegrini N., Simonetti P., Gardana C., Brenna O., Brighenti F., Pietta P., 2000. Polyphenol content and total antioxidant activity of vini novella (young red wines). J. Agric. Food Chem. 48 (3): 732-735.

Pereira G.A., Arruda H.S., Pastore G.M., 2018. Modification and validation of Folin-Ciocalteu assay for faster and safer analysis of total phenolic content in food samples. Braz. J. Food Res. 9 (1): 125-140.

Pérez-Coello M.S., Sanz J., Cabezudo M.D., 1999. Determination of volatile compounds in hydroalcoholic extracts of French and American oak wood. Am. J. Enol. Vitic. 50 (2): 162-165.

Peri C. and Pompei C., 1971. An assay of different phenolic fractions in wines. Am. J. Enol. Vitic. 22 (2): 55-58.

Peterson R.G., 1976. Formation of reduced pressure in barrels during wine aging. Am. J. Enol. Vitic. 27 (2): 80-81.

Petrozziello M., Torchio F., Piano F., Giacosa S., Ugliano M., Bosso A., Rolle L., 2018. Impact of increasing levels of oxygen consumption on the evolution of color, phenolic, and volatile compounds of Nebbiolo wines. Frontiers in Chemistry 6 (137): 1-15, https://doi.org/10.3389/fchem.2018.00137

Pizarro C., Rodríguez-Tecedor S., Esteban-Díez I., Pérez-del-Notario N., Gonzáles-Sáiz J.M., 2014. Experimental design approach to evaluate the impact of oak chips and micro-oxygenation on the volatile profile of red wines. Food Chem. 148: 357-366.

Plane R.A., Mattick I.R., Weirs L.D., 1980. An acidity index for the taste of wines. Am. J. Enol. Vitic. 31 (3): 265-268.

Pons A., Nikolantonaki M., Lavigne V., Shinoda K., Dubourdieu D., Darriet P., 2015. New insights into intrinsic and extrinsic factors triggering premature aging in white wines. *In:* Advances in Wine Research, ACS Symposium Series, Vol. 1203: 229-251.

Pontallier P., 1981. Recherches sur les conditions d'élevage des vins rouges. Rôle des phénomènes oxidatifs. PhD Thesis, University of Bordeaux II.

Pontallier P. and Ribéreau-Gayon P., 1983. Influence de l'aération et du sulfitage sur l'évolution de la matière colorante des vins rouges au cours de la phase d'élevage. Conn. Vigne Vin 17 (2): 105-120.

Porter L.J., Hrstich L.N., Chan B.G., 1985. The conversion of procyanidins and prodelphinidins to cyanidin and delphinidin. Phytochemistry 25 (1): 223-230.

Pour Nikfardjam M.P., 2004. Mikrooxidation – ein Gerätevergleich. Deut. Weinbau 9: 72-74.

Prat-García S, Nevares I, Martínez-Martínez V, Del Alamo-Sanza M., 2020. Customized oxygenation barrels as a new strategy for controlled wine aging. Food Res Int. 131:108982. doi: 10.1016/j.foodres.2020.108982.

Qiu Y., Lacampagne S., Mirabel M., Mietton-Peuchot M., Ghidossi R., 2018. Oxygen desorption and oxygen transfer through oak staves and oak stave gaps: an innovative permeameter. Oeno One 52 (1): DOI: https://doi.org/10.20870/oeno-one.2018.52.1.909

Quijada-Morín N., Regueiro J., Simal-Gándara J., Thomás E., Rivas-Gonzalo J.C., Escribano-Bailón M.T., 2012. Relationship between the sensory-determined astringency and the flavanolic composition of red wines. J. Agric. Food Chem. 60 (50): 12355–12361.

Retali E., 2004. Macération préfermentaire à froid: Application à une vendange de Niellucio. Rev. Fr. d'Œnologie 209: 16-18.

Reynolds A., Cliff M., Girard B., Kopp T.G., 2001. Influence of fermentation temperature on composition and sensory properties of Semillon and Shiraz wines. Am. J. Enol. Vitic. 52 (3): 235-240.

Ribéreau-Gayon P. and Stonestreet E., 1965. Le dosage des anthocyanes dans le vin rouge. Bull. Soc. Chim. France 9: 2649-2652.

Ribéreau-Gayon P. and Stonestreet E., 1966. Le dosage des tanins du vin rouge et la détermination de leur structure. Chim. Anal. 48: 188-196.

Ribéreau-Gayon P. and Milhé J.C., 1970. Recherches technologiques sur les composés phénoliques du vin rouge. I. Influence des différentes parties de la grappe. Conn. Vigne Vin 4: 63-74.

Ribéreau-Gayon P., Sudraud P., Milhé J.C., Canbas A., 1970. Recherches technologiques sur les composés phénoliques du vin rouge. II. Facteurs de dissolution des composés phénoliques. Conn. Vigne Vin 4: 133-144.

Ricci A., Perpinello G.P., Versari A., 2018. Recent advances and applications of pulsed electric fields (PEF) to improve polyphenol extraction and color release during red winemaking. Beverages 4, 18, https://doi:10.3390/beverages4010018

Ristić R., Bindon K., Francis L.J., Herderich M.J., Iland P.G., 2010. Flavonoids and C_{13}-norisoprenoids in Vitis vinifera L. cv. Shiraz: Relationships between grape and wine composition, wine colour and wine sensory properties. Aust. J. Grape Wine Res. 16 (3): 369-388.

Rivas-Gonzalo J.C., Gutierrez Y., Hebrero E., Santos-Buelga C., 1992. Comparison of methods for the determination of anthocyanins in red wines. Am. J. Enol. Vitic. 43 (2): 210-214.

Robichaud J.L. and Noble A.C., 1990: Astringency and bitterness of selected phenolics in wine. J. Sci. Food Agric. 53 (3): 343-353.

Rodrigues A., Ricardo-da-Silva J.M., Lucas C., Laureano O., 2013. Effect of winery yeast lees on Touriga Nacional red wine color and tannin evolution. Am. J. Enol. Vitic. 64 (1): 98-109.

Rodríguez-Bencomo J.J., Ortega-Heras M., Pérez-Magariño S., Gonzáles-Huerta C., Gonzáles-San José M.L., 2008. Importance of chip selection and elaboration process on the aromatic composition of finished wines. J. Agric. Food Chem. 56 (13): 5102-5111.

Rodriguez Montealegre R., Romero Peces J.L., Chacón Vozmediano J., Martínez Gascueña E., García Romero E., 2006. Phenolic compounds in skins and seeds of ten grape (*Vitis vinifera*) varieties grown in a warm climate. J. Food Comp. Anal. 19 (6-7): 687-693.

Rolle L., Siret R., Rio Segade S., Maury C., Gerbi V., Jourjon F., 2012. Instrumental texture analysis parameters as markers of table-grape and wine grape quality: A Review. Am. J. Enol. Vitic. 63 (1): 11-28.

Rosenfeld E., Beauvoit B., Rigoulet M., Salmon J.-M., 2002. Non-respiratory oxygen consumption pathways in anaerobically-grown Saccharomyces cerevisiae: evidence and partial characterization. Yeast 19 (15), 1299-1321.

Ross, C.F., Hinken, C., Weller K., 2007. Efficacy of palate cleansers for reduction of astringency carry-over during repeated ingestions of red wine. J. Sens. Stud. 22 (3): 293-312.

Rousseau Y. and Delteil D., 2000. Présentation d'une méthode d'analyse sensorielle des raisins. Principe, méthode et grille d'interprétation. Rev. Fr. d'Œnologie 183: 10-13.

Rubio-Bretón R., Garde-Cerdán T., Martínez J., 2018. Use of oak fragments during the aging of red wines. Effect on the phenolic, aromatic, and sensory composition of wines a a function of the contact time with the wood. Beverages 4, 102; https.doi:10.3390/beverages4040102

Sablayrolles, J.M., 1990. Besoins en oxygène lors des fermentations œnologiques. Rev. Fr. d'Œnologie 124: 77-79.

Sacci K.I., Bisson L.F., Adams D.O., 2005. A review of the effect of winemaking techniques on phenolic extraction in red wines. Am. J. Enol. Vitic. 56 (3): 197-206.

Sack M., Eing C., Stängle R., Wolf A., Müller G., Sigler J., Stukenbrock L., 2009. Electric measurement of the electroporation efficiency of mash from wine grapes. IEEE Transactions on Dielectrics and Electrical Insulation 16 (5): 1329-1337.

Sacks G.L., Howe P.A., Standing M., Danilewicz J.C., 2020. Free, bound, and total sulfur dioxide (SO_2) during oxidation of wines. Am. J. Enol. Vitic. 71 (4): 266-277.

Salinas M.R., Alonso G.L., Navarro G., Pardo F., Jimeno J., Huerta M.D., 1996. Evolution of the aromatic composition of wines undergoing carbonic maceration under different aging conditions. Am. J. Enol. Vitic. 47 (2): 134-144.

Salmon J.M., 2006. Interactions between yeast, oxygen and polyphenols during alcoholic fermentations: practical implications. Food Sci. Technol. 39 (9): 959-65.

Salmon J.-M., Fornairon-Bonnefond C., Mazauric J.-P., Moutounet M., 2000. Oxygen consumption by wine lees: impact on lees integrity during wine ageing. Food Chem. 71 (4): 519-528.

Salmon J.M, Fornairon-Bonnefond C., Mazauric J.P., 2002. Interactions between wine lees and polyphenols: Influence on oxygen consumption capacity during simulation of wine aging. J Food Sci. 67 (5): 1604-1609.

Sarneckis C.J., Dambergs R.G., Jones P., Mercurio M., Herderich M.J., Smith P.A., 2006. Quantification of condensed tannins by precipitation with methyl cellulose: development and validation of an optimised tool for grape and wine analysis. Aust. J. Grape Wine Res. 12 (1): 39-49.

Sarni-Manchado P., Cheynier V., Moutounet M., 1999 a. Interaction of grape seed tannins with salivary proteins. J. Agric. Food Chem. 47 (1): 42-47.

Sarni-Manchado P., Deleris A., Avallone S., Cheynier V., Moutounet M., 1999 b. Analysis and characterization of wine condensed tannins precipitated by proteins used as fining agent in enology. Am. J. Enol. Vitic. 50 (1): 81-86.

Saucier C., Bourgeois G., Vitry C., Roux D., Glories Y., 1997 a. Characterization of (+)-catechin-acetaldehyde polymers : A model for the colloidal state of wine polyphenols. J. Agric. Food Chem. 45 (4) : 1045-1049.

Saucier C., Little D., Glories Y., 1997 b. First evidence of acetaldehyde-flavanol condensation products in red wine. Am. J. Enol. Vitic. 48 (3): 370-373.

Schmidtke L.M., Clark A.C., Scollary G.R., 2011. Micro-oxygenation of red wine: Techniques, applications, and outcomes. Crit. Rev. Food Sci. Nutr. 51 (2): 115-131.

Schneider V., 1998. Herausforderung Rotwein. Deut. Weinmagazin 3: 32-37.

Schneider V., 1999. Oenologische Tannine im Rotwein. Holzton, aber kein Barrique. Deut. Weinmagazin 22, 34-38.

Schneider V., 2001. Mikrooxidation junger Rotweine, III. Einfluss unterschiedlicher Verfahren und Intensitäten. Deut. Weinmagazin 05: 41-46.

Schneider V., 2004. Einsatz und Wirkung von Tannin in der Rotweinbereitung. Deut. Weinmagazin 13: 12-16.

Schneider V., 2005 a. Dornfelder: Welche Inhaltsstoffe bestimmen seine Qualität? Deut. Weinmagazin 02: 12-15.

Schneider V., 2005 b. Aufnahme von Sauerstoff im Keller und bei der Abfüllung. Der Winzer 12: 6-9.

Schneider V., 2009 a. Extraktionskinetik des Tannins bei der Maischegärung. Deut. Weinmagazin 16/17: 34-38.

Schneider V., 2009 b. Glättende Schönung: Schönungsmittel zur Minderung der Adstringens von Rotwein. Deut. Weinmagazin 18: 14-17.

Schneider V., 2019. White wine enology. Advanced winemaking strategies for fine white wines, chapter 2. Board and Bench Publishing, San Francisco 2019.

Schneider V., Müller J., Schmidt D., 2016. Oxygen consumption by postfermentation yeast lees: Factors affecting its rate and extent under enological conditions. Food Technol. Biotechnol. 54 (4): 395-402.

Schneider V., Schmitt M., Kroeger R., 2017. Wine screw cap closures: The next generation. Grapegrower & Winemaker 638: 50-52.

Schneider V. and Troxell S., 2018. Acidity management in musts and wines. Board and Bench Publishing, San Francisco 2018.

Scrimgeour N. and Wilkes E., 2014. Closure trials show volatile sulfur compounds formation can still cause a stink. Australian & New Zealand Grapegrower & Winemaker 602: 62-67.

Scudamore-Smith P.D., Hooper R.L., McLaran E.D., 1990. Color and phenolic changes of Cabernet Sauvignon wine made by simultaneous yeast/bacterial fermentation and extended pomace contact. Am. J. Enol. Vitic. 41 (1): 57-67.

Serrano M., Vannier A.-C., Ribéreau-Gayon P., 1988. Clarification des vins par filtration en flux tangentiel (ultrafiltration). Incidence sur la composition chimique et les qualités organoleptiques. Évolution des produits au cours du vieillissement. Conn. Vigne Vin 22 (1): 49-71.

Sheridan M.K. and Elias R.J., 2015. Exogenous acetaldehyde as a tool for modulating wine color and astringency during fermentation. Food Chem. 177: 17-22.

Sidari R., Postorino S. Caparello A., Caridi A., 2007. Evolution during wine aging of colour and tannin differences induced by wine starters. Ann. Microbiol. 57 (2): 197-201.

Silva M. A., Julien M., Jourdes M., Teissedre P.-L., 2011. Impact of closures on wine post-bottling development: A review. Eur. Food Res. Technol. 233: 905-914.

Singleton V.L., 1972. Effects on red wine quality of removing juice before fermentation to simulate variation in berry size. Am. J. Enol. Vitic. 23 (3): 106-113.

Singleton V.L., 1987. Oxygen with phenols and related reactions in musts, wines, and model systems: Observations and practical reactions. Am. J. Enol. Vitic. 38 (1): 69-77.

Singleton V.L. and Rossi J.A., 1965. Colorimetry of total phenolics with phosphomolybdic-phosphotungstic acid reagents. Am. J. Enol. Vitic. 16 (3): 144-158.

Singleton V.L. and Draper D.E., 1964. The transfer of polyphenolic compounds from grape seeds into wines. Am. J. Enol. Vitic. 15 (1): 34-40.

Singleton V.L. and Trousdale E.K., 1992. Anthocyanin-tannin interactions explaining differences in polymeric phenols between white and red wines. Am. J. Enol. Vitic. 43 (1): 63-70.

Singleton V.L., Orthofer R., Lamuela-Raventos R.M., 1999. Analysis of total phenols and other oxidation substrates and antioxidants by means of Folin-Ciocalteu reagent. Methods Enzymol. 299: 152-178.

Sito S., Horvatiček B., Devrnja A., Obad N., 2013. Utjecaj načina filtriranja na organoleptička svojstva vina. Glasnik zaštite bilja 1: 64-69.

Slinkard K. and Singleton V.L., 1977. Total phenol analysis: Automatation and comparison with manual methods. Am. J. Enol. Vitic. 28 (1): 49-55.

Smith A.K., June H., Noble A.C., 1996. Effects of viscosity on the bitterness and astringency of grape seed tannins. Food Qual. Pref. 7 (3-4): 161-166.

Somers T.C., 1971. The polymeric nature of wine pigments. Phytochemistry 10 (9): 2175-2186.

Somers T.C., 1998. The wine spectrum, 1st edition. Winetitles, Adelaide 1998.

Somers T.C. and Evans M.E., 1974. Wine quality: Correlations with colour density and anthocyanin equilibria in a group of young red wines. J. Sci. Food. Agric. 25: 1369-1379.

Somers T.C. and Evans M.E., 1977. Spectral evaluation of young red wines: Anthocyanin equilibria, total phenolics, free and molecular SO_2, “chemical age”. J. Sci. Food Agric. 28: 279-287.

Somers T.C. and Ziemelis G., 1980. Gross interference by sulphur dioxide in standard determinations of wine phenolics. J. Sci. Food Agric. 31 (6): 600-610.

Somers T.C. and Ziemelis G., 1985. Spectral evaluation of total phenolic composition in Vitis vinifera: Grapes and wine. J. Sci. Food Agric. 36 (12): 1275-1284.

Somers T.C. and Evans M.E., 1986. Evolution of red wines. I. Ambient influences on colour composition during early maturation. Vitis 25 (1): 31-39.

Somers T.C. and Wescombe L.G., 1987. Evolution of red wines. II. An assessment of the role of acetaldehyde. Vitis 26 (1): 27-36.

Somers T.C. and Pocock K.F., 1990. Evolution of red wines. III. Promotion of the maturation phase. Vitis 29 (2): 109-121.

Sowalsky R.A. and Noble A.C., 1998. Comparison of the effects of concentration, pH and anion species on astringency and sourness of organic acids. Chem. Senses 23 (3): (343-349).

Spillman P.J., Pollnitz A.P., Liacopoulos D., Skouroumounis G.K. Sefton M.A., 1997. Accumulation of vanillin during barrel aging of white, red, and model wines. J. Agric. Food Chem. 45 (7): 2584-2589.

Springer L.F. and Sacks G.L., 2014. Protein-precipitable tannin in wines from Vitis vinifera and interspecific hybrid grapes (Vitis ssp.): Differences in concentration, extractability, and cell wall binding. J. Agric. Food Chem. 62 (30): 7515-7523.

Sun B., Spranger I., Roque-do-Vale F., Leandro C., Belchior P., 2001. Effect of different winemaking technologies on phenolic composition in Tinta Miúda red wines. J. Agric. Food Chem. 49 (12): 5809-5816.

Suriano S., Ceci G., Tamborra T., 2012. Impact of different winemaking techniques on polyphenolic compounds of Nero di Troia wine. Ital. Food Bev. Technol. 70: 5-15.

Taladrid D., Lorente L., Bartolomé B., Moreno-Arribas M.V., Laguna L., 2018. An integrative salivary approach regarding palate cleanser in wine tasting. J. Texture Stud., 50 (1): 75-82.

Tamborra P., 1993. Antociani monomeri e polimeri dei vini rossi durante l'invecchiamento. Riv. Vitic. Enol. 2: 61-74.

Tao, J., Dykes, S. I., and Kilmartin, P. A. (2007). Effect of SO_2 concentration on polyphenol development during red wine micro-oxygenation. J. Agric. Food Chem. 55 (15): 6104–6109.

Tesnière C. and Flanzy C., 2011. Carbonic maceration wines : Characteristics and winemaking process. Adv. Food Nutr. Res. 63: 1-15.

Timberlake C.F. and Bridle P., 1976. Interactions between anthocyanins, phenolic compounds, and acetaldehyde and their significance in red wines. Am. J. Enol. Vitic. 27 (3): 97-105.

Towey J.P. and Waterhouse A.L., 1996 a. The extraction of volatile compounds from French and American oak barrels in Chardonnay during three successive vintages. Am. J. Enol. Vitic. 47 (2): 163-172.

Towey J.P. and Waterhouse A.L., 1996 b. Barrel-to-barrel variation of volatile oak extractives in barrel-fermented Chardonnay. Am. J. Enol. Vitic. 47 (1): 17-20.

Tupinier C. and Gerbeaux V., 2002. Attention aux vins non filtrés. Bourgogne Aujourd'hui 50: 61-63.

Ulbricht M., Ansorge W., Danielzik I., König M., Schuster O., 2009. Fouling in microfiltration of wine: The influence of the membrane polymer on adsorption of polyphenols and polysaccharides. Sep. Purif. Technol. 68 (3): 335-342.

Valade M., Tribaut-Sohier I., Brunner D., Pierlot C., Moncomble D., Tusseau D., 2006. Les apports d'oxygène en vinification et leurs impacts sur les vins. 1ère partie. Le Vignéron Champanois 8: 17-28.

Valade M., Tribaut-Sohier I., Brunner D., Laurent M., Moncomble D., Tusseau D., 2007. Les apports d'oxygène en vinification et leurs impacts sur les vins: Le cas particulier du Champagne (2ème partie). Rev. Fr. d'Œnologie 222: 17-28.

Valentová H., Škrovánková S., Panovská Z., Pokorny J., 2001. Determination of astringent taste in model solutions and beverages. Czech J. Food Sci. 19 (5): 196-200.

Valentová H., Skrovánková S., Panovská Z., Pokorny J., 2002. Time-intensity studies of astringent taste. Food Chem. 78 (1): 29-37.

Vela E., Hernandez-Orte P., Franco-Luesma E., Ferreira V., 2018. Micro-oxygenation does not eliminate hydrogen sulfide and mercaptans from wine; it simply shifts redox and complex-related equilibria to reversible oxidized species and complexed forms. Food Chem. 243: 222-230.

Vernhet A., Carrillo S., Poncet-Legrand C., 2014. Condensed tannin changes induced by autoxidation: Effect of the initial degree of polymerization and concentration. J. Agric. Food Chem. 62 (31): 7833-7842.

Versari A., Barbanti D., Potentini G., Parpinello G.P., Galassi S., 1999. Preliminary study of the interaction of gelatin-red wine components. Ital. J. Food Science 11 (3): 231-239.

Versari A., du Toit W., Parpinello G.P., 2013. Oenological tannins: a review. Aust. J. Grape Wine Research 19 (1): 1-10.

Vidal J.-C., Dufourcq T., Boulet J.-C., Moutounet M., 2001. Les apports d'oxygène au cours des traitements des vins. Bilan des observations sur site. 1ère partie. Rev. Fr. d'Œnologie 190: 24-31.

Vidal J.-C., Boulet J.-C., Moutounet M., 2003. Les apports d'oxygène au cours des traitements des vins. Bilan des observations sur site. 2ème partie. Rev. Fr. d'Œnologie 201: 32-38.

Vidal J.-C., Boulet J.-C., Moutounet M., 2004 a. Les apports d'oxygène au cours des traitements des vins. Bilan des observations sur site. 3ème partie. Rev. Fr. d'Œnologie 205: 25-33.

Vidal S., Cartalade D., Souquet J.-M., Fulcrand H., Cheynier V., 2002. Changes in proanthocyanidin chain length in wine-like solutions. J. Agric. Food Chem. 50 (8): 2261-2266.

Vidal S., Leigh F., Noble A., Kwiatkowski M., Cheynier V., Waters E., 2004 b. Taste and mouth-feel properties of different types of tannin-like polyphenolic compounds and anthocyanins in wine. Anal. Chim. Acta 513 (1): 57-65.

Vidal S., Francis L., Williams P., Kwiatkowski M., Gawel R., Cheynier V., Waters E., 2004 c. The mouth-feel properties of polysaccharides and anthocyanins in a wine-like medium. Food Chem. 85 (4): 519-525.

Vidal S., Courcoux P. Francis L., Kwiartkowski M., Gawel R., Williams P., Waters E., Cheynier V., 2004 d. Use of an experimental design approach for evaluation of key wine components on mouth-feel perception. Food Qual. Pref. 15 (3): 209-217.

Vidal J.-C. and Moutounet M., 2006. Suivi de l'oxygène des phases gazeuse et liquide de bouteilles de vin à l'embouteillage et en conservation. J. Int. Sci. Vigne Vin 40 (1): 35-45.

Vidal L., Antúnez L., Giménez A., Ares G., 2016. Evaluation of palate cleansers for astringency evaluation of red wines. J. Sens. Stud. 31 (2) : 93-100.

Vidal J.-C., Caillé S., Samson A., Salmon J.-M., 2017. Impact of eight closures in controlled industrial conditions on the shelf life of two (red and rosé) wines. OENO One, 51 (4) : 387-399.

Vivas N., 1999. Les oxydations et les réductions dans les moûts et les vins. Éditions Féret, Bordeaux 1999.

Vivas N. and Glories Y., 1996. Role of oak wood ellagitannins in the oxidation process of red wines during aging. Am. J. Enol. Vitic. 47 (1): 103-107.

Vivas N., Vivas de Gaulejac N., Nonier M.F., 2003 a. Sur l'estimation et la quantification des composes phénoliques des vins. Bull. de l'OIV 865-866: 281-303.

Vivas N., Debeda H., Menil F., Vivas de Gaulejac N., Nonier M.F., 2003 b. Mise en évidence du passage de l'oxygène au travers des bouelles constituant les barriques par l'utilisation d'un dispositif original de mesure de la porosité du bois. Premiers résultats. Sci. Aliment. 23: 655-678.

Wang J., Huo S., Zhanag X., Lui Y., Fan W., 2016. Effect of different pre-fermentation treatments on polyphenols, color, and volatile compounds on three wine varieties. Food Sci. Biotechnol. 25: 735–743.

Waterhouse A.L. and Laurie V.F., 2006. Oxidation of wine phenolics: A critical evaluation and hypotheses. Am. J. Enol. Vitic. 57 (3): 306-313.

Waterhouse A., Frost S., Ugliano M., Cantu A. R., Currie B. L., Anderson M., Chassy A.W., Vidal S., Diéval J.-B., Aagaard O., Heymann H., 2016. Sulfur dioxide-oxygen consumption ratio reveals differences in bottled wine oxidation. Am. J. Enol. Vitic. 67 (4): 449-459.

Waters E. J., Peng Z., Pocock K. F., Williams P. J., 1996. The role of corks in oxidative spoilage of white wines. Aust. J. Wine Grape Res, 2 (3): 191-197.

Weber F., Greve K., Durner D., Fischer U., Winterhalter P., 2013. Sensory and chemical characterization of phenolic polymers from red wine obtained by gel permeation chromatography. Am. J. Enol. Vitic. 64 (1): 15-25.

Weiand J., 2009. Spätburgunder. Was bewirken Gärtemperatur und önologische Maßnahmen? Deut. Weinmagazin 19: 25-29.

Weiand J., 2010. Gärtemperatur bei Rotwein. Deut. Weinmagazin 16/17: 10-16.

Weiand J., 2014. Sauerstoffeinsatz während der Maischegärung. Deut. Weinbau 16-17: 36-39.

Wightman J.D., Price S.F., Watson B.T., Wrolstad R.E., 1997. Some effects of processing enzymes on anthocyanins and phenolics in Pinot noir and Cabernet Sauvignon wines. Am. J. Enol. Vitic. 48 (1): 39-48.

Wildenradt H.L. and Singleton V.L., 1974. The production of aldehydes as a result of oxidation of polyphenolic compounds and its relation to wine aging. Am. J. Enol. Vitic. 25 (2): 119-126.

Wilker K.L. and Gallander J.F., 1988. Comparison of Seyval blanc wine aged in barrels and stainless steel tanks with oak chips. Am. J. Enol. Vitic. 39 (1): 38-43.

Yang H. H.-L. and Lawless H.T., 2005. Descriptive analysis of divalent salts. J. Sens. Stud. 20 (2): 97-113.

Yokotsuka K. and Singleton V.L., 1995. Interactive precipitation between phenolic fractions and peptides in wine-like model solutions: Turbidity, particle size, and residual content as influenced by pH, temperature and peptide concentration. Am. J. Enol. Vitic. 46 (3): 329-338.

Yokotsuka K., Sato M., Ueno N., Singleton V.L., 2000. Colour and sensory characteristics of Merlot red wines caused by prolonged pomace contact. J. Wine Res. 11 (1): 7-18.

Zamora F., Luengo G., Margalef P., Magriña M., Arola L., 1994. Nota: Effecto del sangrado sobre el color y la composición en compuestos fenólicos del vinho tinto. Rev. Esp. Cienc. Tecnol. Aliment. 34: 663-671.

Zimman A., Joslin W.S., Lyon M.L., Meier J., Waterhouse A.W., 2002. Maceration variables affecting phenolic composition in commercial-scale Cabernet Sauvignon winemaking trials. Am. J. Enol. Vitic. 53 (2): 93-98.

Zini S., Siliani P., Vgnoli P., Rosi I., 2002. Un nuovo ceppo selezionato di Saccharomyces cerevisiae nella fermentazione di uve Sangiovese. Enologo 38 (3): 95-100.

Zini S., Canuti V., Siliani A., Bertuccioli M., 2003. Criomacerazione in Toscana: Esperienze sul Sangiovese. Ind. Bev. 32 (1): 16-23.

Zironi R., Buiatti S., Celotti E., 1992. Evaluation of a new colourimetric method for the determination of catechins in musts and wines. Vitic. Enol. Sci. 47 (1): 1-7.

Zou H., Kilmartin PA., Inglis M.J., Frost A., 2008. Extraction of phenolic compounds during vinification of Pinot noir wine examined by HPLC and cyclic voltammetry. Aust. J. Grape Wine Res. 8 (3): 163-174.

^

INDEX

www.ingramcontent.com/pod-product-compliance
Lightning Source LLC
LaVergne TN
LVHW081938200926
848484LV00009B/2462

* 9 7 8 1 9 3 5 8 7 9 1 5 2 *